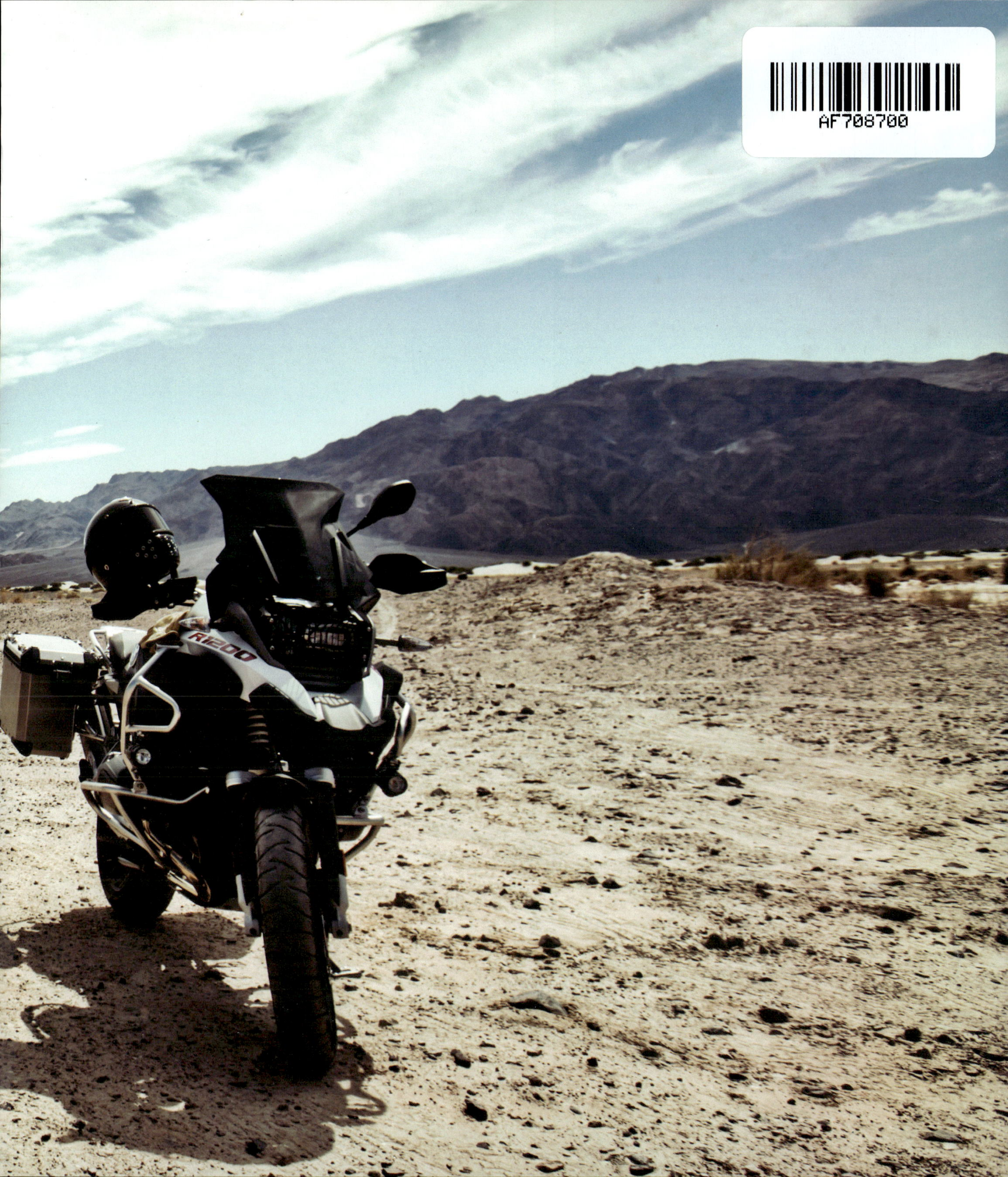

AETHER

HAVANA, CUBA

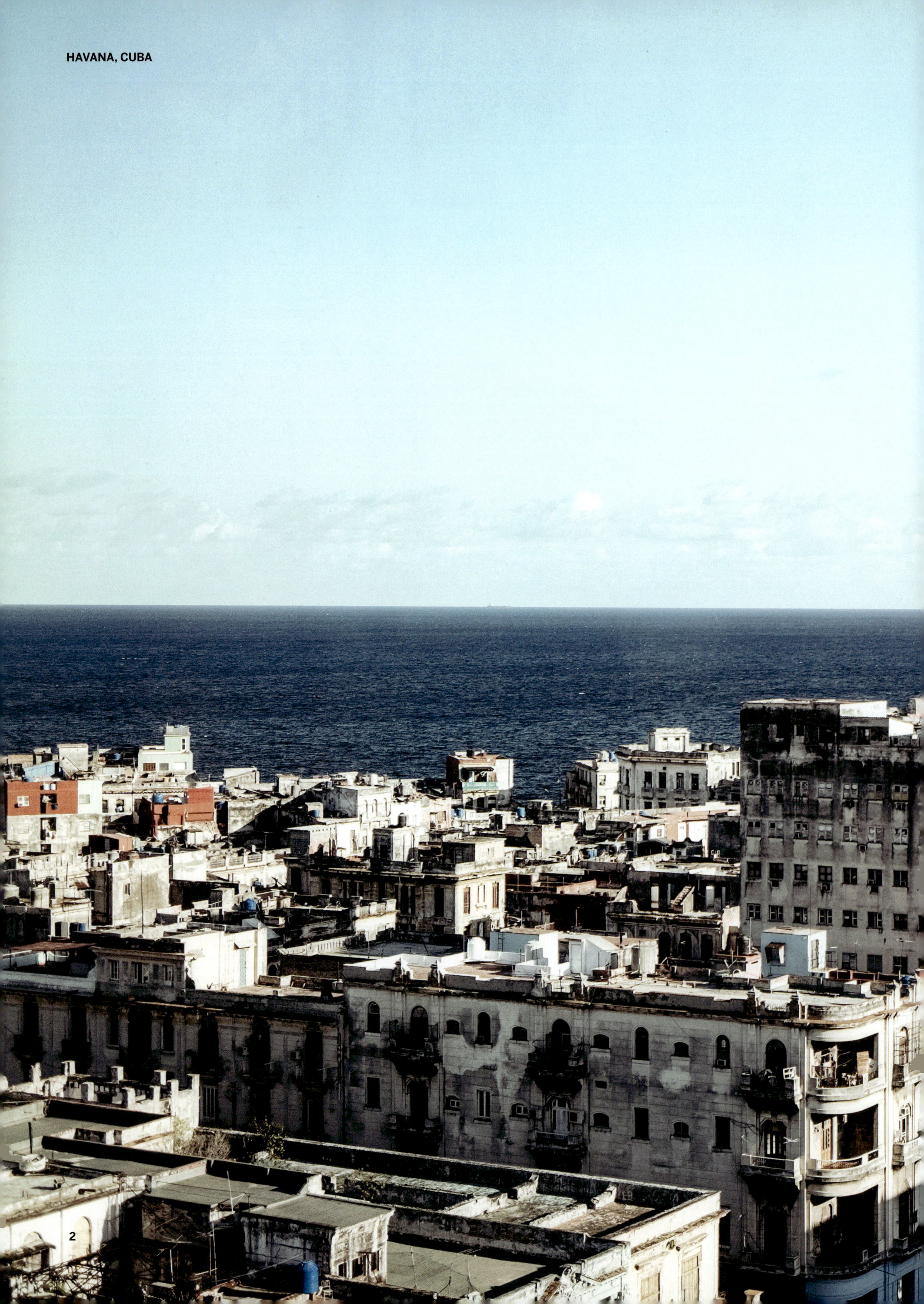

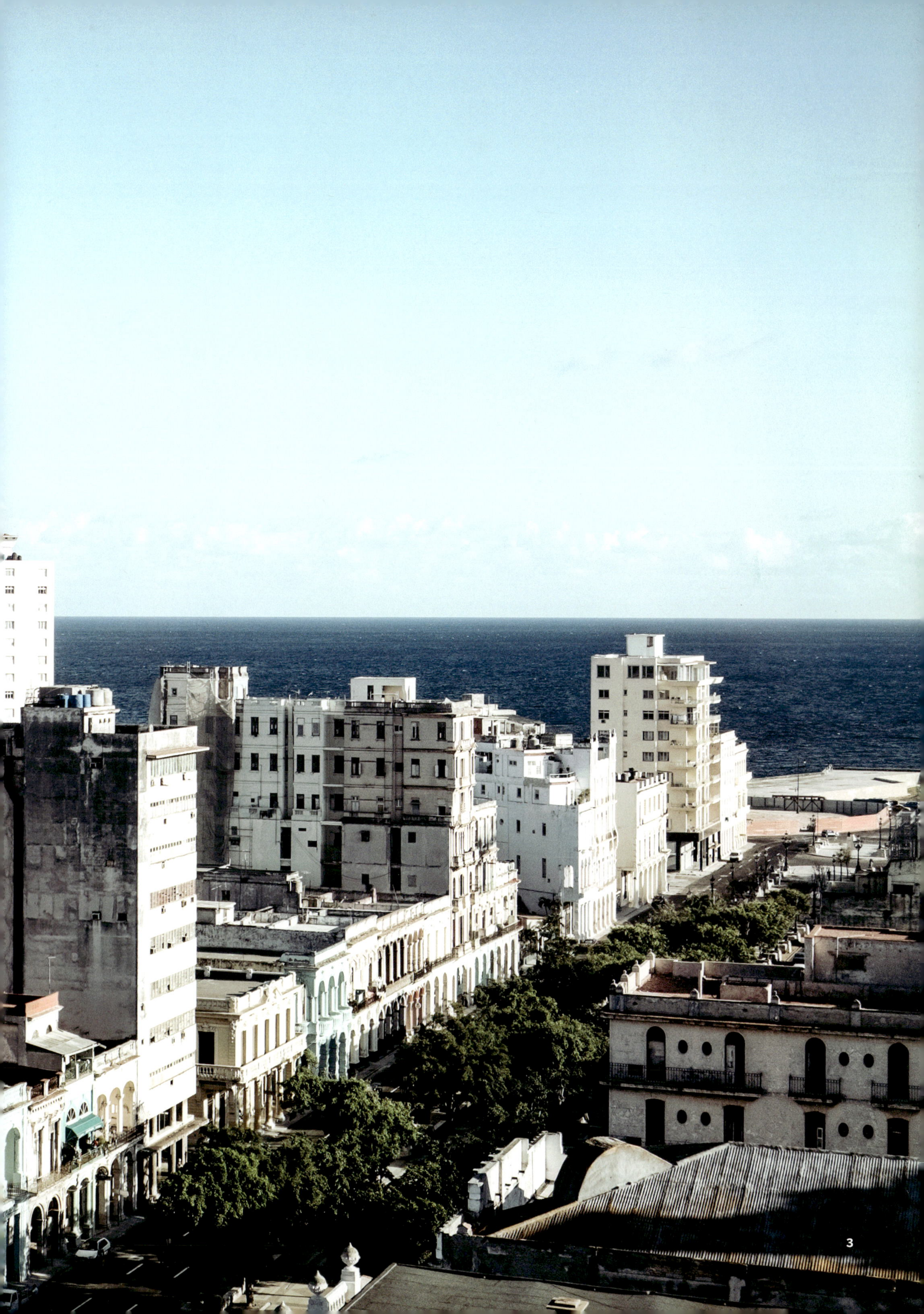

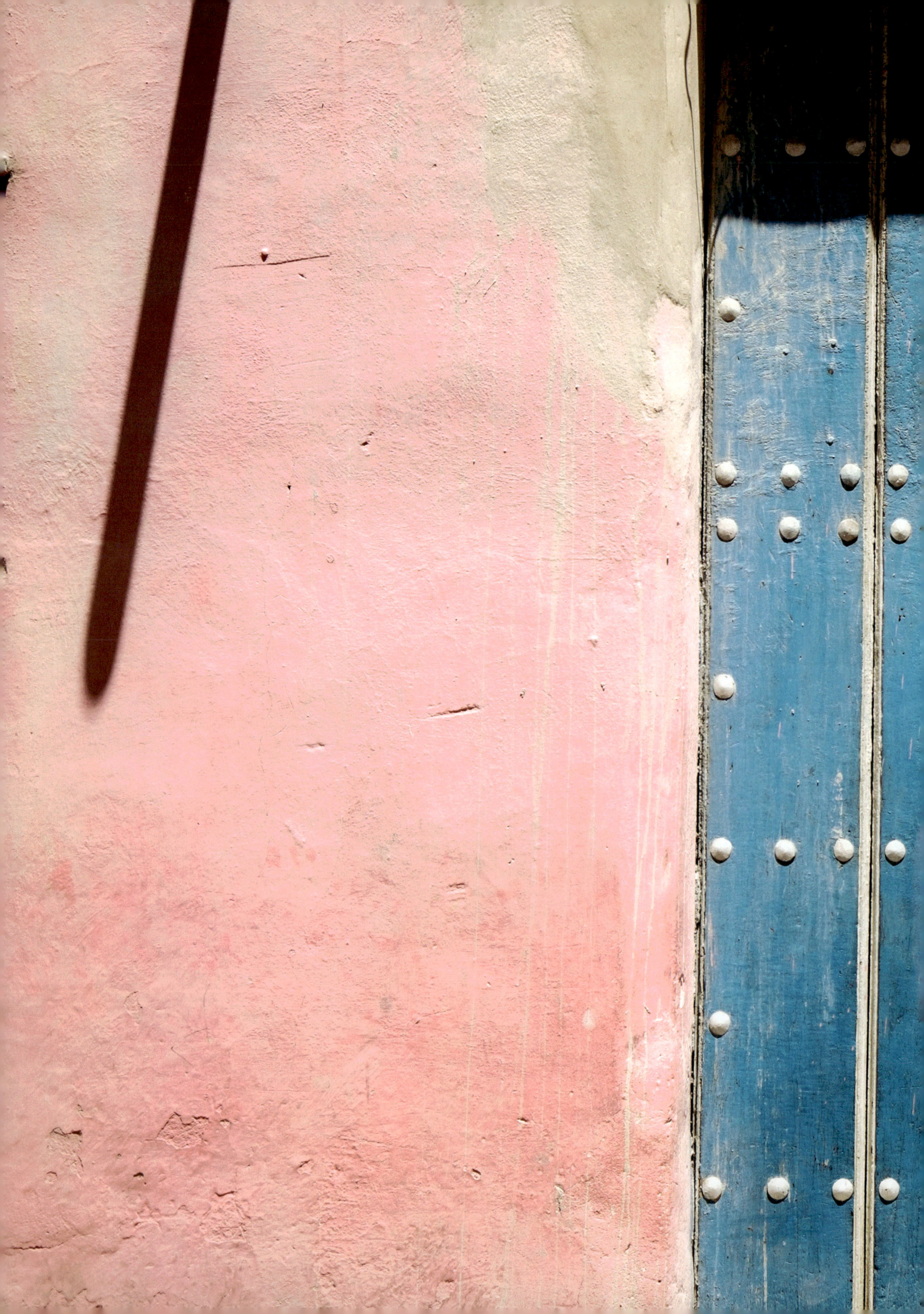

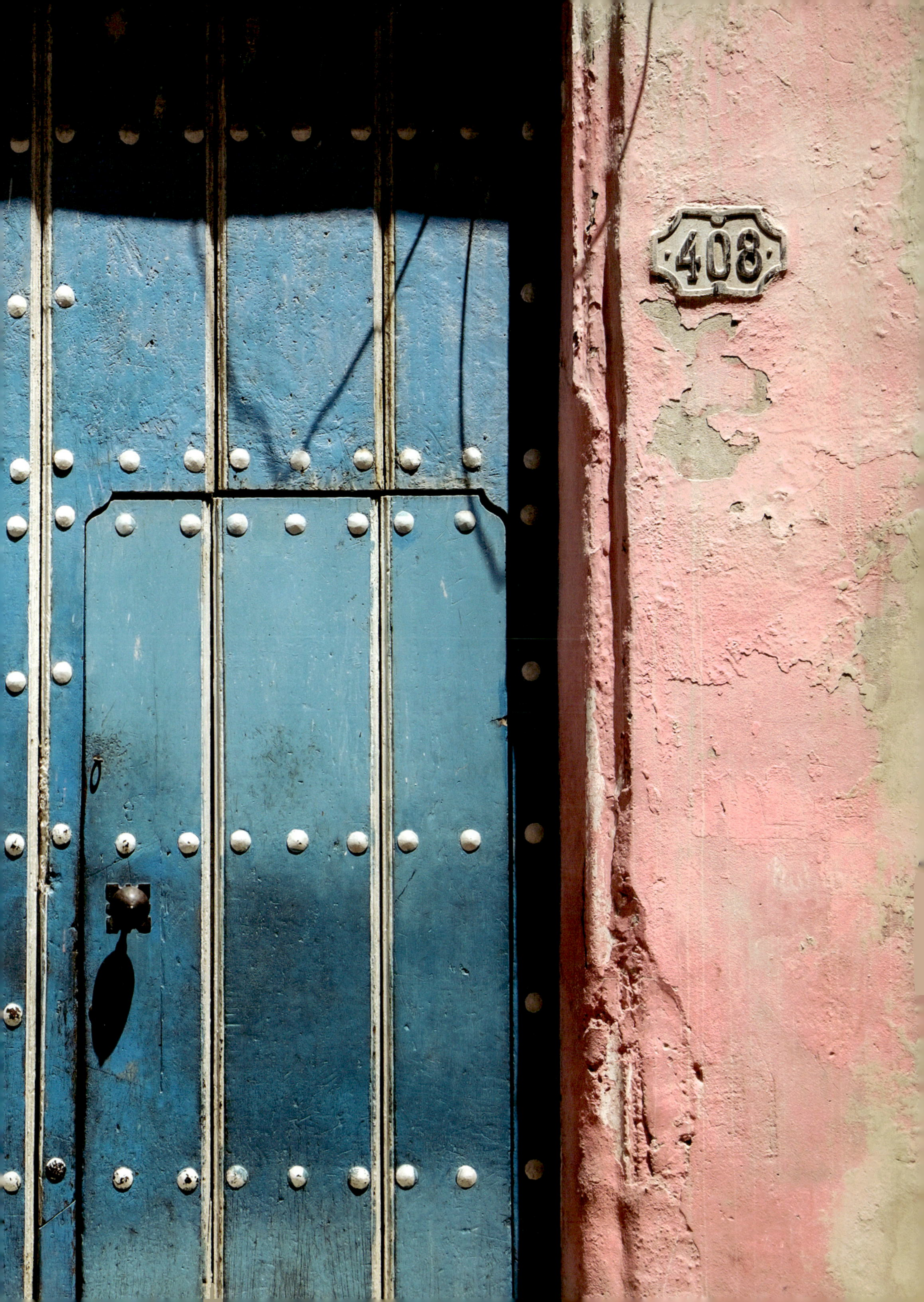
408

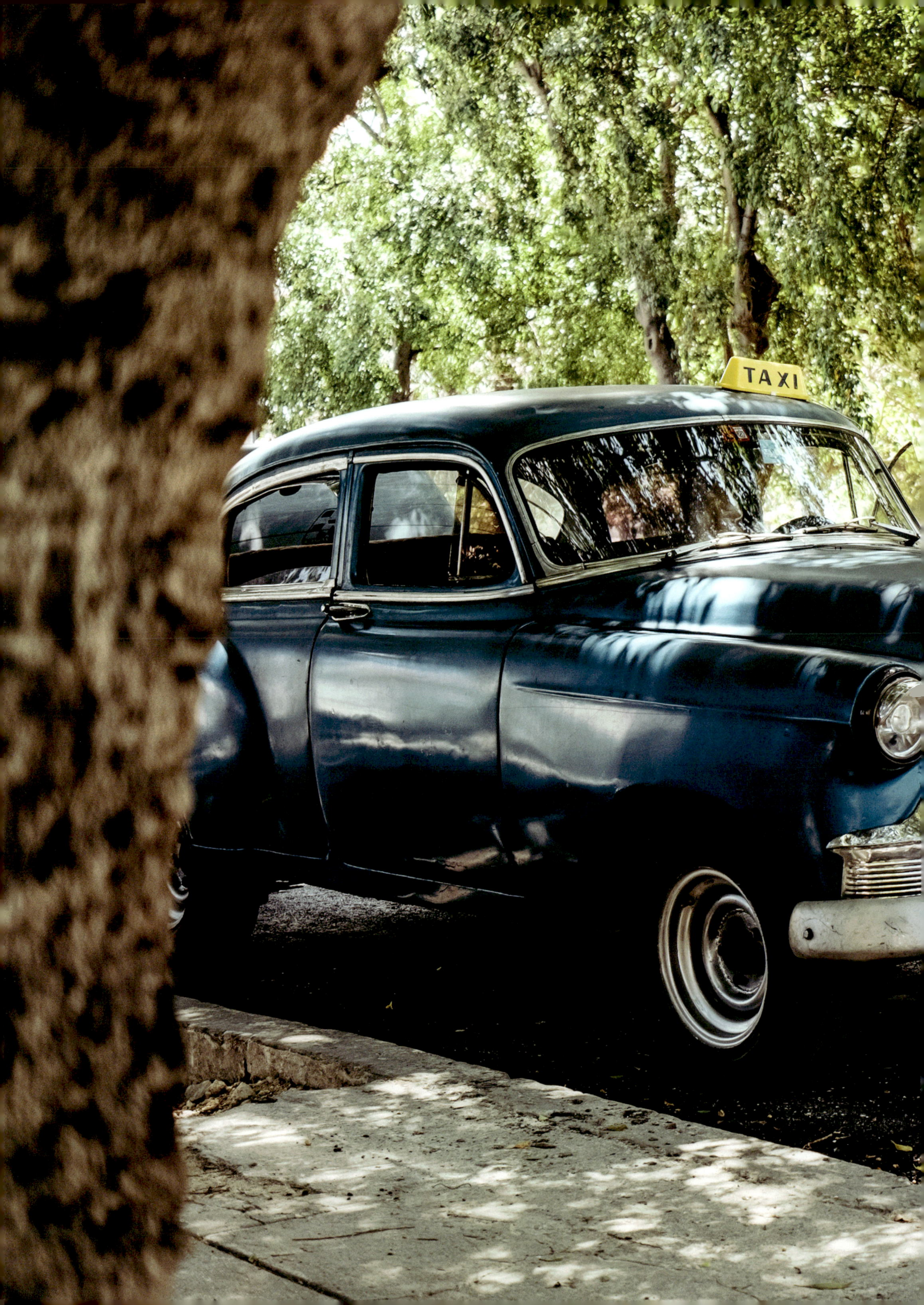
TAXI

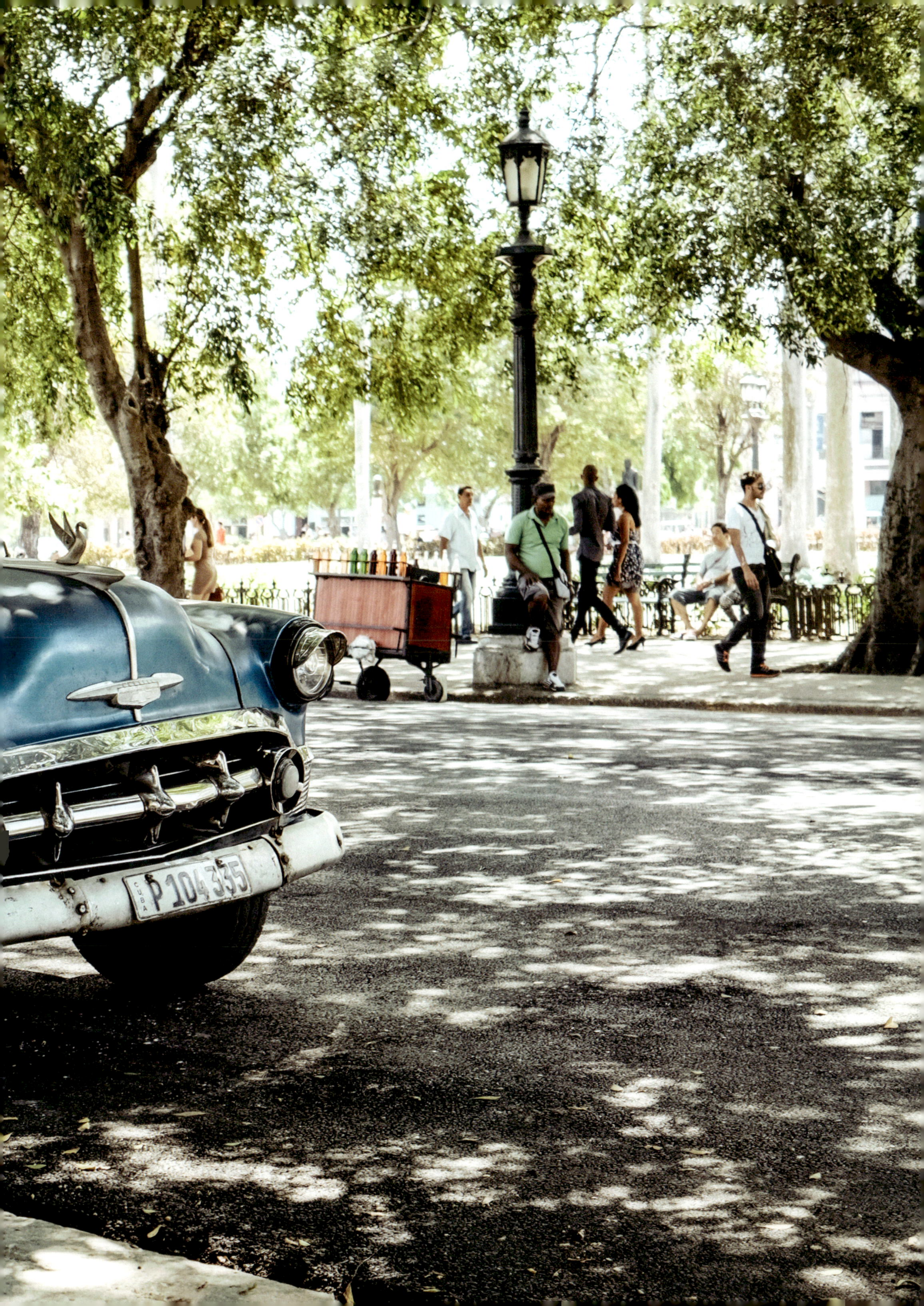
CUBA
P 104 335

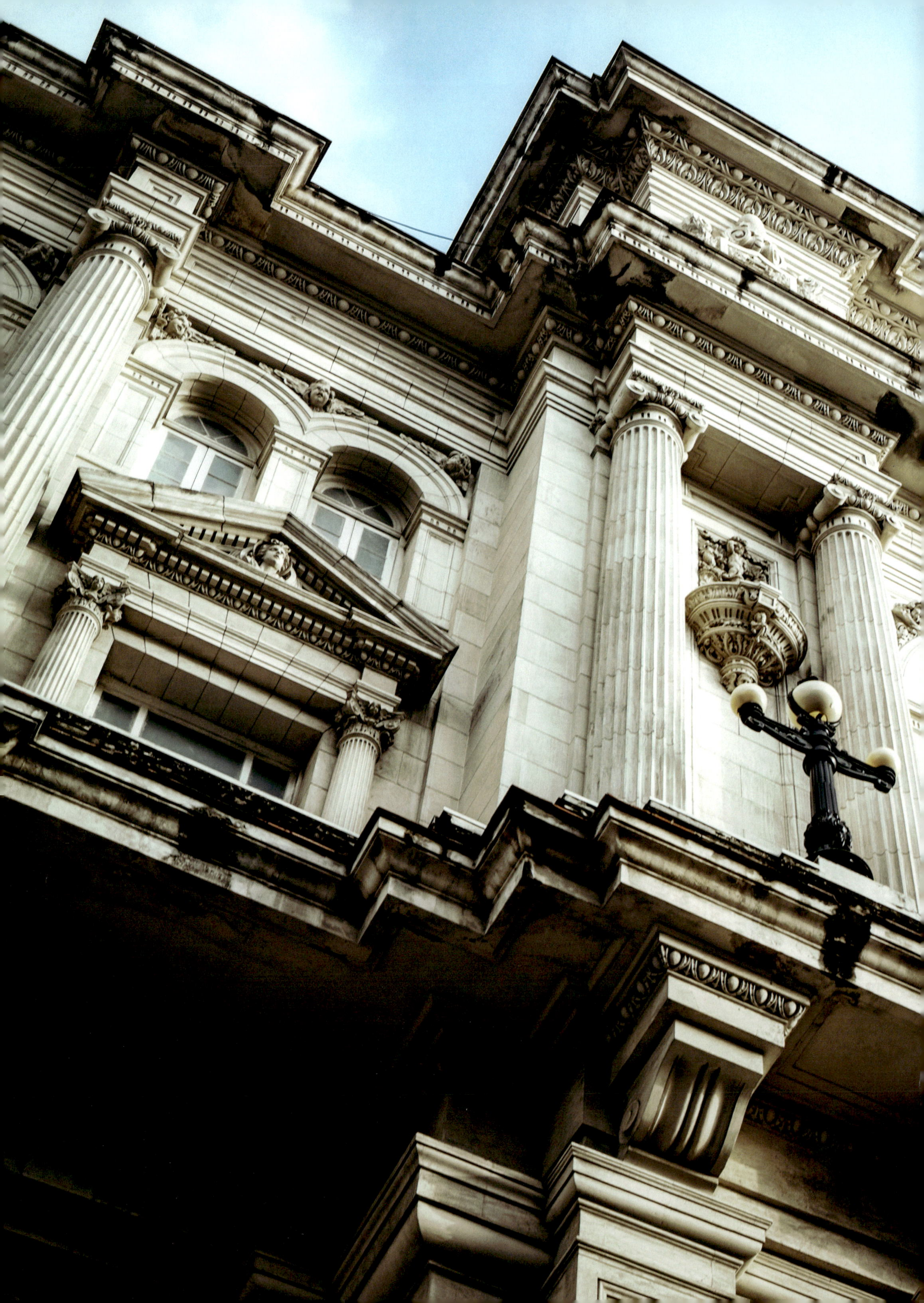

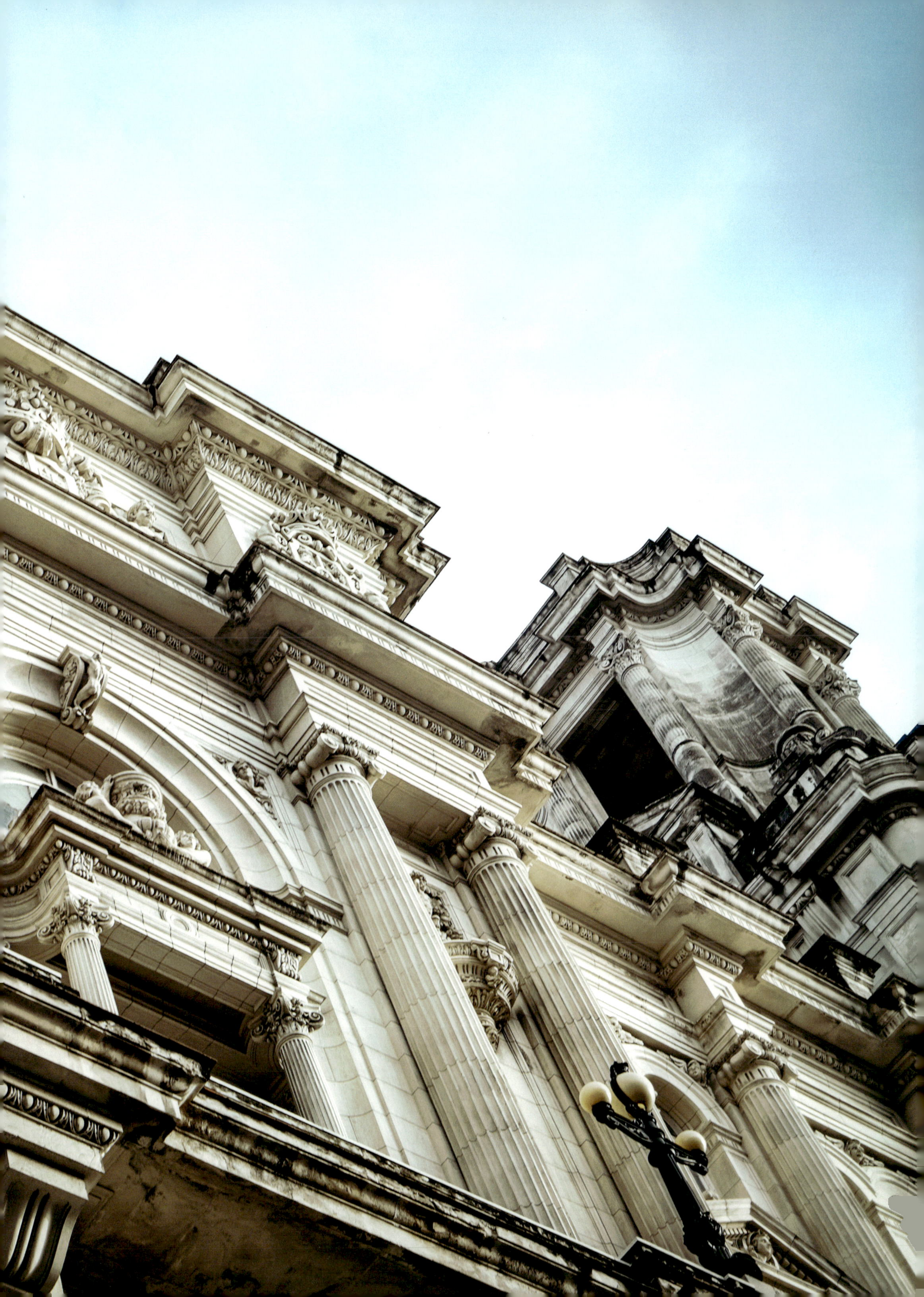

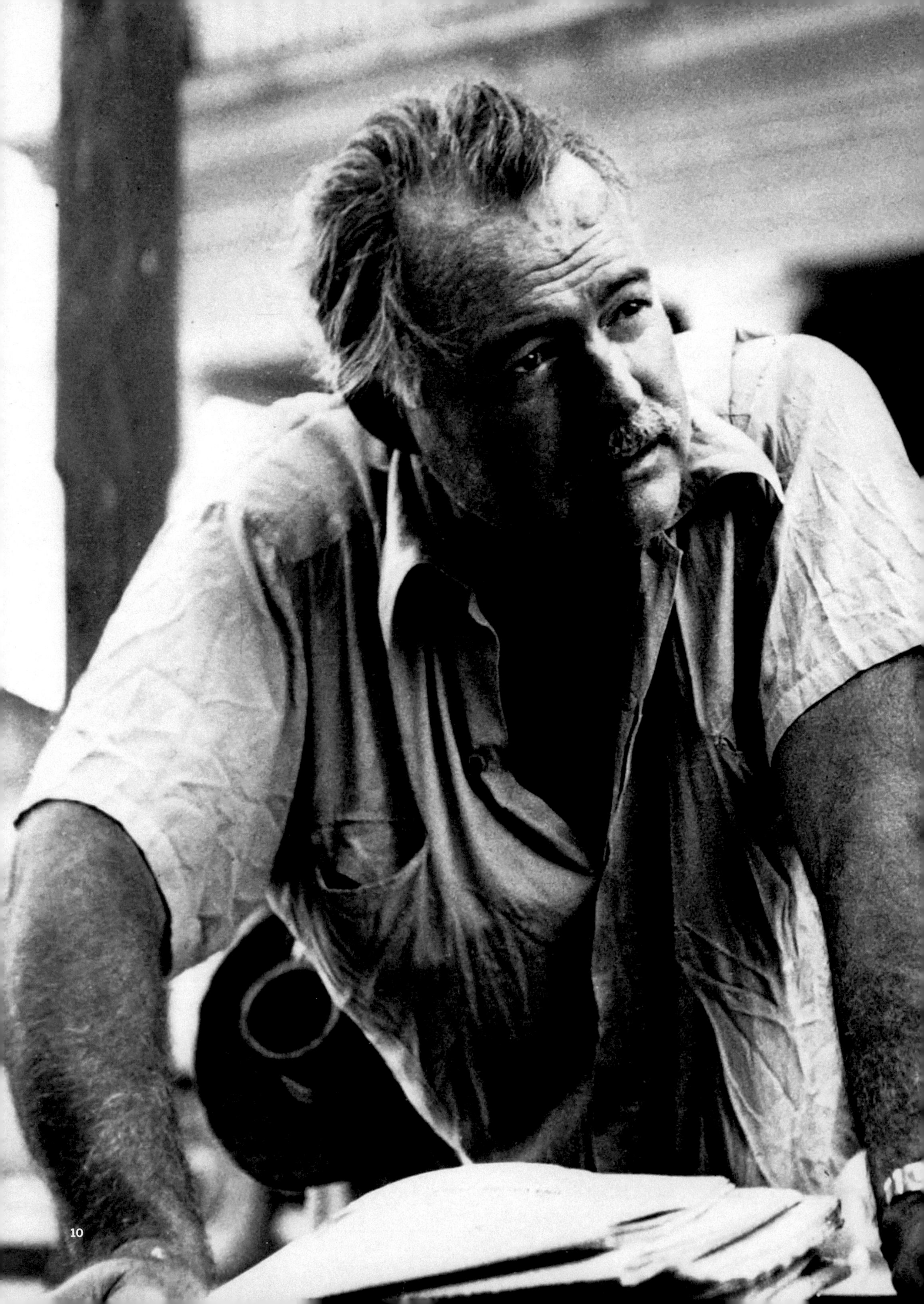

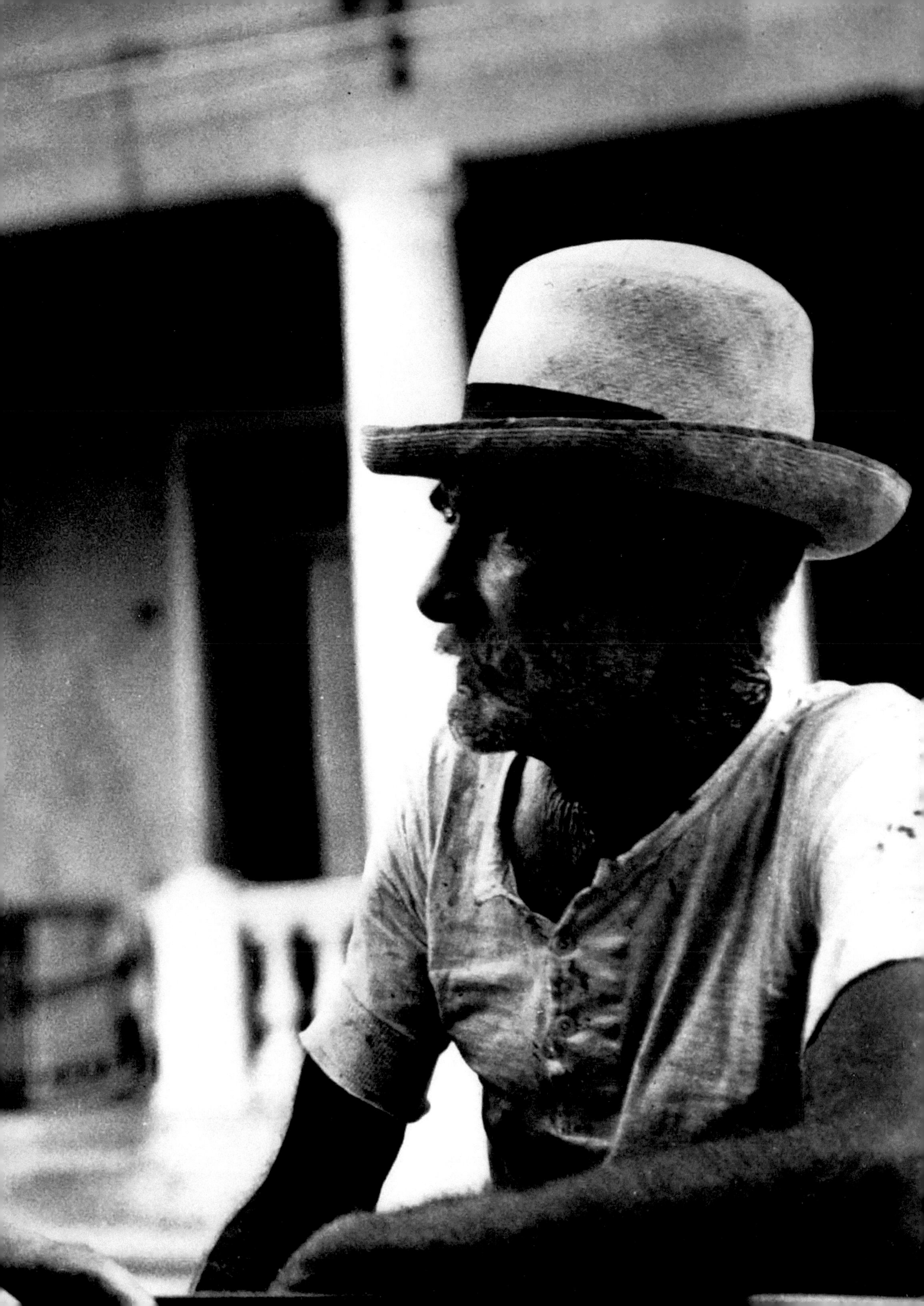

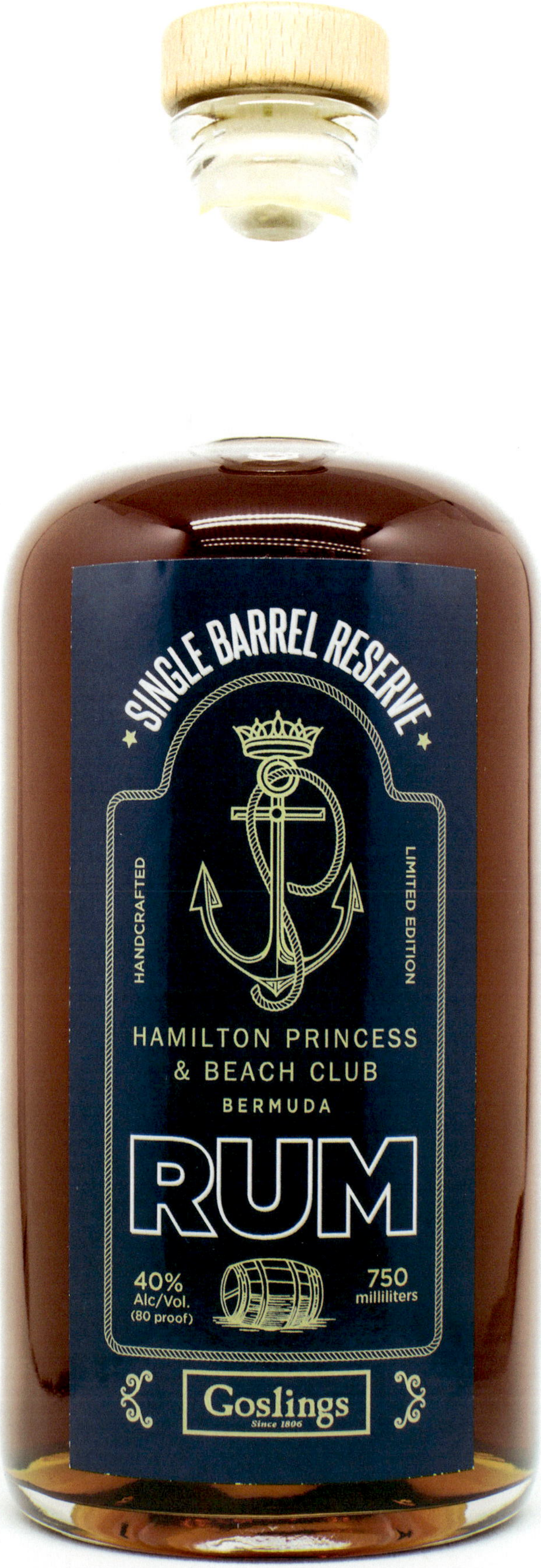

GOSLINGS X HAMILTON PRINCESS SINGLE BARREL RUM / $200

A resort having a signature drink is nothing new. Having a signature rum is unique. This dark, smooth rum is distilled by Bermuda's most storied rum producer, and is exclusive to the island's Hamilton Princess hotel. So you'll more than likely have to brave a trip to the island in order to get a bottle for yourself.

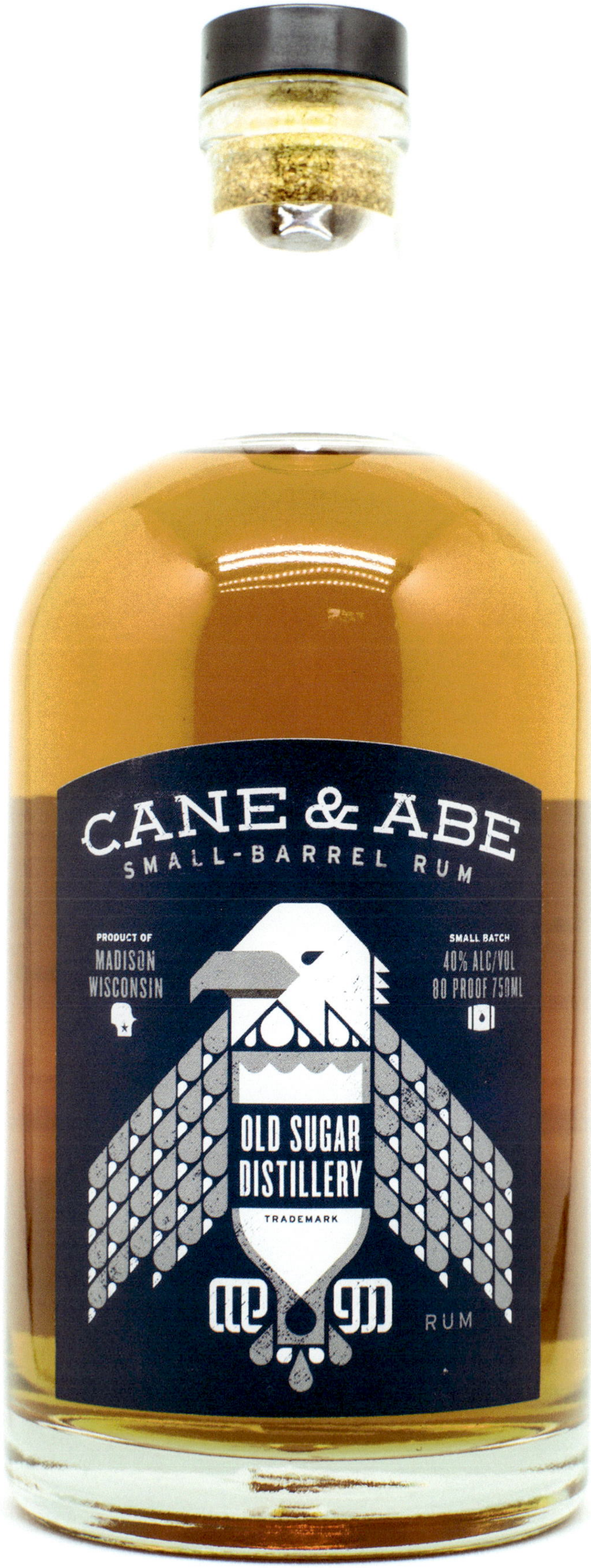

OLD SUGAR DISTILLERY CANE & ABE RUM / $25

Named for Wisconsin's 8th infantry mascot during the Civil War, this Madison-produced spirit is made using Hawaiian-grown dark brown cane sugar. It's fermented and distilled in-house, aged in a mix of both new and used small American Oak barrels, and is left unsweetened, giving it an unexpected flavor.

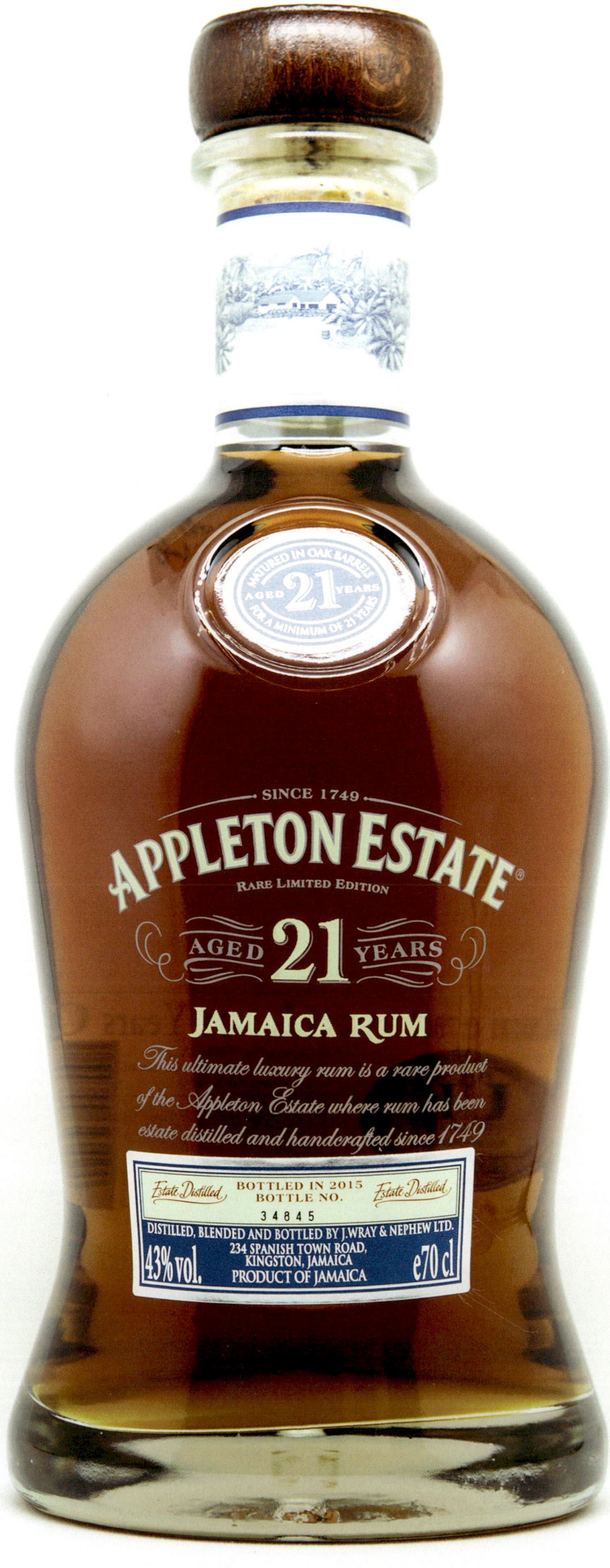

APPLETON ESTATE 21 YEAR OLD RUM / $150

Distilled in small batches using sugarcane grown in Jamaica's Nassau Valley — a place with a unique soil composition claimed by only two other spots on Earth — and water from a limestone spring, this smooth sipping rum is aged in used Number One Select barrels acquired straight from Jack Daniels.

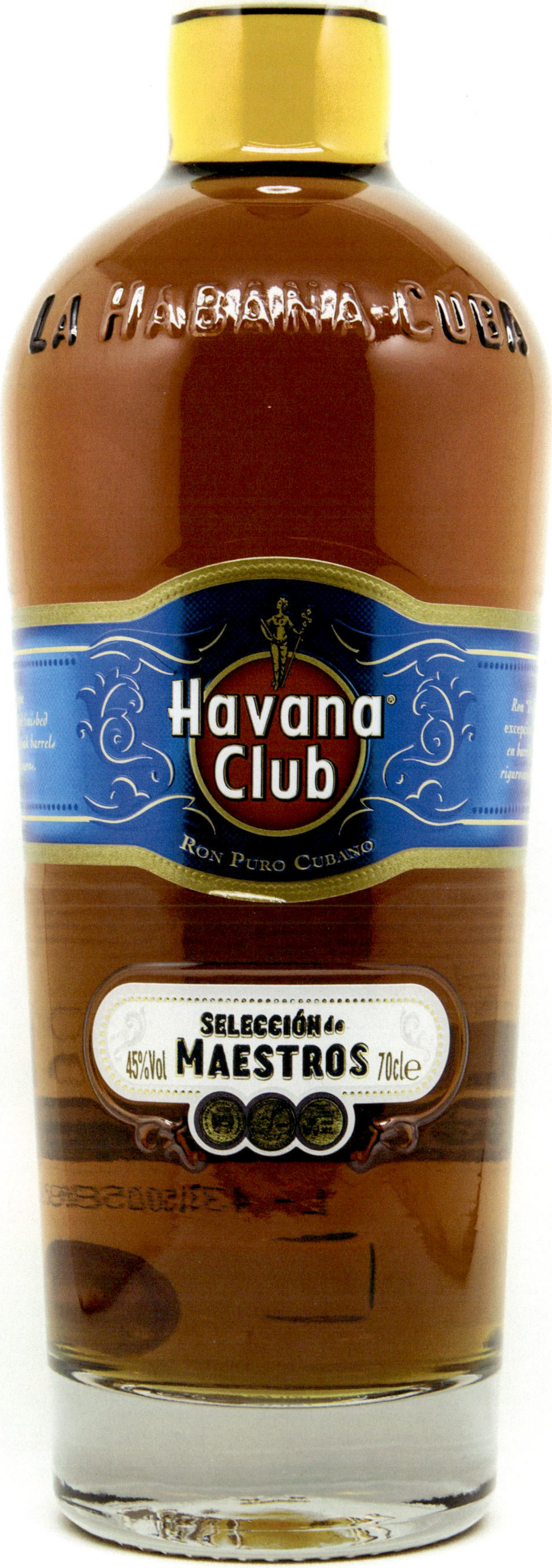

HAVANA CLUB SELECCIÓN DE MAESTROS RUM / $50

The collective knowledge of Cuba's master rum-makers is the definitive mark of this exceptional rum. They first meet to select the brand's best aged rums, then blend, age some more, and meet again to ensure maturation is complete, before the final mix is made and bottled straight from the barrel.

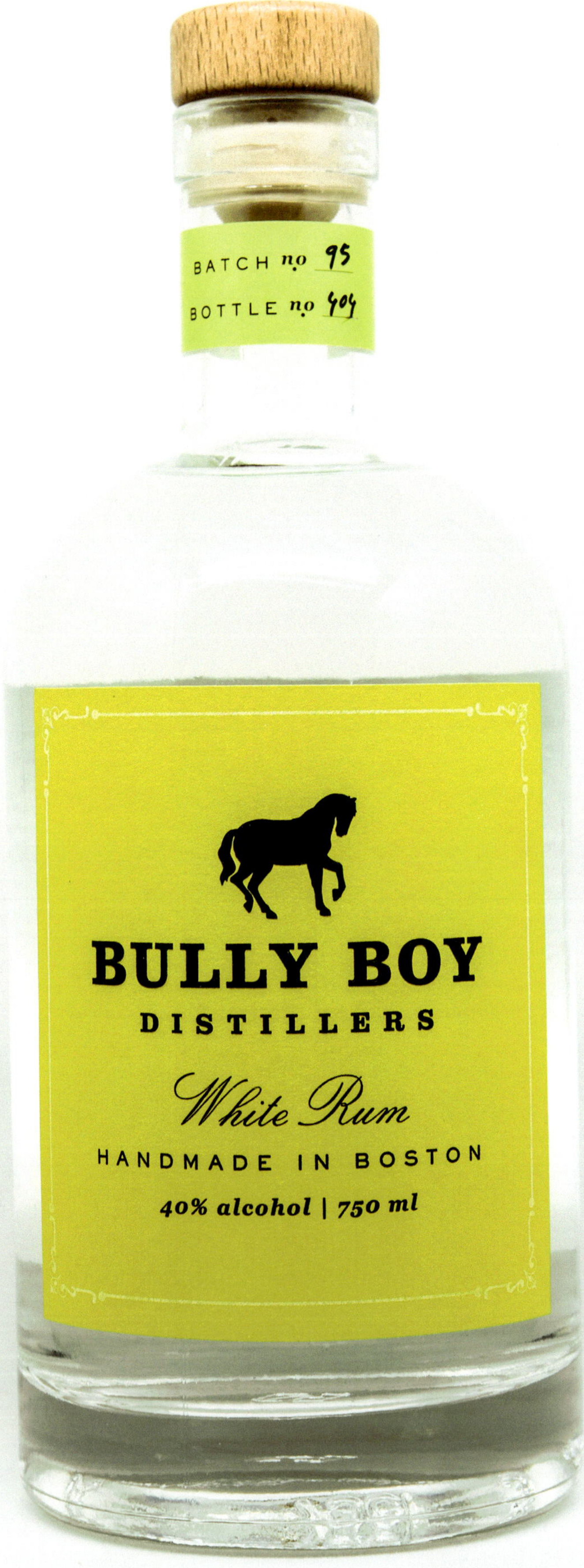

BULLY BOY WHITE RUM / $25

During Colonial times, New England was home to over 150 distilleries, most of which were making rum. This spirit draws from that heritage, born in small batches using blackstrap molasses, with samples blended at different times — like cognac.

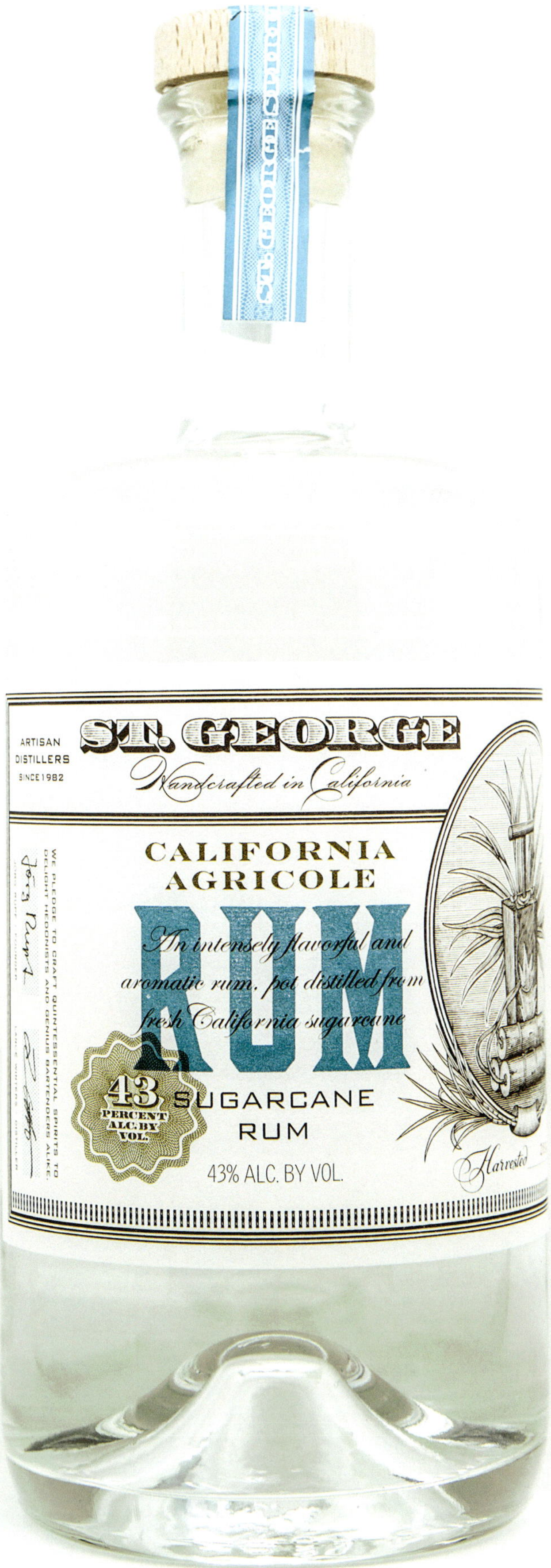

ST. GEORGE CALIFORNIA AGRICOLE RUM / $50

Rarely does a spirit reflect its origins so boldly. This rum is made by hand-pressing sugarcane from the Imperial Valley then fermenting and distilling the juice in small batches, lending it an earthy quality that's enjoyable by itself or in cocktail form.

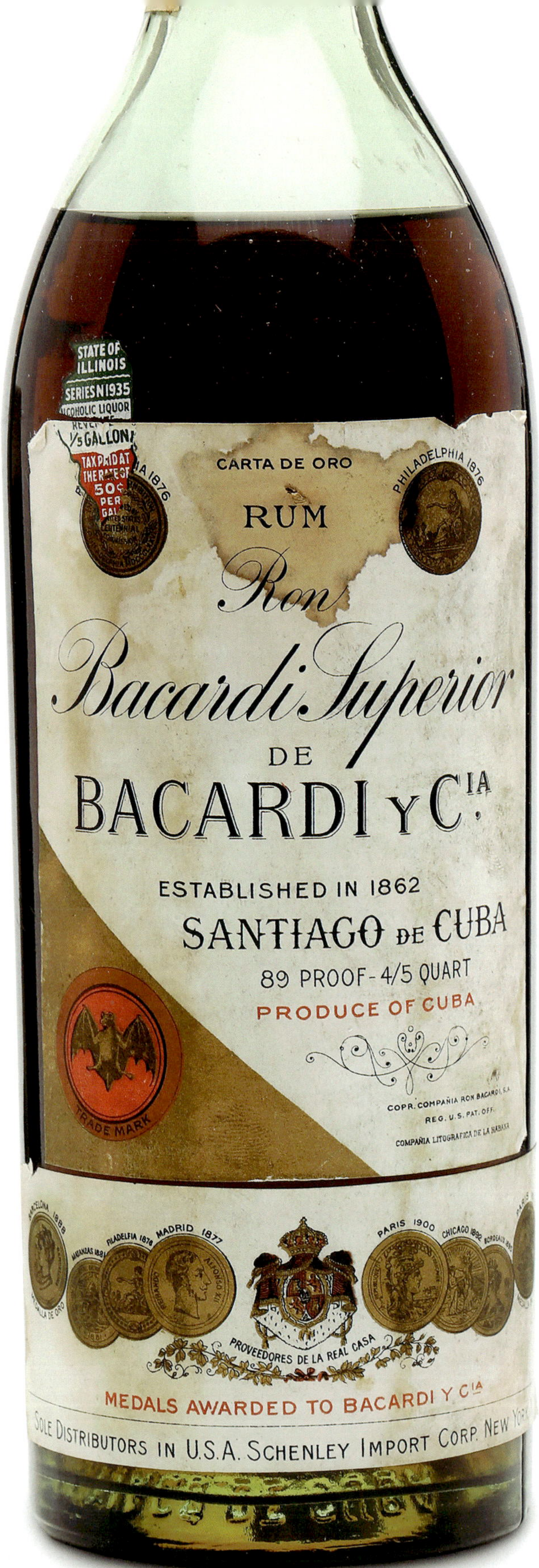

BACARDI SUPERIOR CARTA DE ORO RUM / $3,000

When Fidel Castro gained power, he decided he wanted the Bacardi distillery for himself — leading the family to flee to the U.S. and move production to Puerto Rico and Mexico. Thus, this one liter bottle from the 1930s is one of the few remaining to have been produced in Cuba, giving the gold rum a story worthy of its complex flavor.

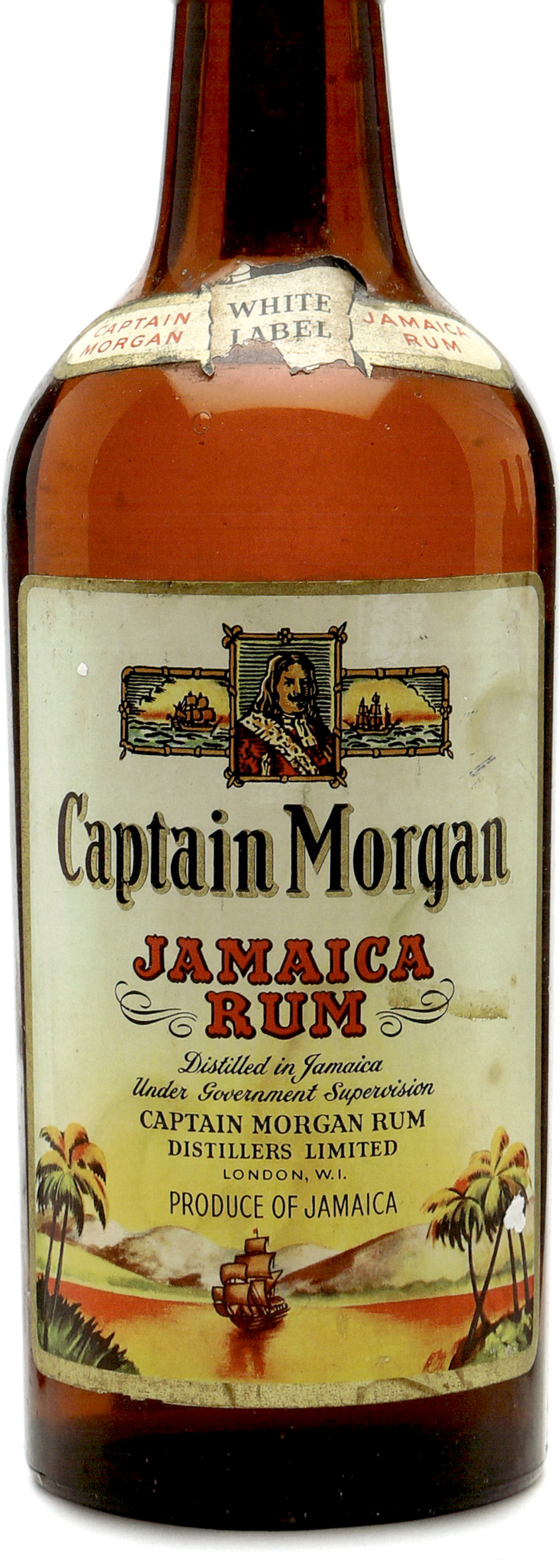

1970'S CAPTAIN MORGAN RUM / $325

Long before he struck his signature pose, the Captain was confined to a more reasonable portion of the label. The rest was devoted to the drink's origin — this bottle was distilled in Jamaica, unlike the current mix of rums from multiple sources — and the spirit of Sir Henry Morgan, who died nearly 300 years before this bottling.

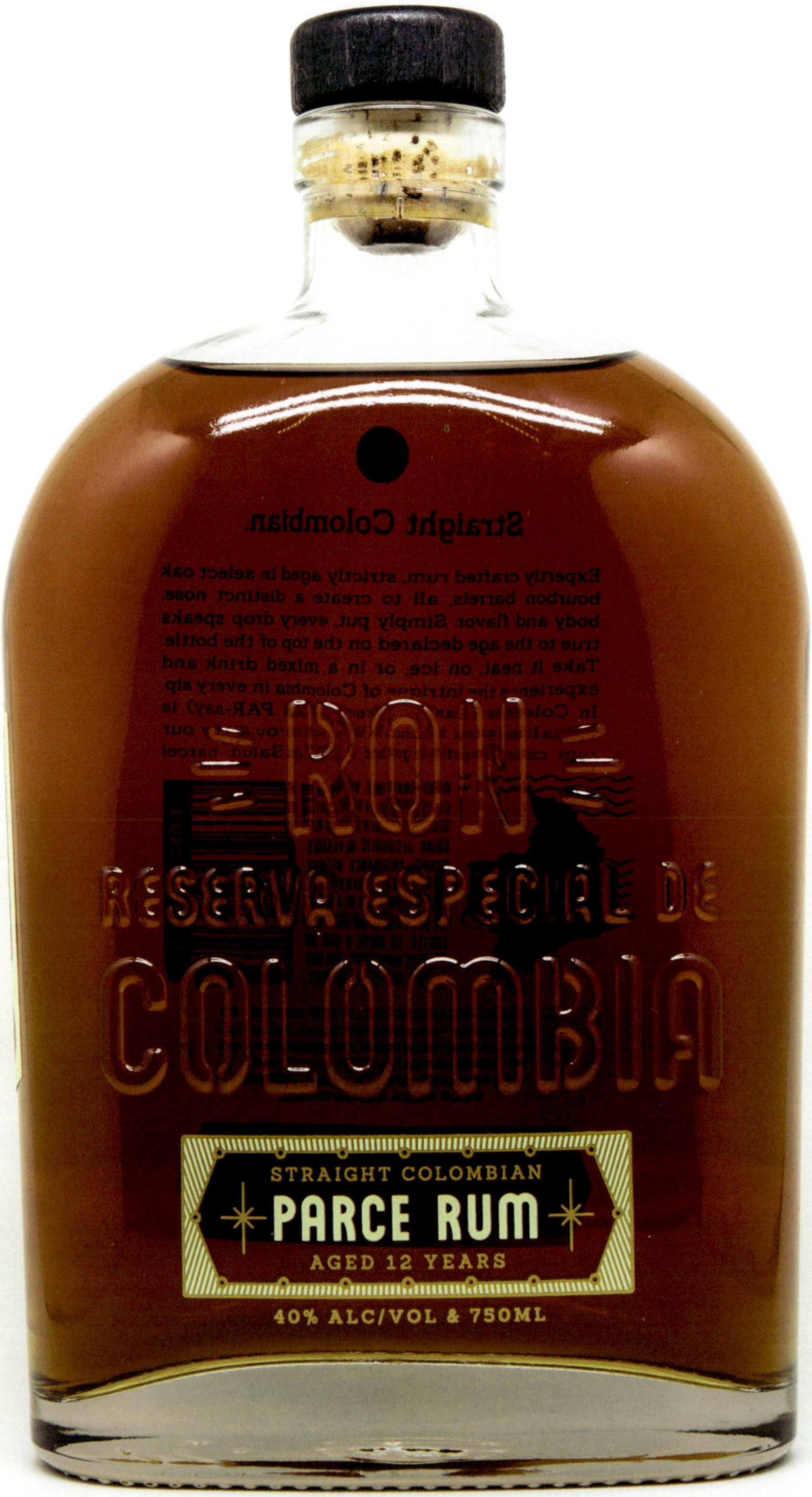

PARCE 12 YEAR COLOMBIAN RUM / $60

The bottle looks great on a shelf. Yet this all-natural elixir could come in a can and be just as good. Named after the Colombian word for "good friend", this aged rum is ideal for sharing at social gatherings, with a semi-sweet flavor, long finish, and enticing aroma that make it all too easy to pour another glass.

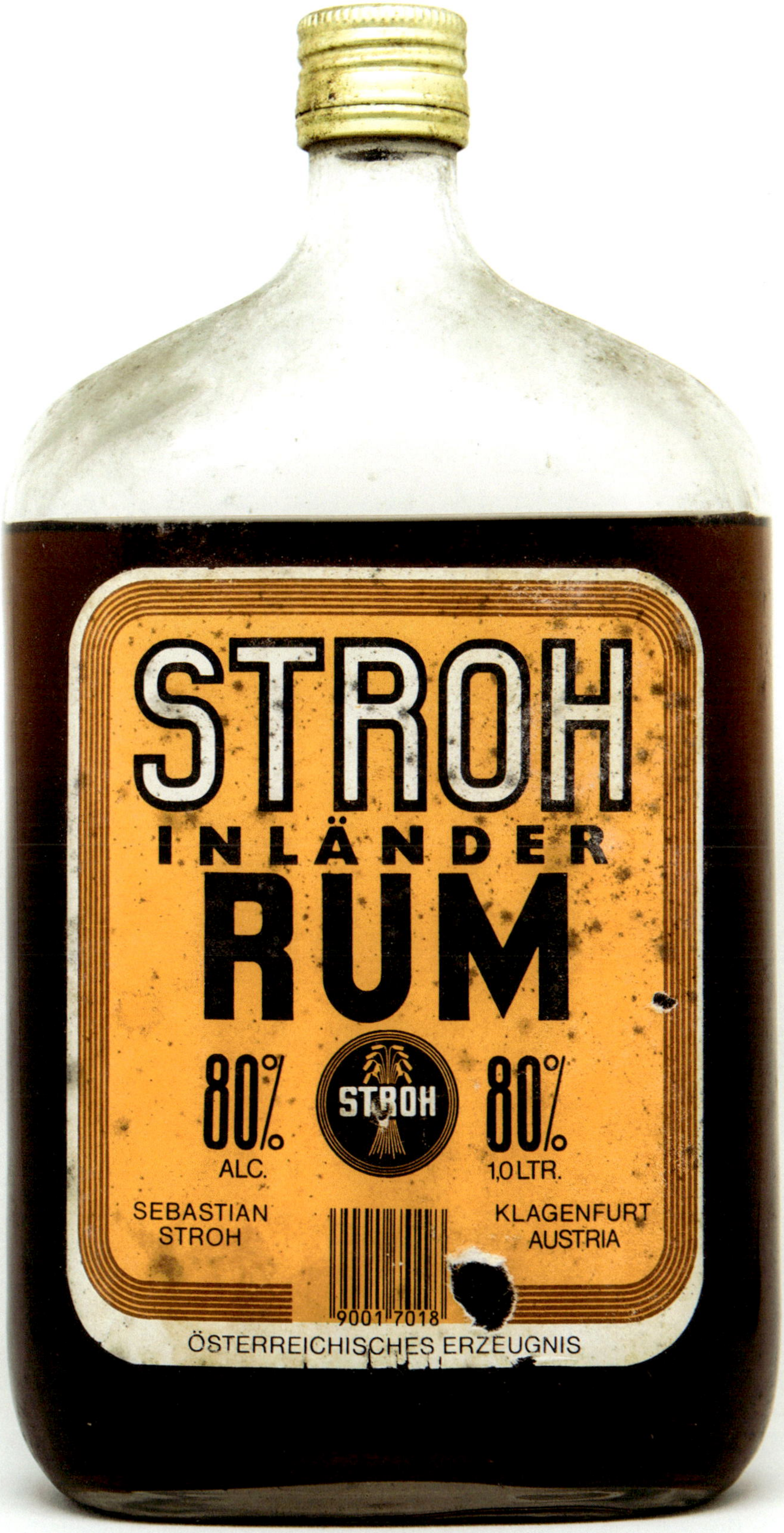

1980'S STROH INLANDER RUM / $160

Austria isn't a place you'd expect to produce rum. Yet Stroh has been making their "inlander" rum since the 1800s. With no access to the tropics, they instead add flavors and dyes to an ethanol base. Rarely enjoyed straight because of the high alcohol content, this 80% blend is instead used for punches and drinks.

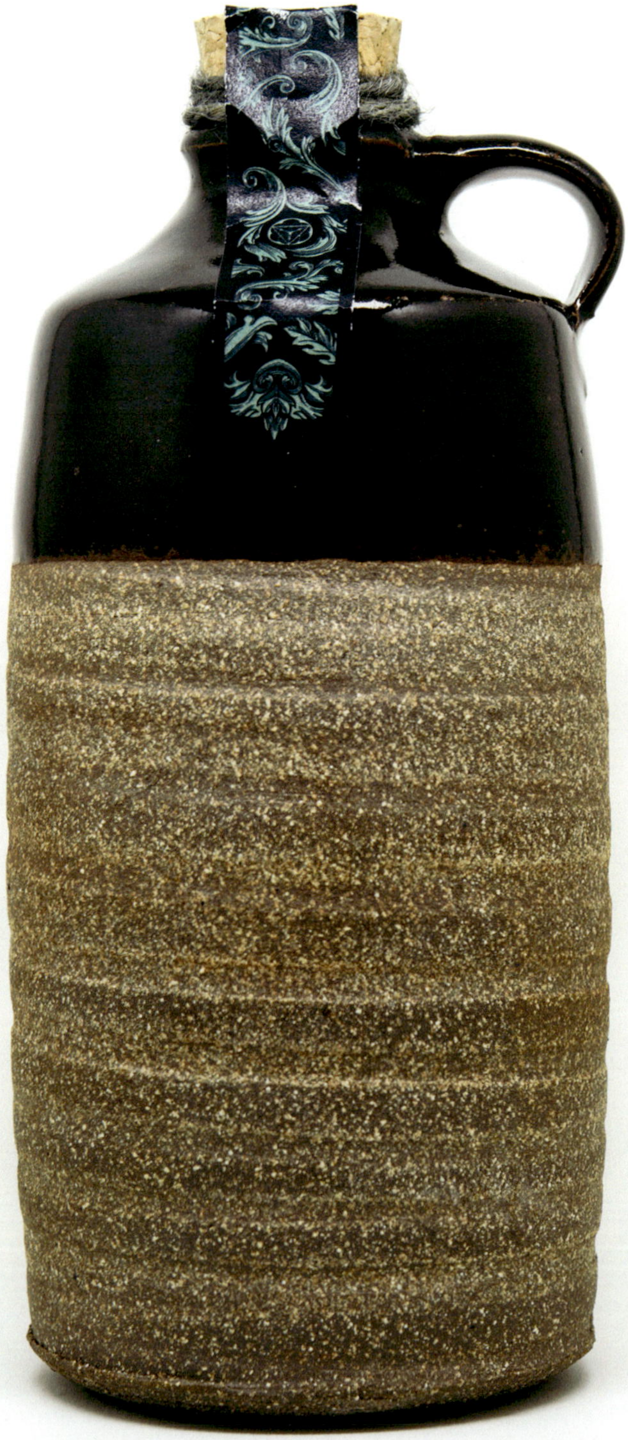

NEW DAWN 18 RUM SOLERA CASK STRENGTH FLAGON / $135

Crafted in the Dominican Republic using the solera process, this rum ages 18 years before it begins its journey across the sea. It's moved from the DR to England on a engineless, sailing cargo ship, then transferred to a horse and cart before arriving at a brewery to be emptied into its distinctive handmade ceramic flagon.

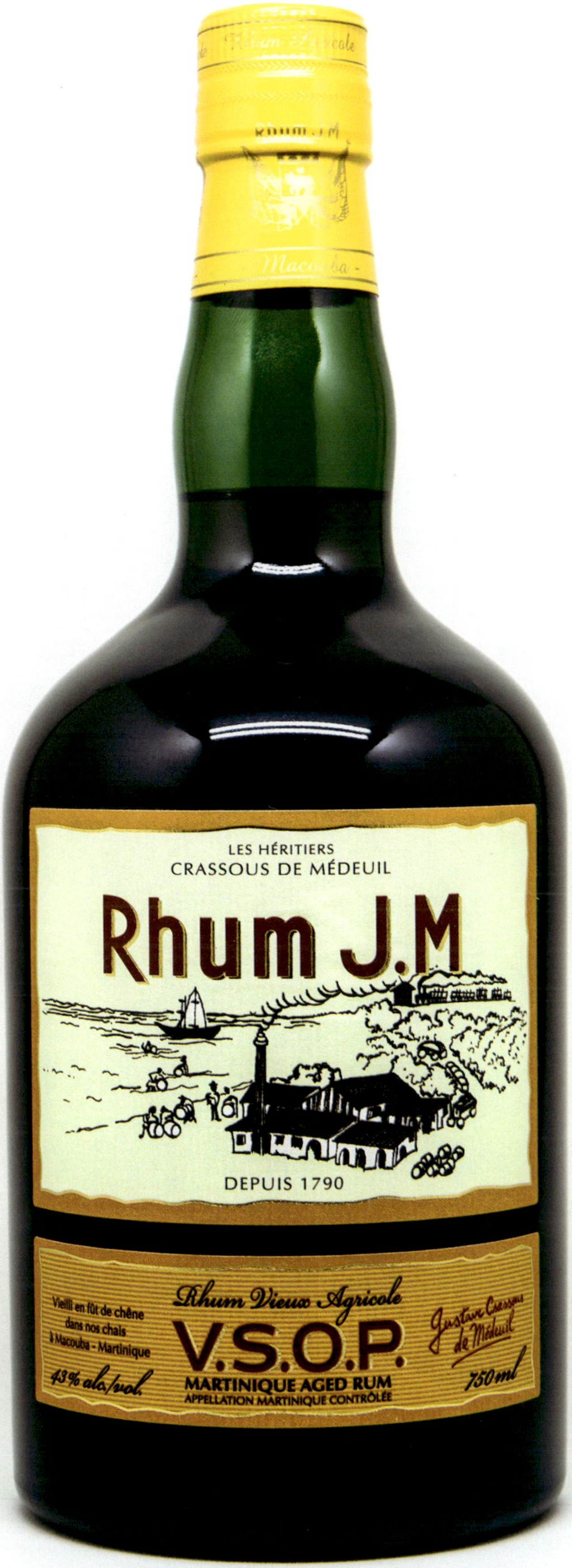

RHUM J.M VIEUX AGRICOLE VSOP RUM / $50

Unique is an understatement. This Martinique-produced rum is made from sugarcane that's grown on the slope of a semi-active volcano — imparting a rich flavor — and cut less than an hour before it's turned into juice and immediately fermented. Distillation and three years of aging come after, resulting in a highly drinkable spirit.

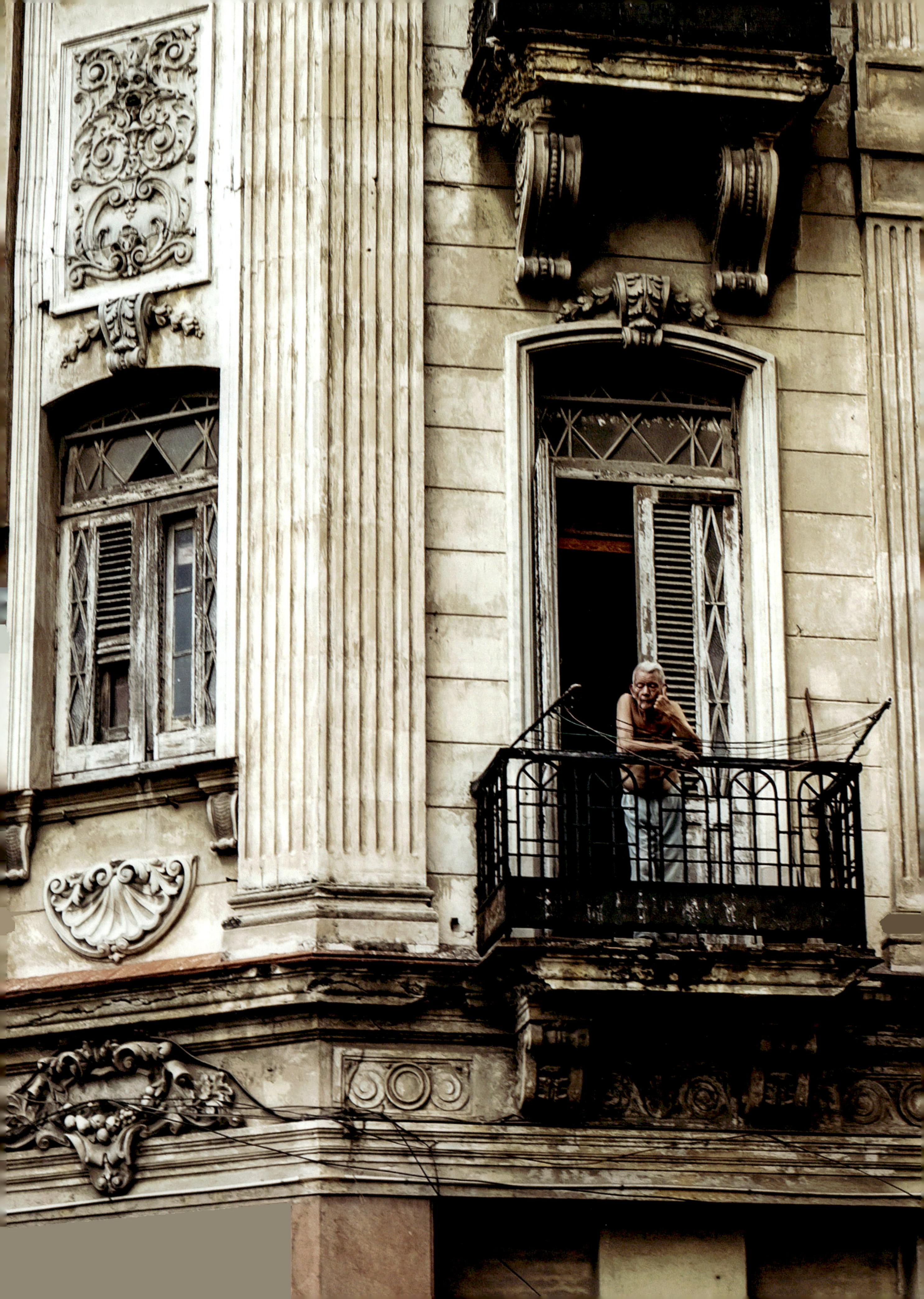

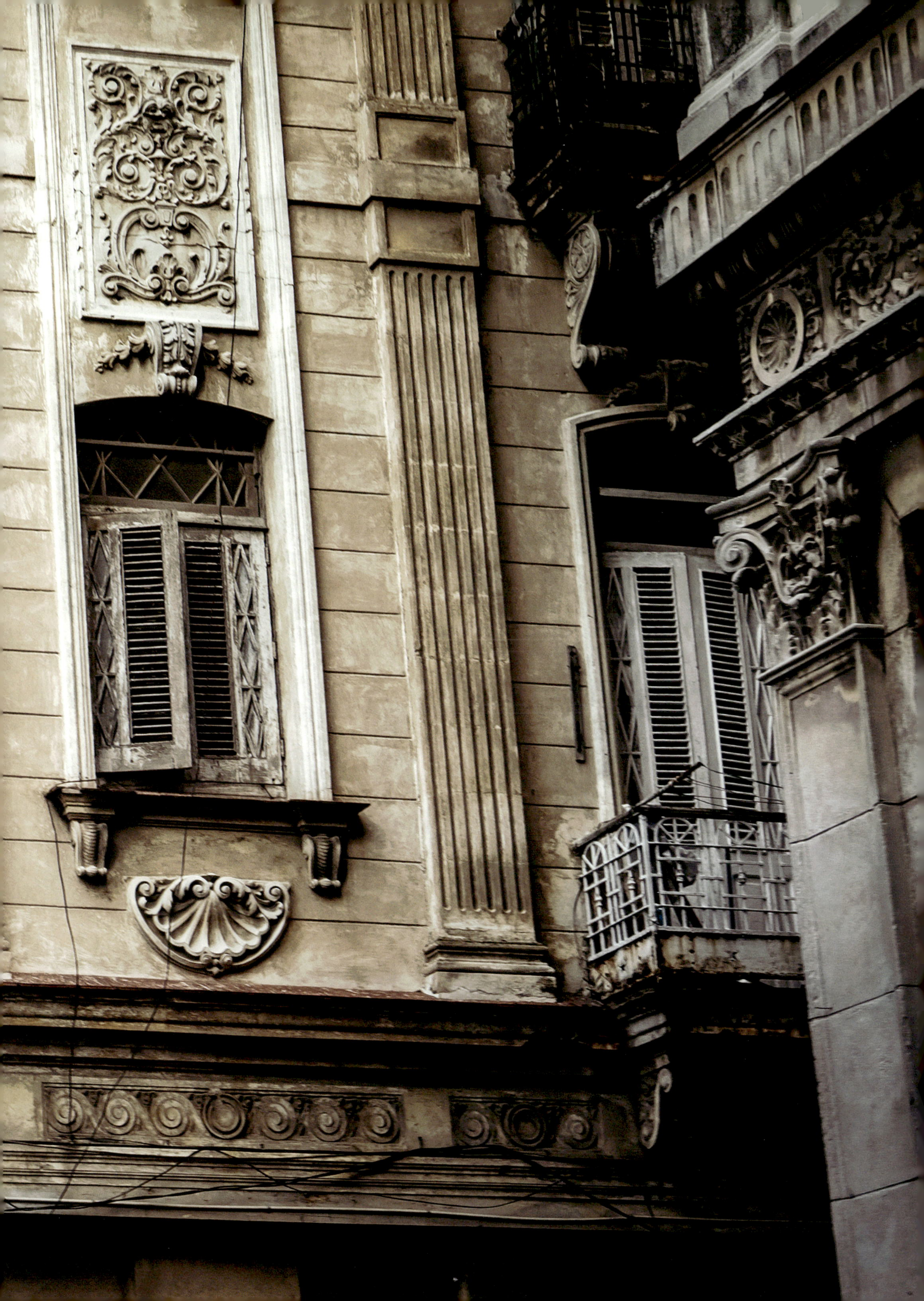

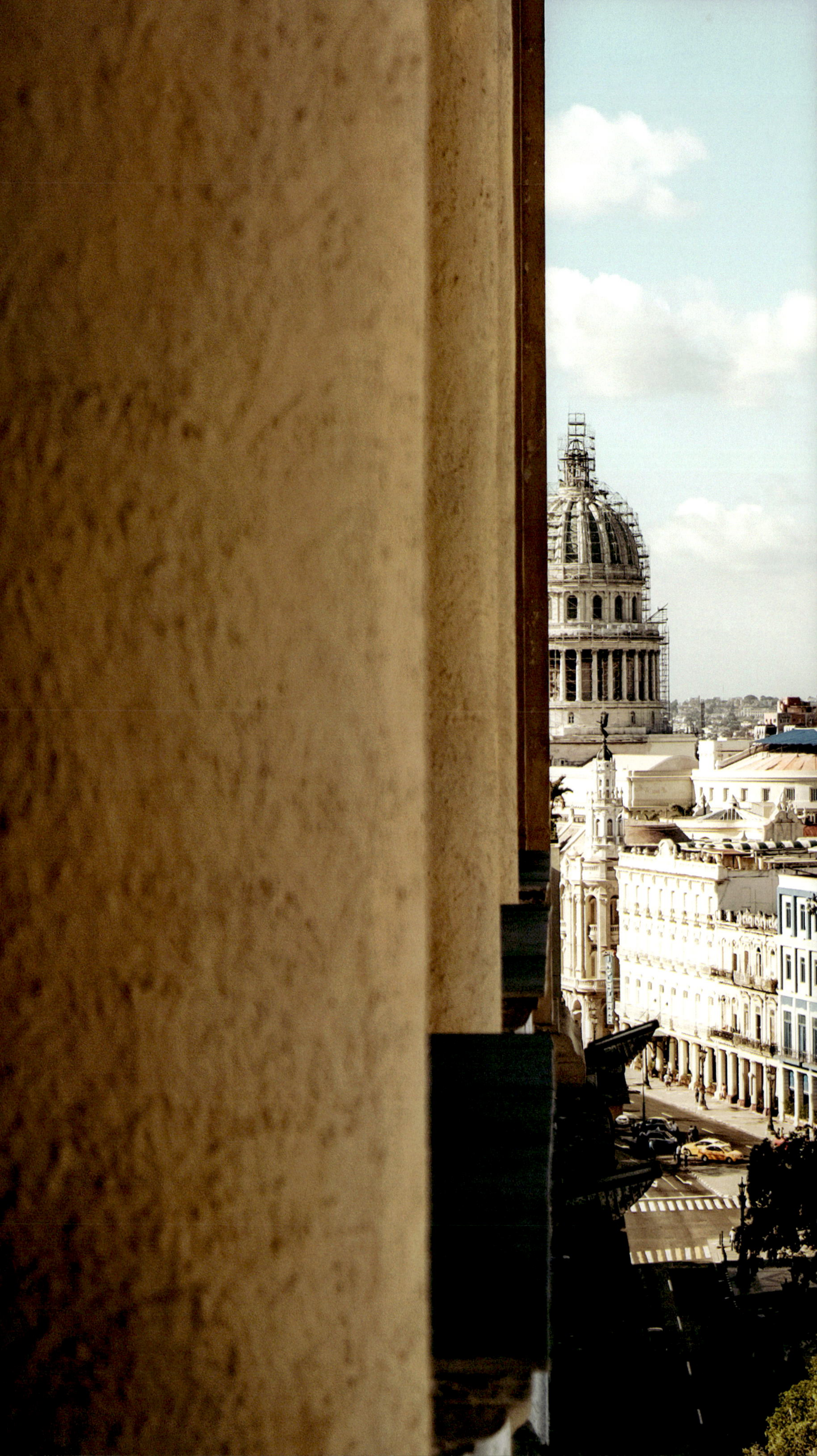

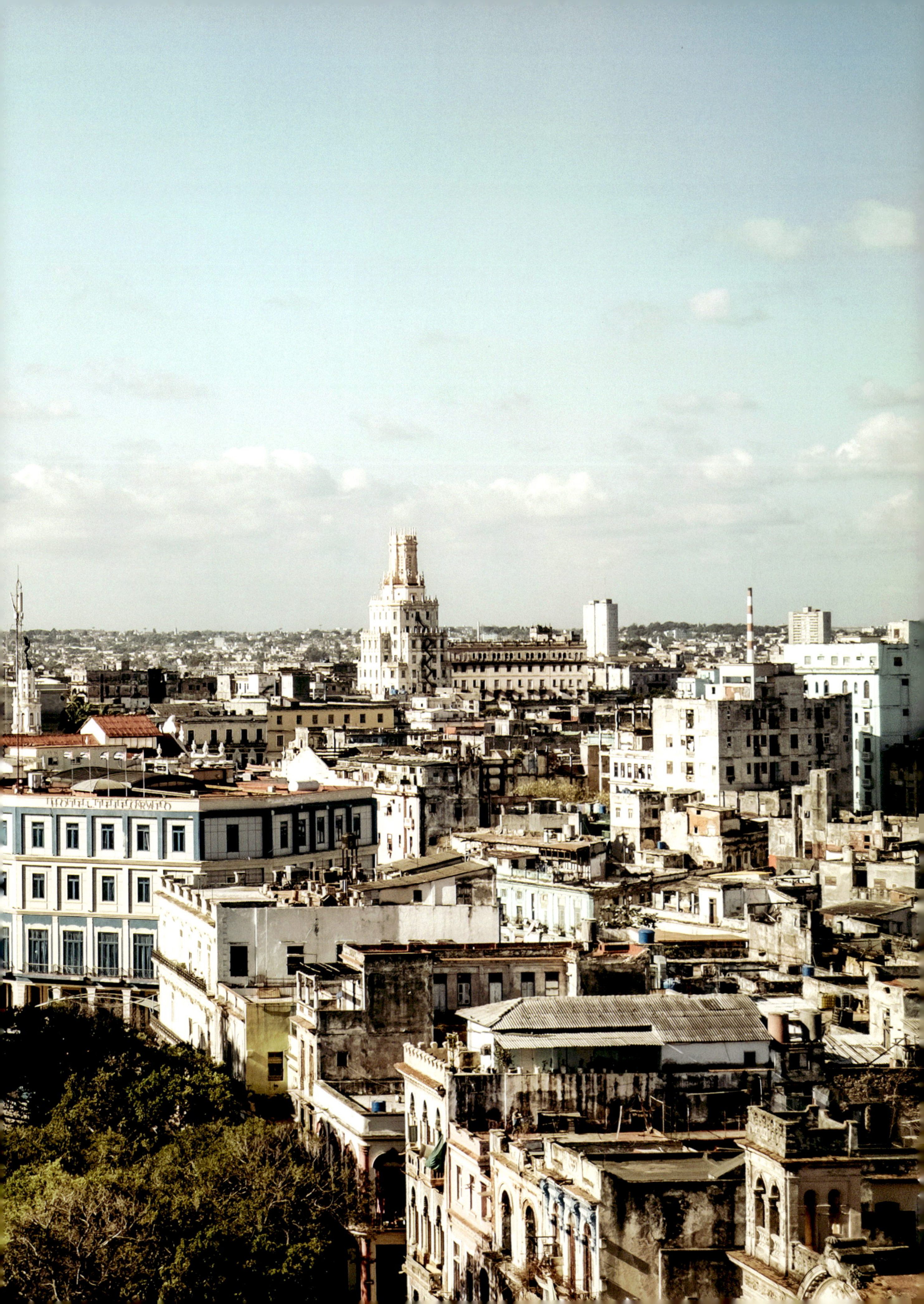

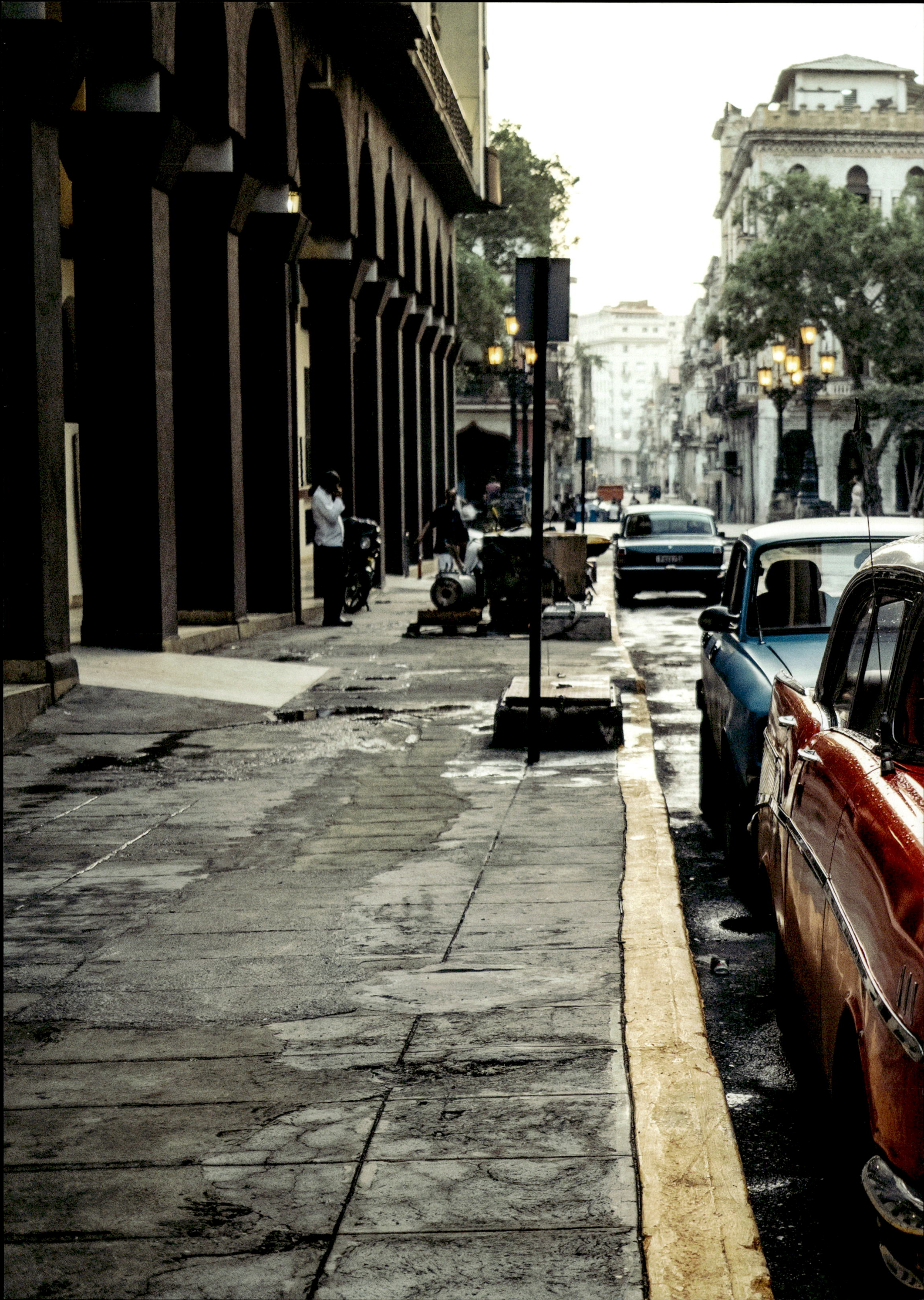

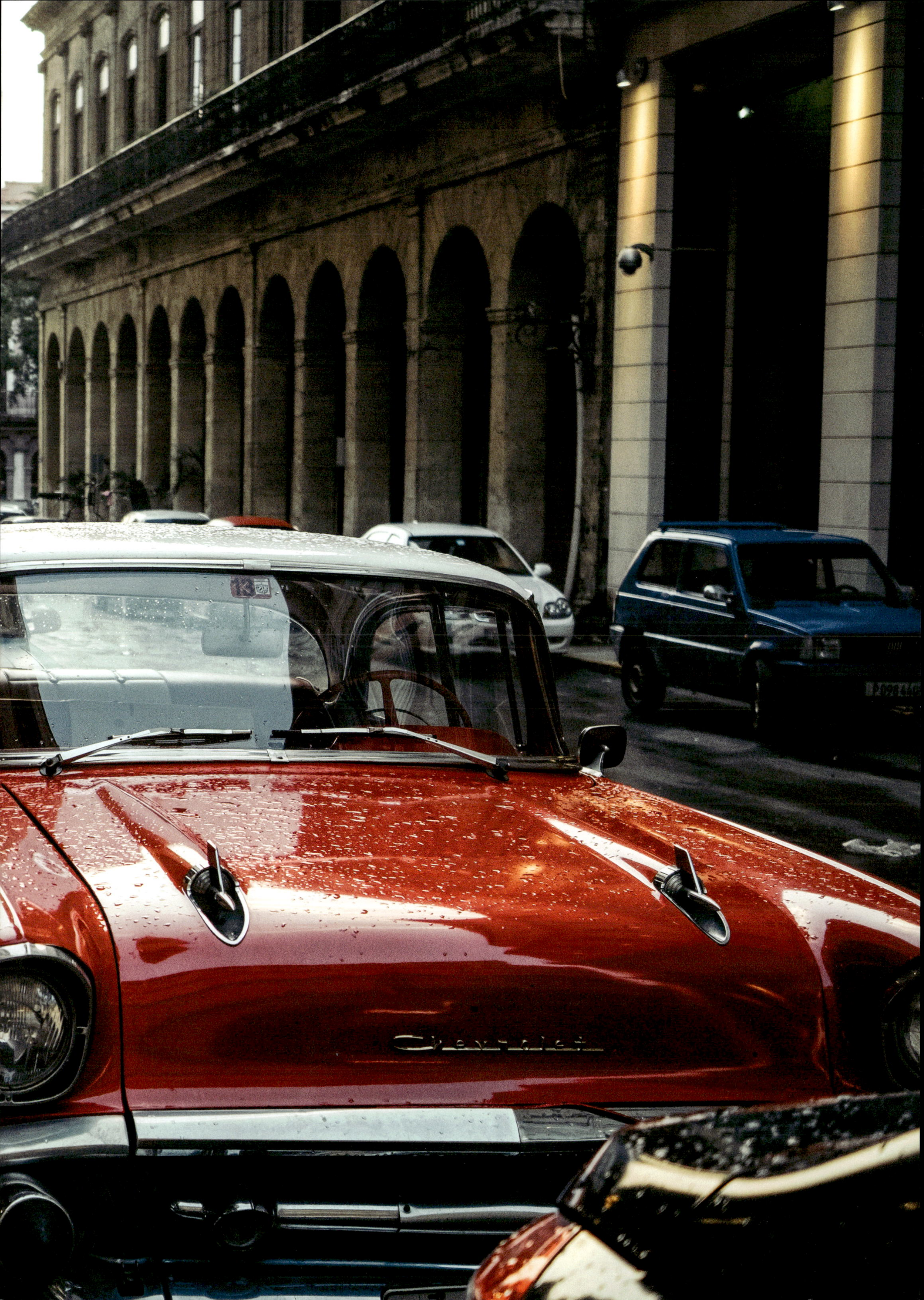
Chevrolet

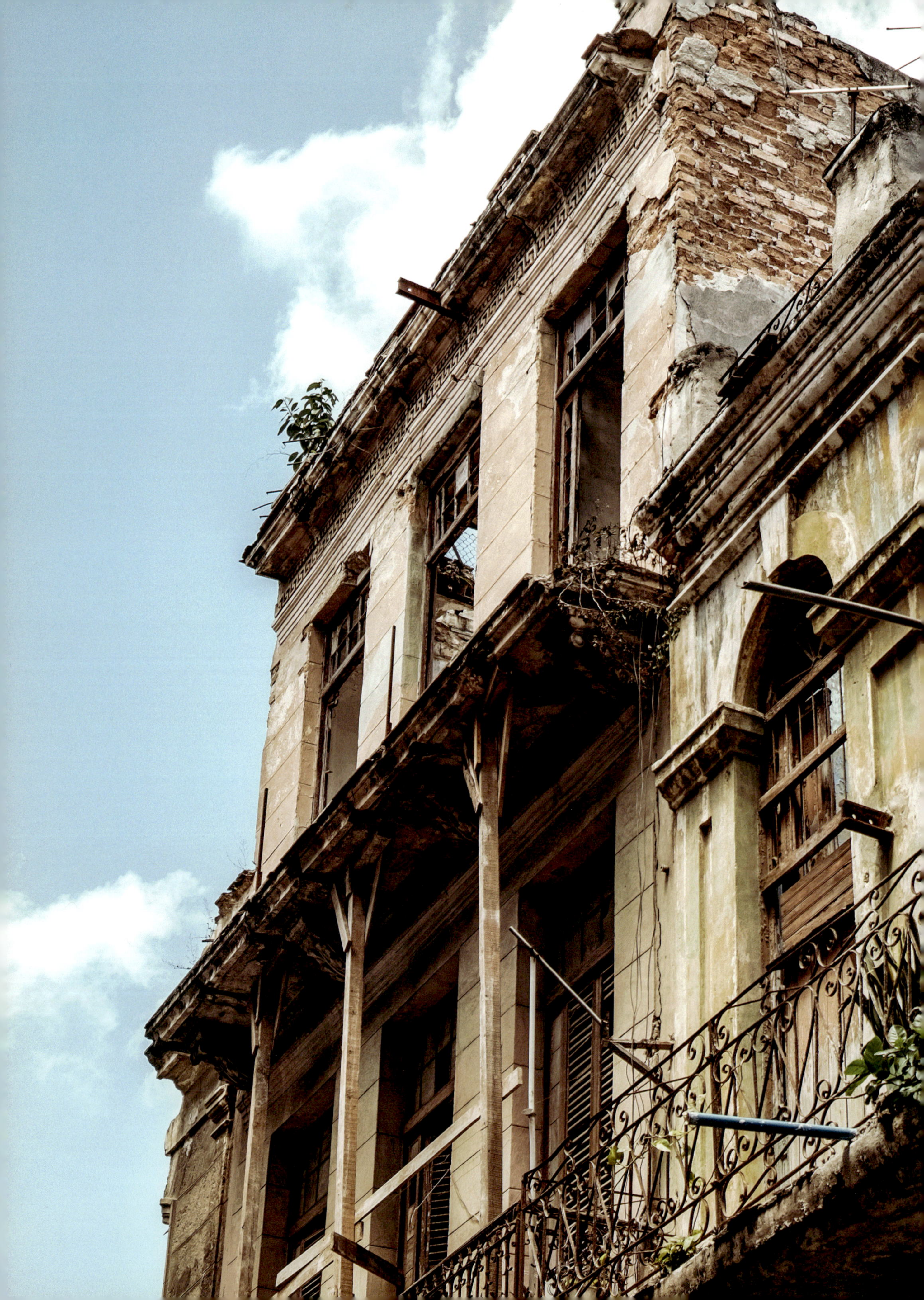

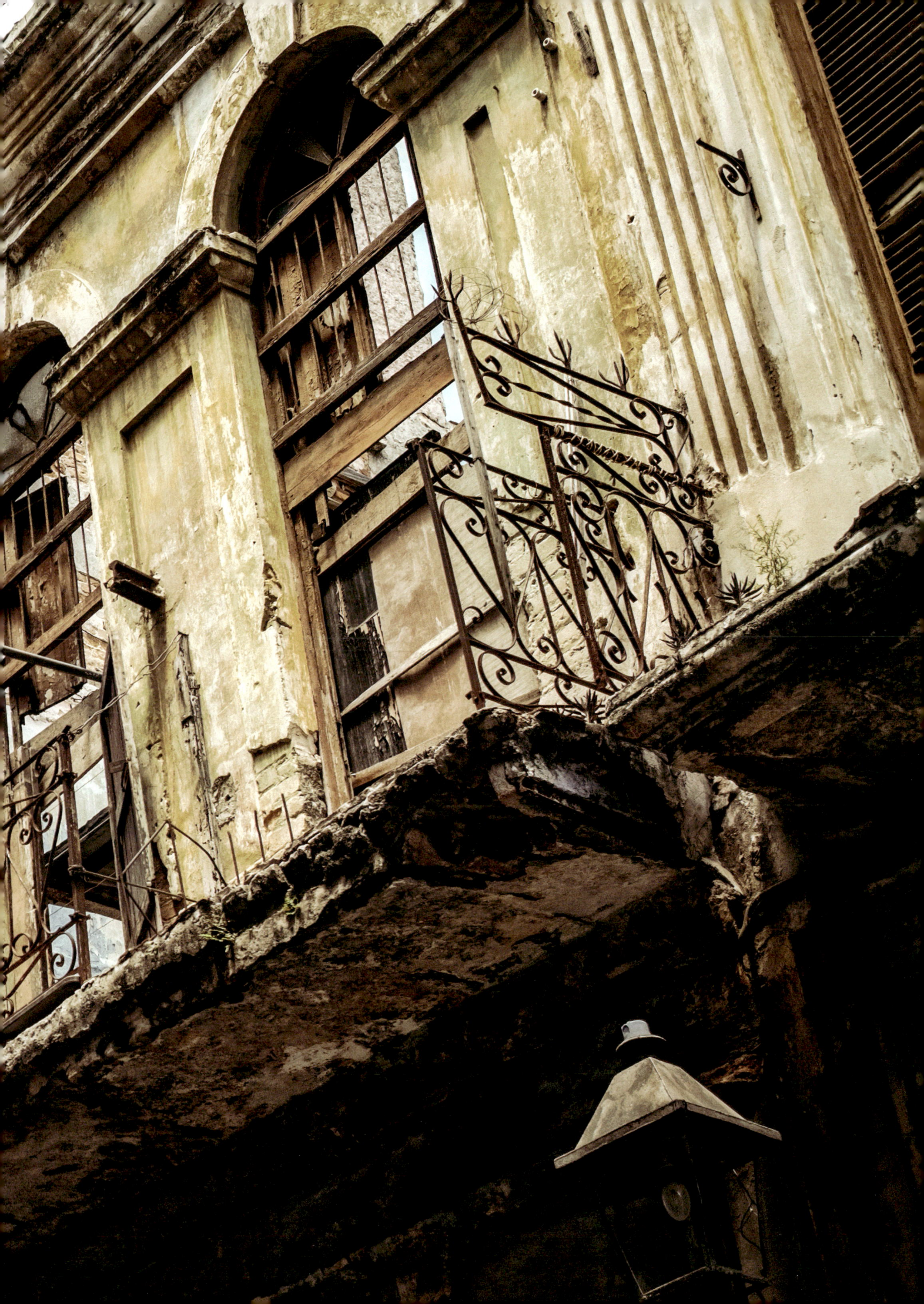

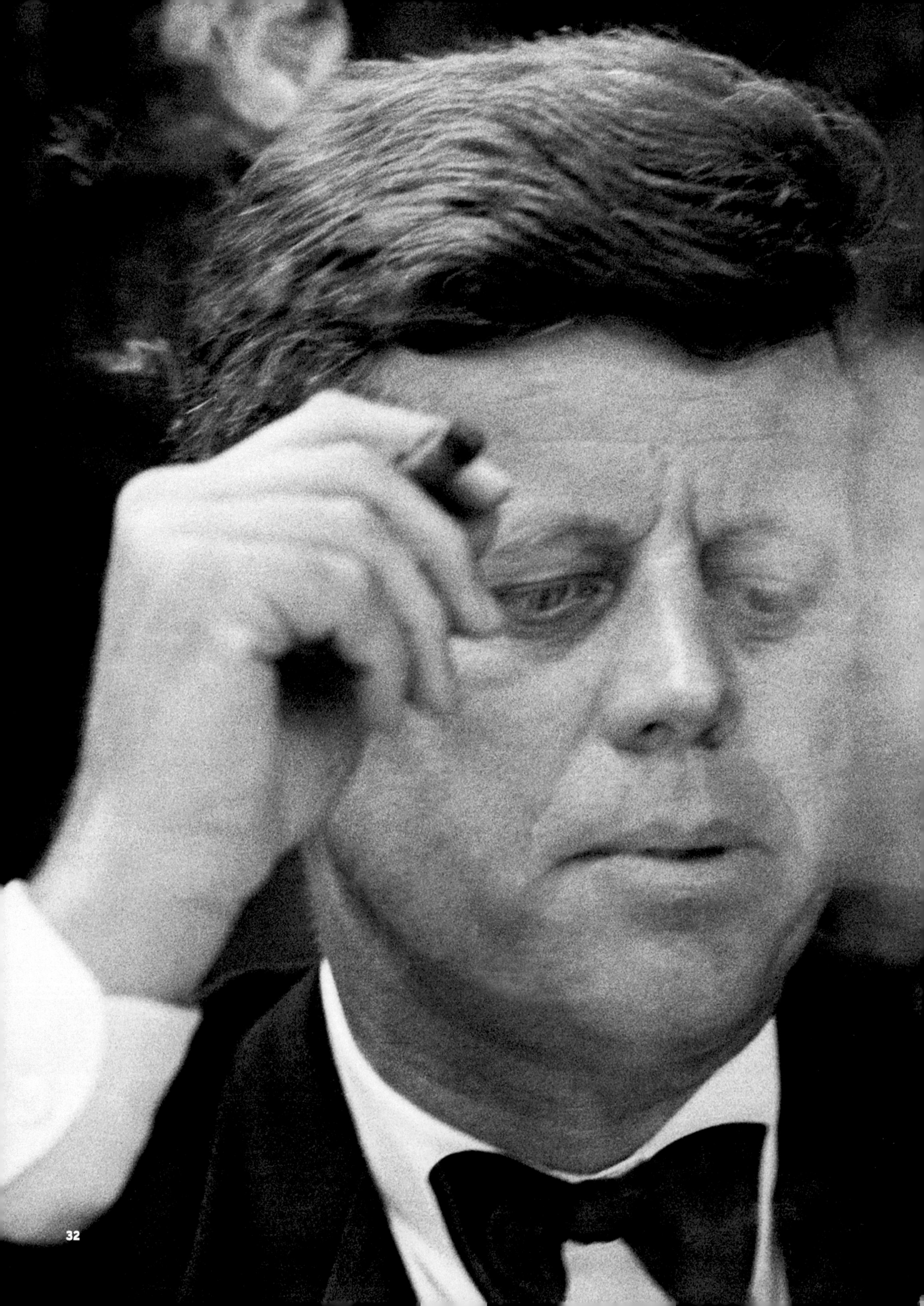

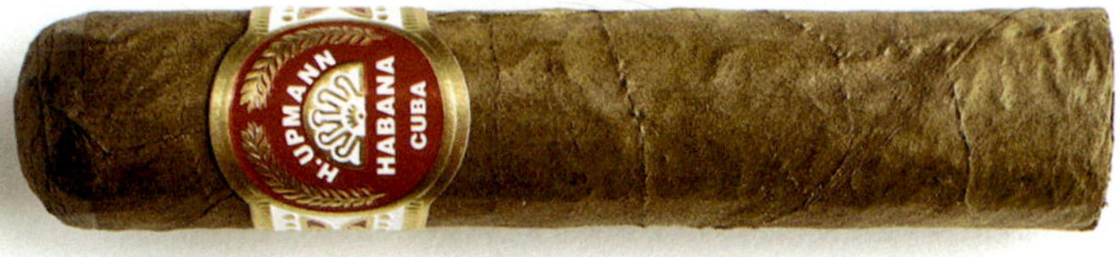

H. UPMANN HALF CORONA / $4

Measuring only 3.5 inches long and 44 on the ring gauge, it isn't big. Its flavor is. Ideal for times when you don't have an hour or more to spend but still want a solid smoke, it's reminiscent of JFK's preferred cigar.

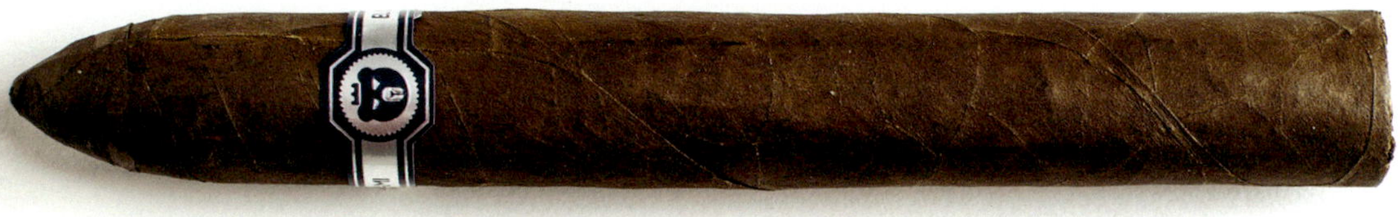

WARPED CIGARS EL OSO PAPA / $11

Each one of these premium belicosos — a shape defined by its conical end that helps direct both aroma and flavor — is rolled by a single Cuban-trained category 9 roller in a family-owned factory in Little Havana, Miami.

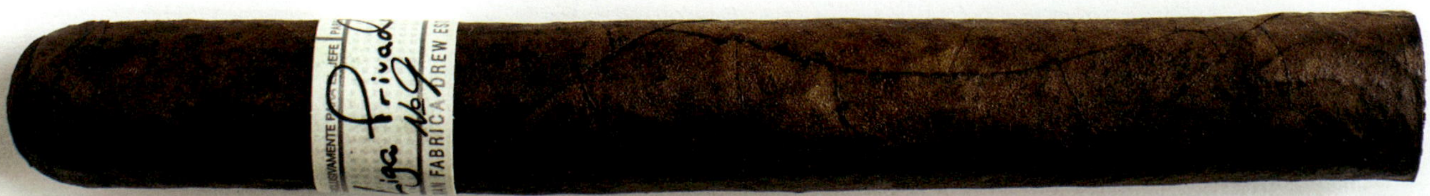

DREW ESTATE LIGA PRIVADA NO. 9 CORONA DOBLE / $16

A bold mix of three Cuban-seed tobaccos, this cigar was originally crafted for the company's own humidors. The blend gives it a robust flavor, while its Connecticut Broadleaf Oscuro wrapper gives it a dark color and slight sweetness.

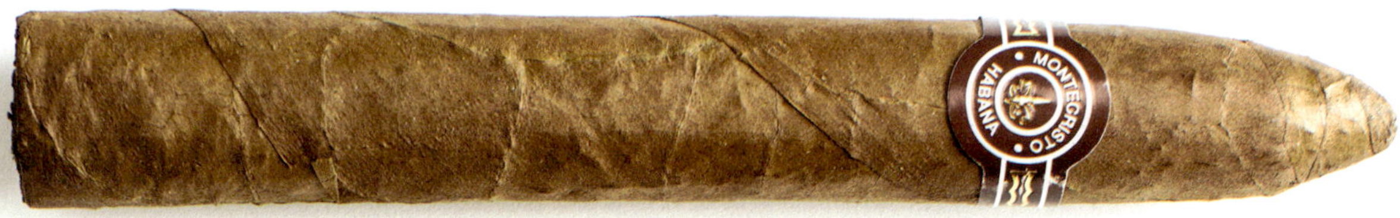

MONTECRISTO NO. 2 / $11

Named Cigar of the Year in 2013, this is arguably the world's most famous torpedo, with a spicy, approachable flavor that pleases both serious aficionados and newcomers equally. Easily found in shops around Havana.

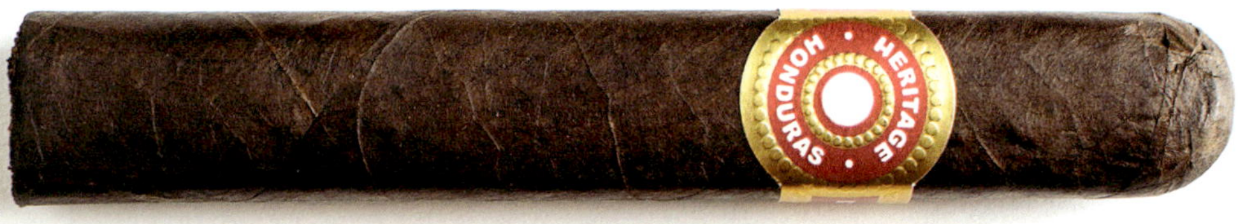

DUNHILL HERITAGE BOX PRESSED ROBUSTO / $10

With a lineage dating back to Alfred Dunhill's 1900s tobacco shop in London, this full-bodied cigar pays homage to the company's earliest Cuban-made offerings with a mix of tobaccos from Central America.

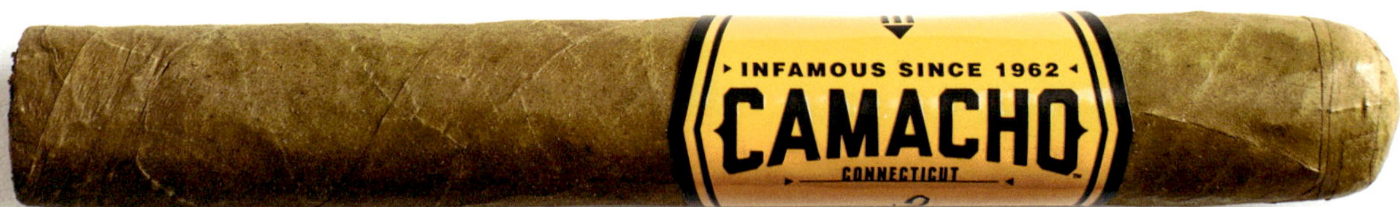

CAMACHO CONNECTICUT TORO / $8

Filled with a blend of Honduran and Dominican tobaccos, this intense, bold, medium-sized smoke gets its name from its high-grade wrapper, which is rolled in Connecticut but grown in Ecuador.

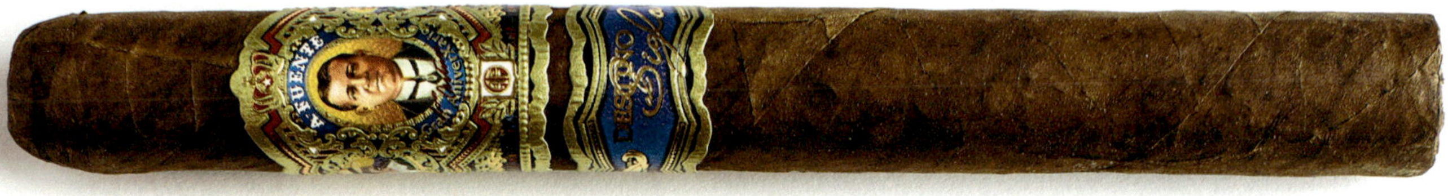

DON ARTURO GRAN ANIVERXARIO 2013 CHURCHILL / $70

Created by Carlito Fuente to honor his grandfather Arturo, the filler of this massive smoke is made up of tobaccos from the company's other labels, while the wrapper comes from the family's Chateau de la Fuente farm in the Dominican Republic.

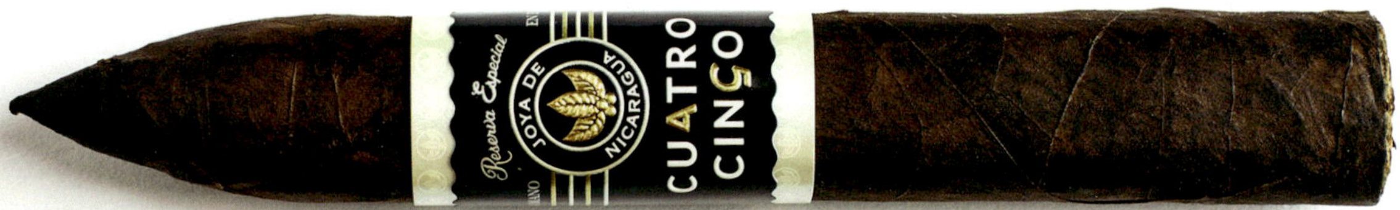

JOYA DE NICARAGUA CUATRO CINCO RESERVA ESPECIAL TORPEDO / $13

Created to celebrate the company's 45th anniversary — hence the name — the dark wrapper of this smoke is aged in vintage white oak barrels, making it well suited to the dark cocoa-based, medium-to-full flavor profile.

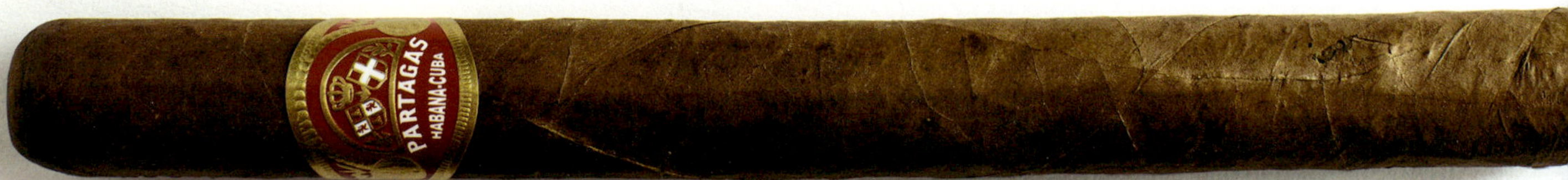

PARTAGÁS LUSITANIA / $13

At over 7.5 inches long, the Partagás Lusitania takes at least two hours to smoke. Good thing it's fantastic. Widely considered among the best in the world, its complex flavors move from mild to full the longer you smoke.

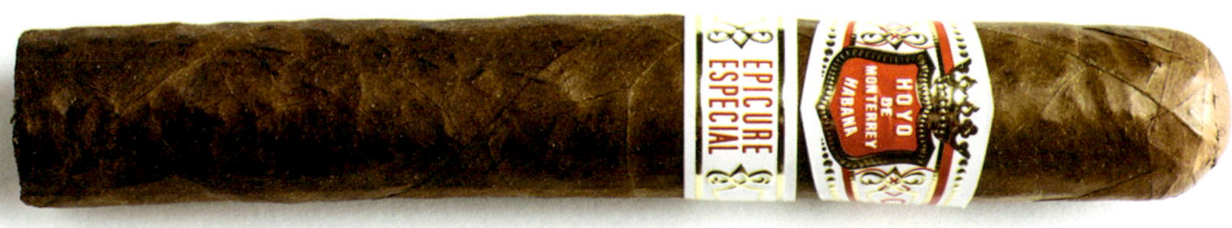

HOYO DE MONTERREY EPICURE ESPECIAL / $10

This lengthier version of the company's beloved Epicure No. 2 is best when packed in a tube, a protection only afforded when the rich yet approachable smoke is purchased in a three-pack. Also available as singles.

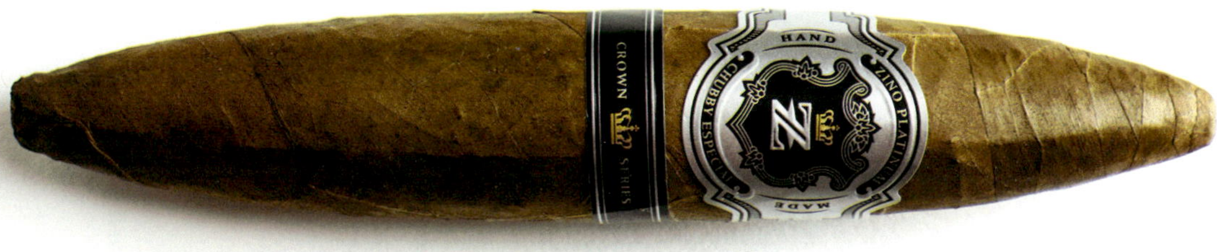

ZINO PLATINUM CROWN SERIES CHUBBY ESPECIAL TUBO / $40

Named after Davidoff's founder, both the Ecuadorian wrapper and Connecticut binder of this football-shaped cigar are aged for two years, then filled with aged, long-leaved tobaccos from Nicaragua and Peru.

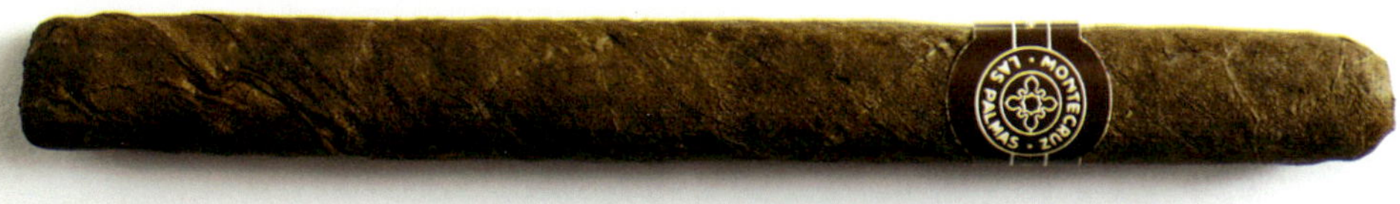

DUNHILL MONTECRUZ C.1970 JUNIORS / $40

Decades old, these were among the first new cigar brands created following the Cuban revolution. They were made in the Canary Islands, and remain as smooth and mild as they were in the days of bell bottoms.

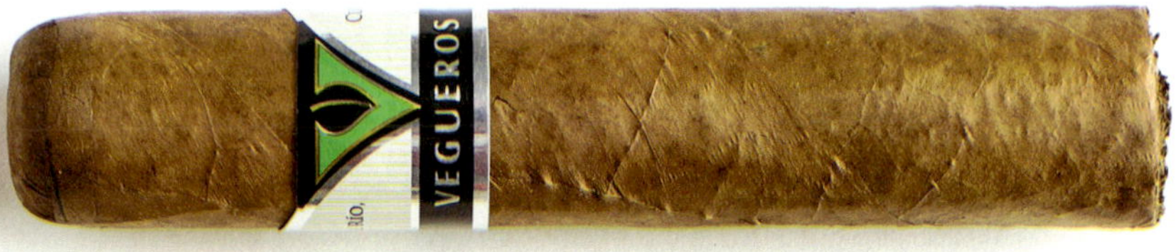

VEGUEROS ENTRETIEMPO / $5

Lesser known than its contemporaries, the Vegueros Entretiempo is a full-flavored petit robusto that provides an enjoyable yet unpretentious smoke, fitting for a brand created to pay homage to the farmers of Pinar del Río.

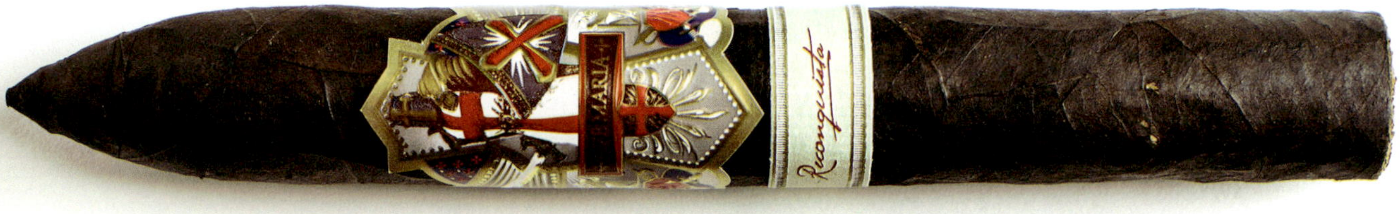

AVE MARIA RECONQUISTA / $20

Arriving in an individual coffin box that gently lifts the cigar for display as you open it, its Cuban-seed Nicaraguan tobaccos are wrapped inside a Brazilian Habano Oscuro wrapper so rare, it's the only cigar in the world to use it.

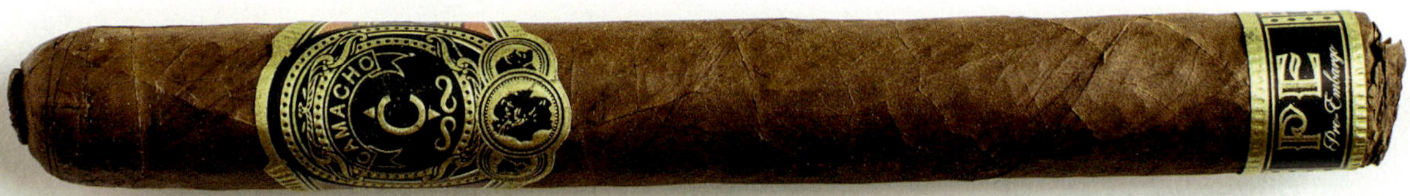

CAMACHO PRE-EMBARGO TORO / $25

It's not a true Cubano. It might be better. Filled with aged Cuban tobacco sourced from one of the four remaining bales of pre-embargo in the world, it's finished with a Corojo wrapper from 1999, one of the first ever grown outside the island nation.

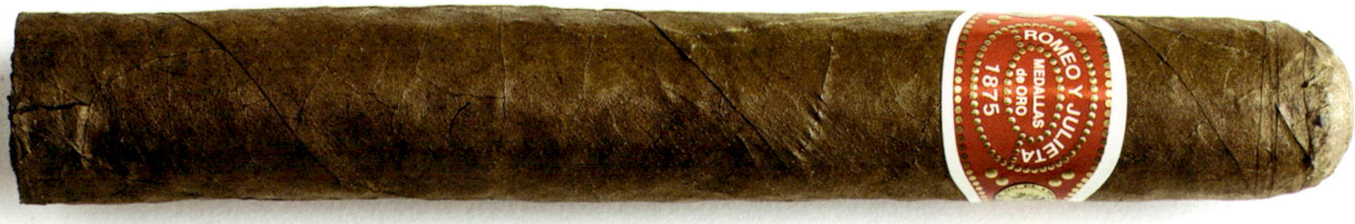

ROMEO Y JULIETA GG CONMERATIVO LIMITED RESERVE C.2002 / $35

These cigars were excellent when they were released. They're even better after 14 years of aging. The wrapper is from Cameroon, and they're filled with a blend of Dominican, Nicaraguan, and Peruvian long leaves.

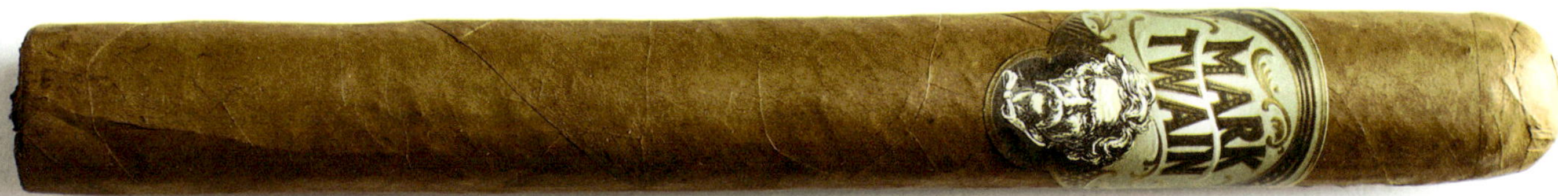

MARK TWAIN NO. 1 / $4

Don't let the inexpensive price fool you. These surprisingly enjoyable cigars are easy on the wallet, yet offer a smooth, robust smoke with a Nicaraguan binder, filler, and a Connecticut Shade wrapper commonly found on more expensive options.

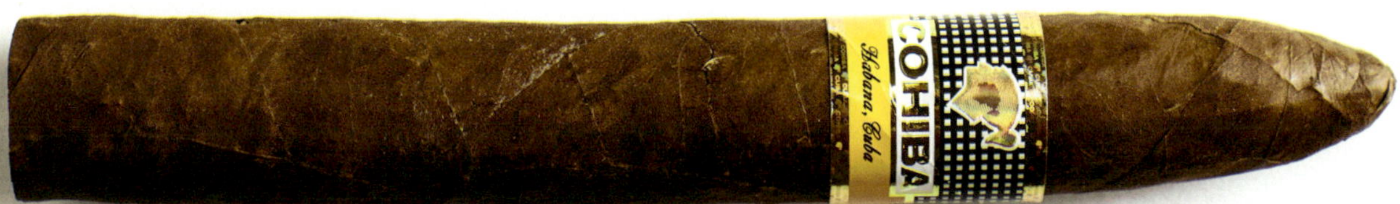

COHIBA PIRAMIDES EXTRA TUBOS / $26

The only production pyramid from the legendary Cuban brand — it was originally founded as a private brand to supply Castro — this cigar offers a unique size and a premium experience. Hand-rolled from the finest leaves Vuelta Abajo has to offer.

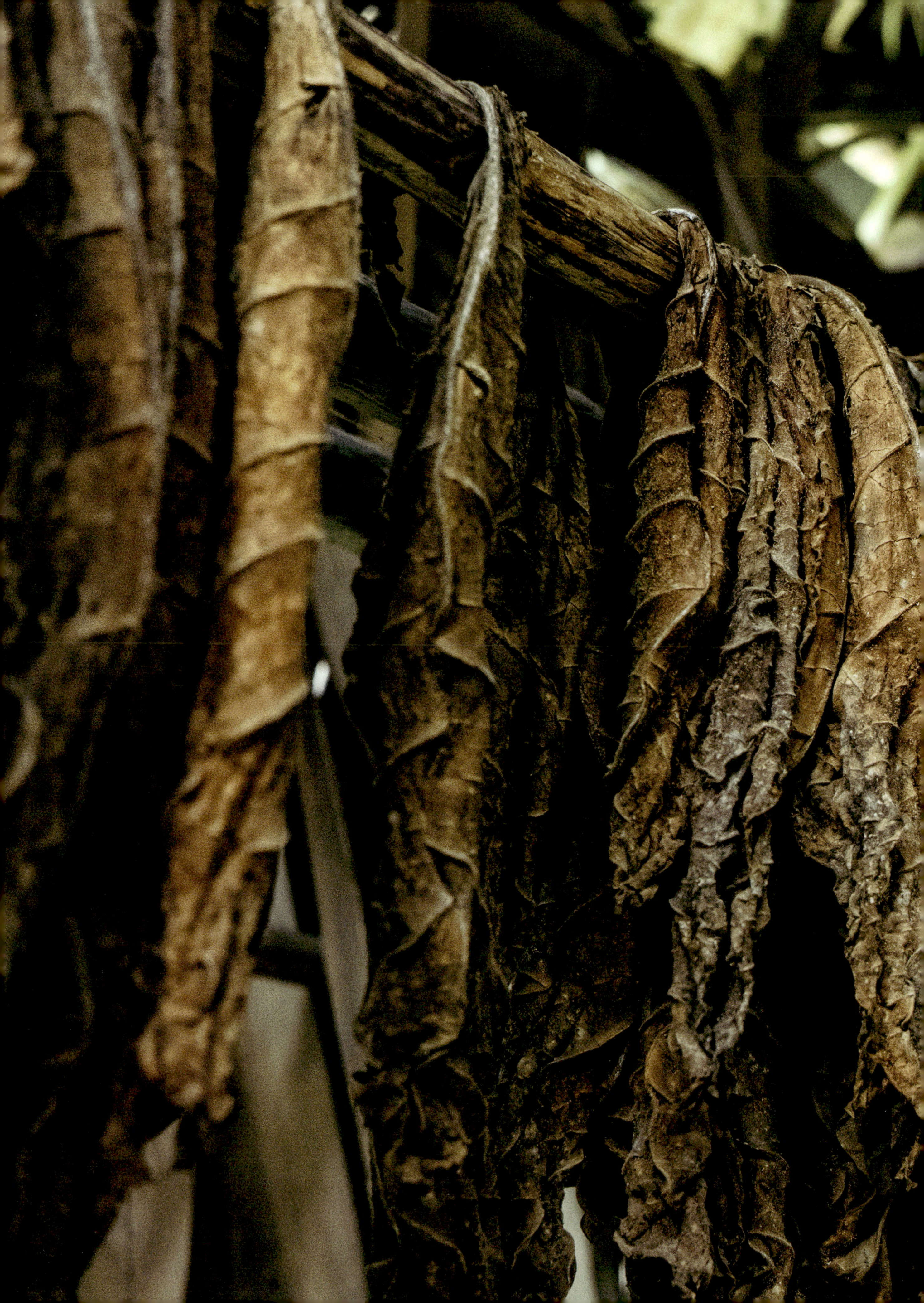

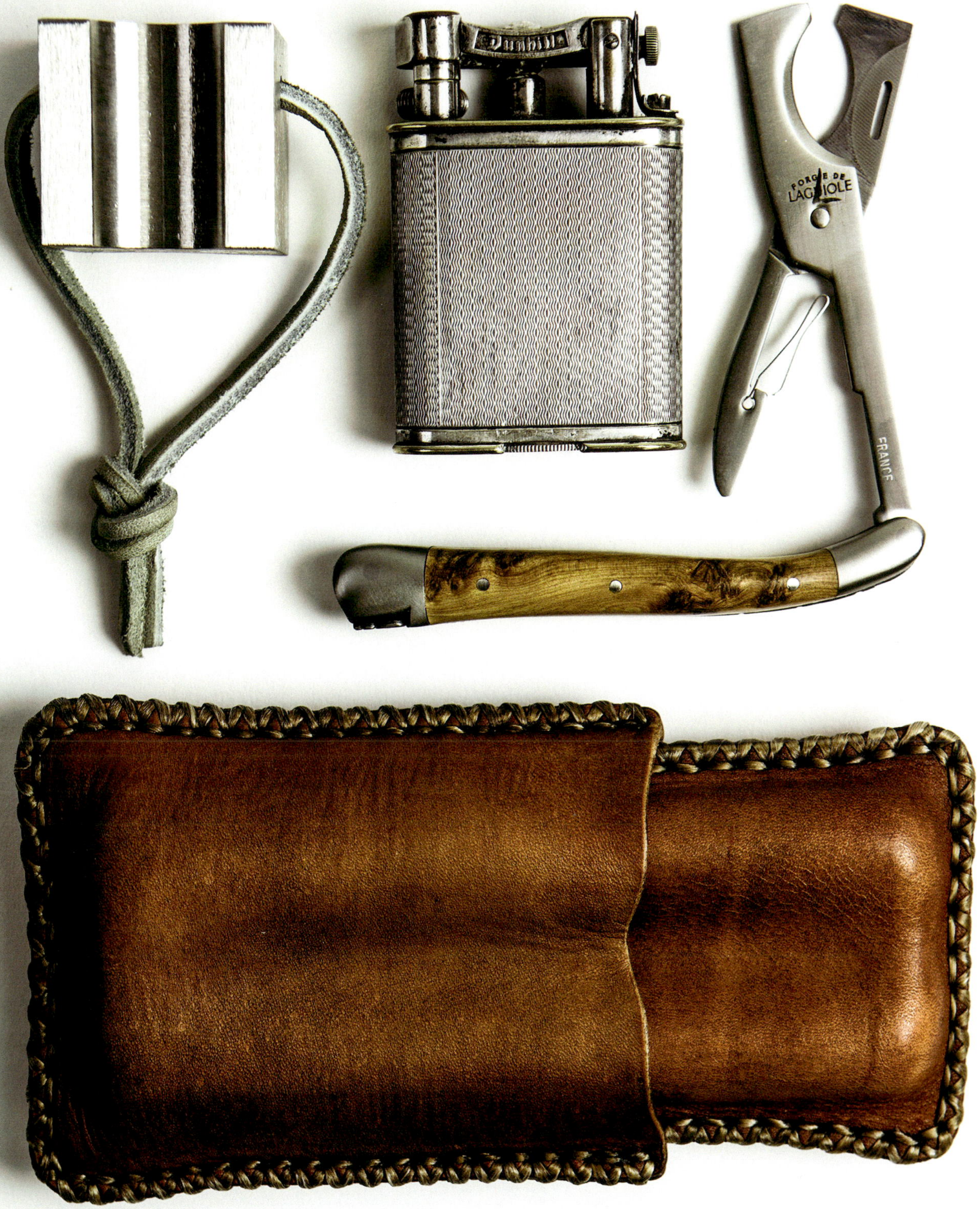

WICKED ANVIL BRIK CIGAR REST / $40
VINTAGE DUNHILL SERVICE LIGHTER / $350
FORGE DE LAGUIOLE CIGAR CUTTER / $230
HALLETT PEAK CASTAWAY CIGAR CASE / $75

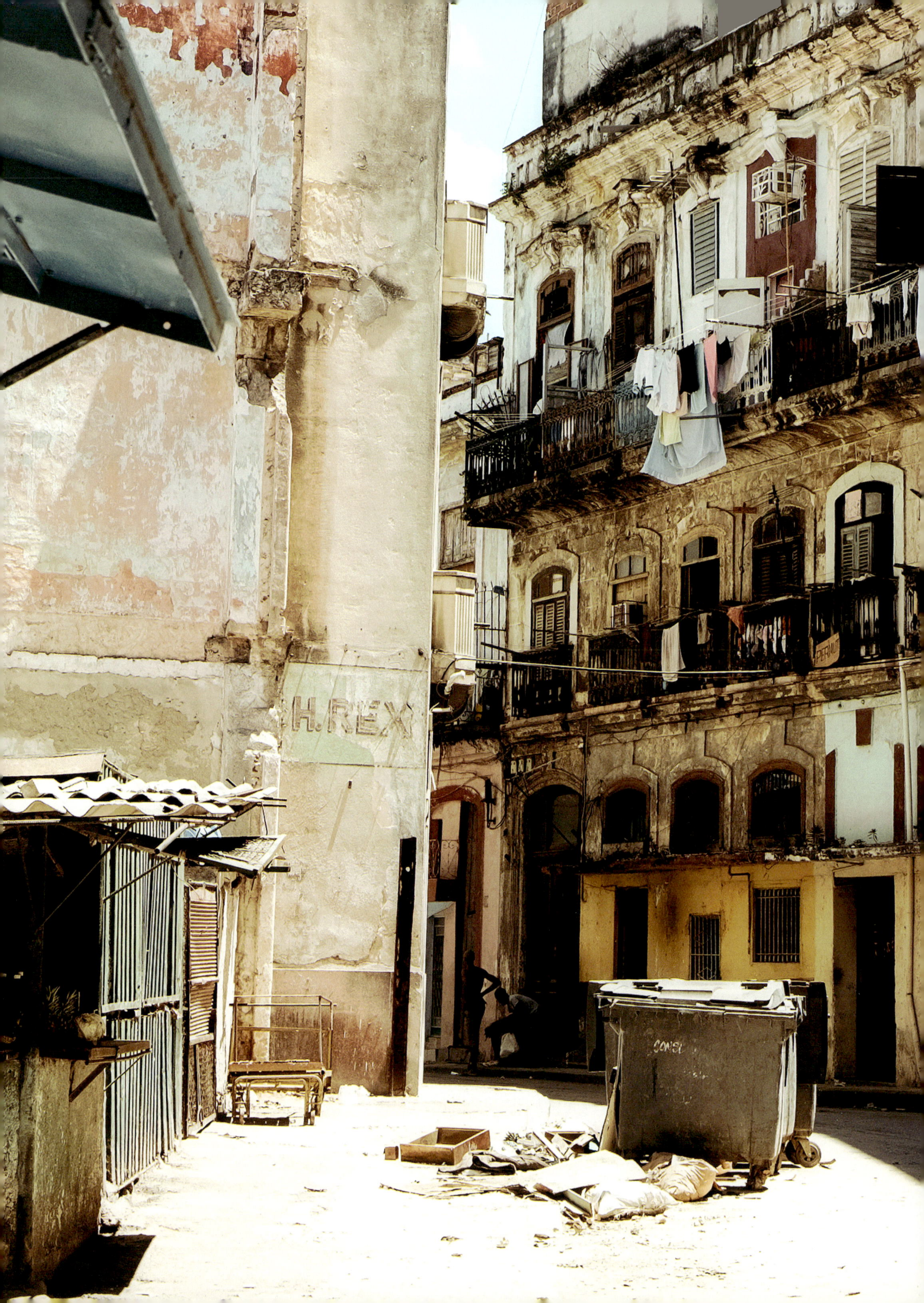
H.REX

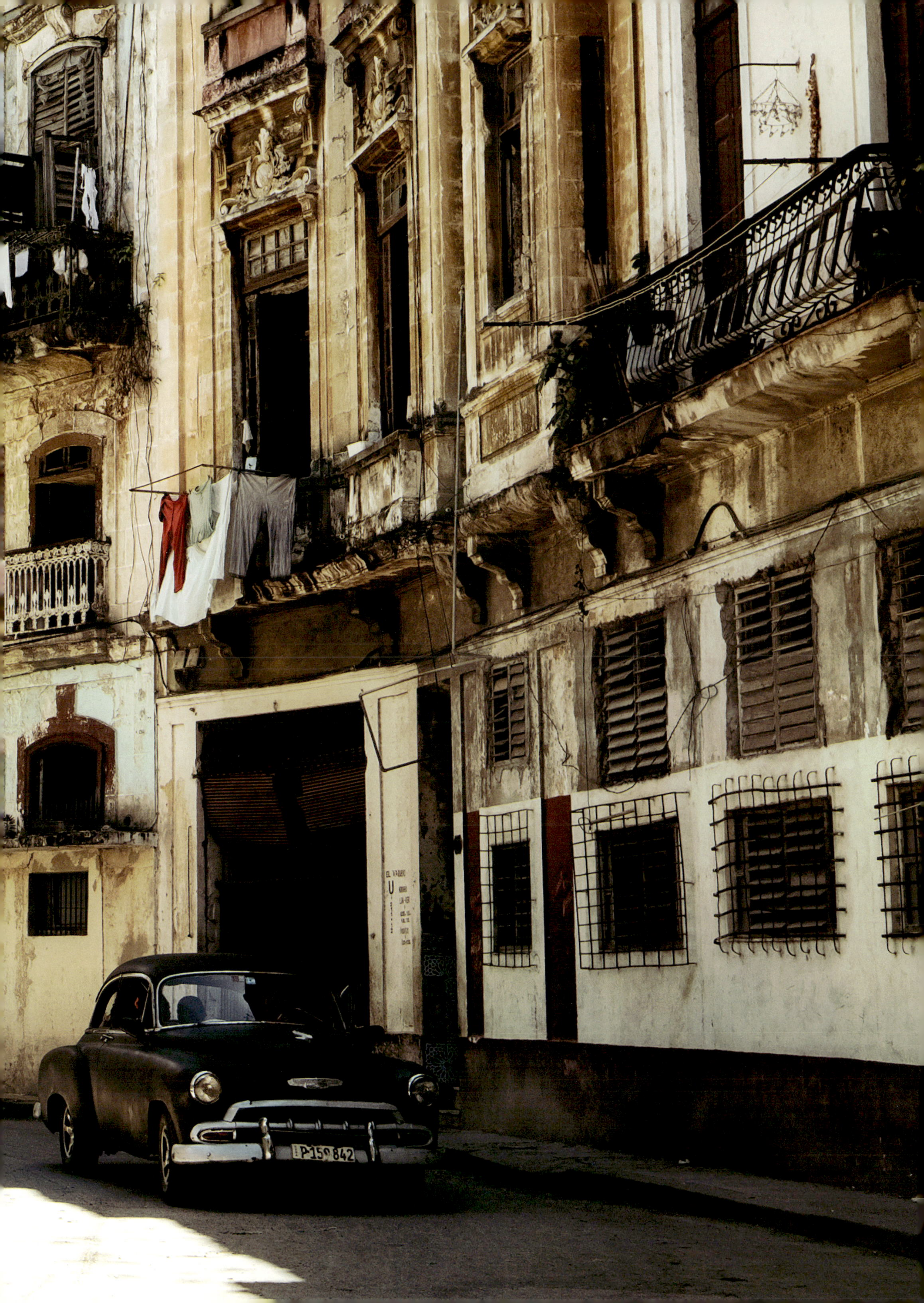

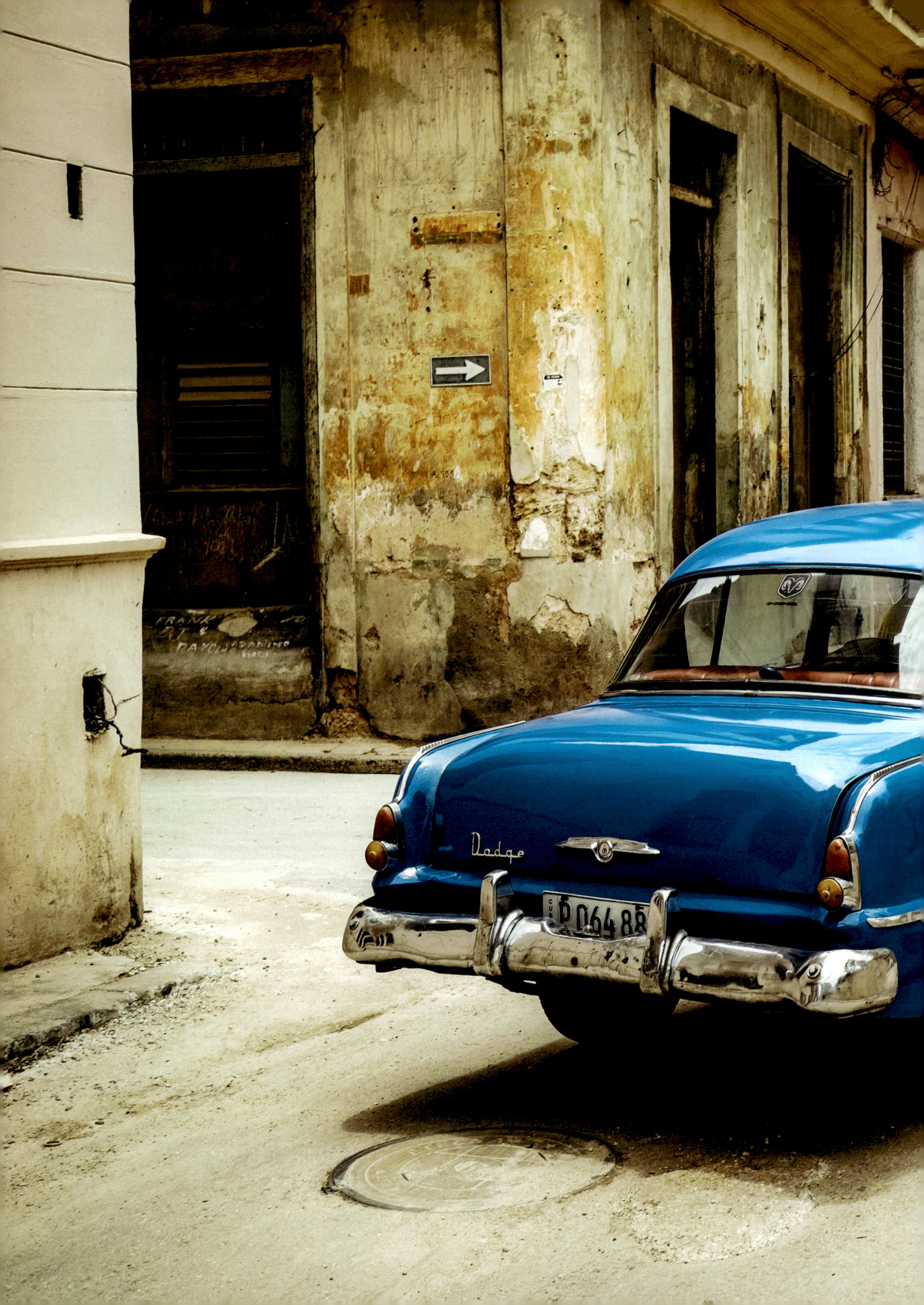
Dodge
P064 88

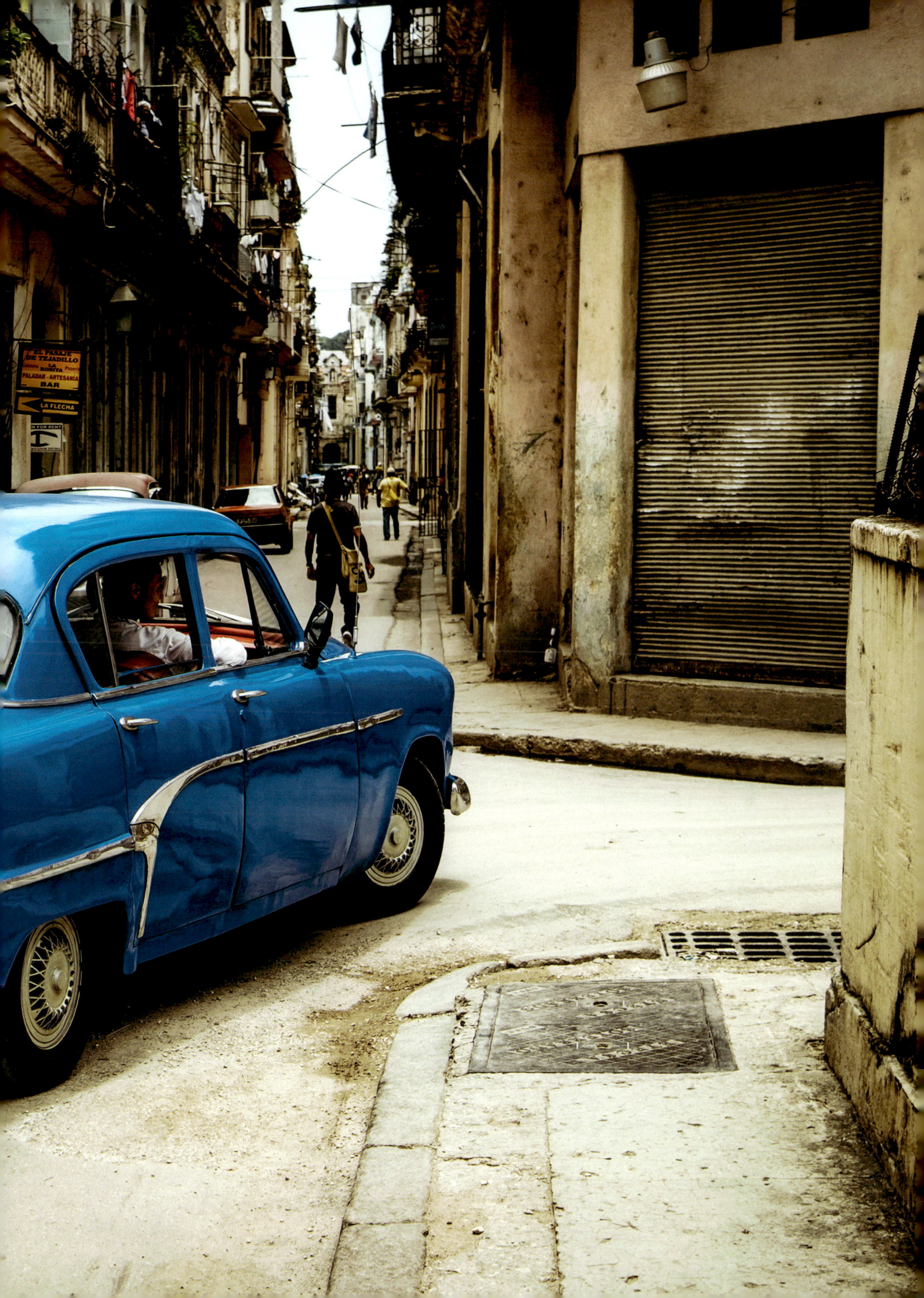
EL PASAJE
DE TEJADILLO
LA
ROSITA
PALADAR - ARTESANÍA
BAR
LA FLECHA
FOR RENT

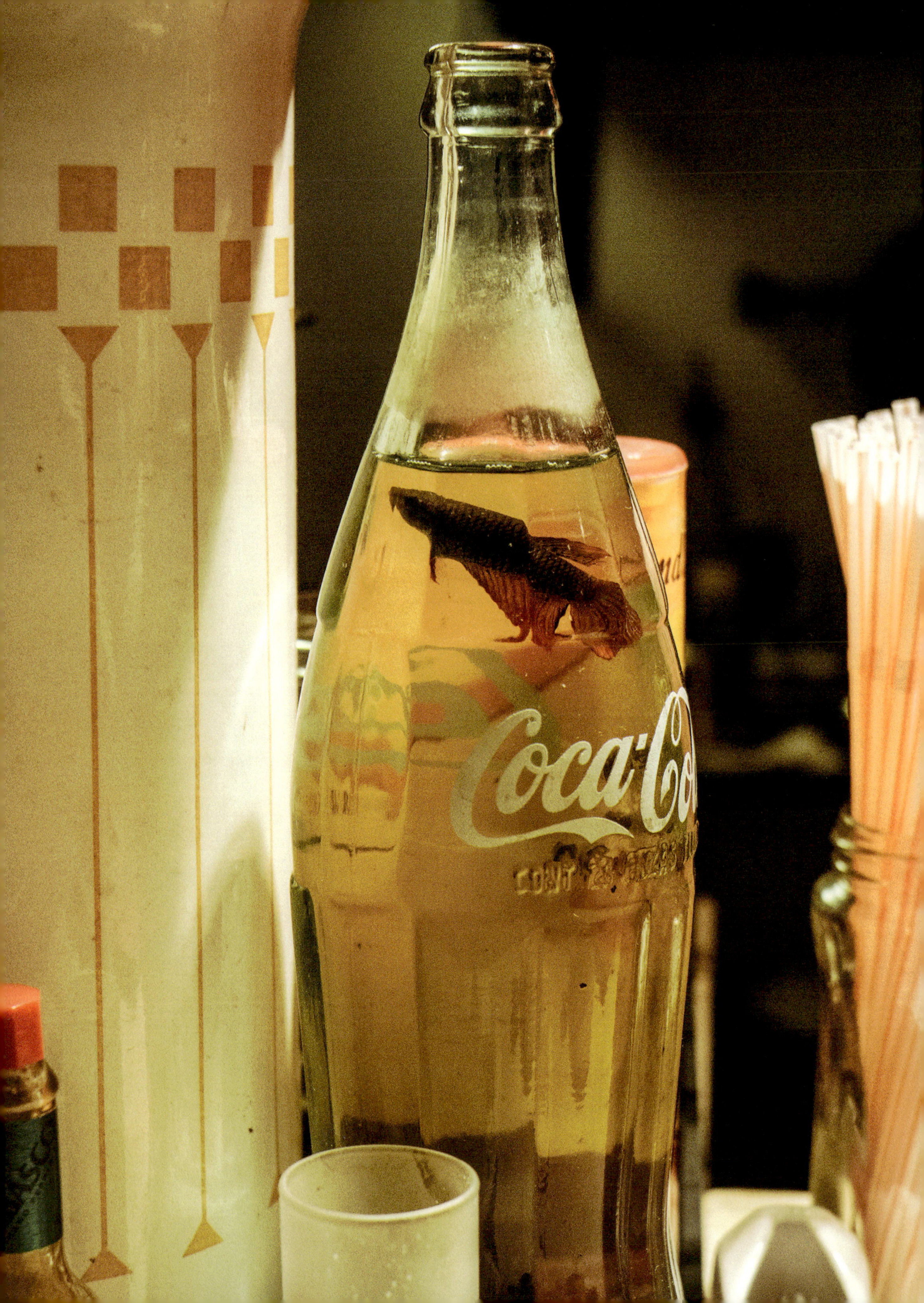

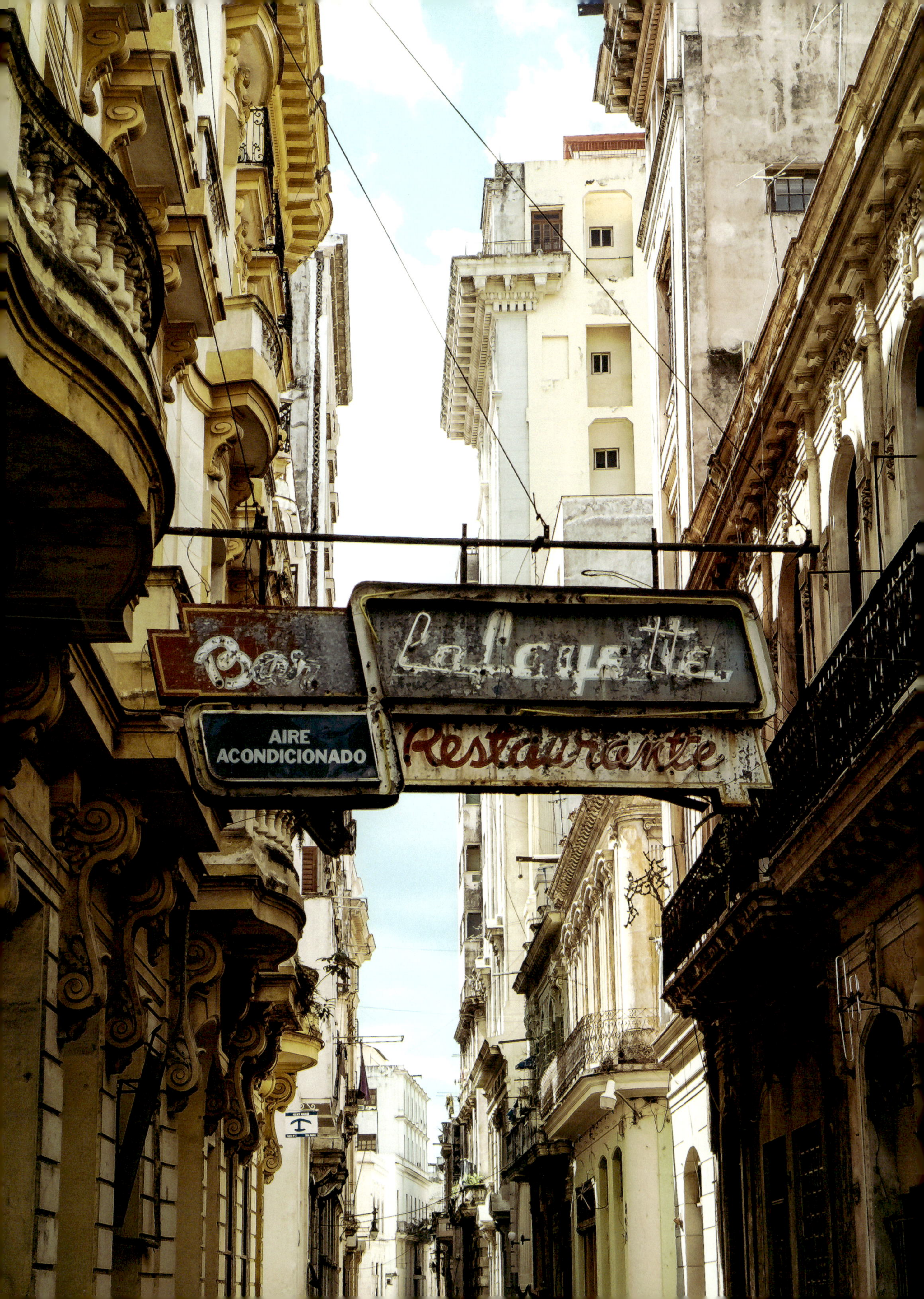
Bar
AIRE
ACONDICIONADO
Restaurante

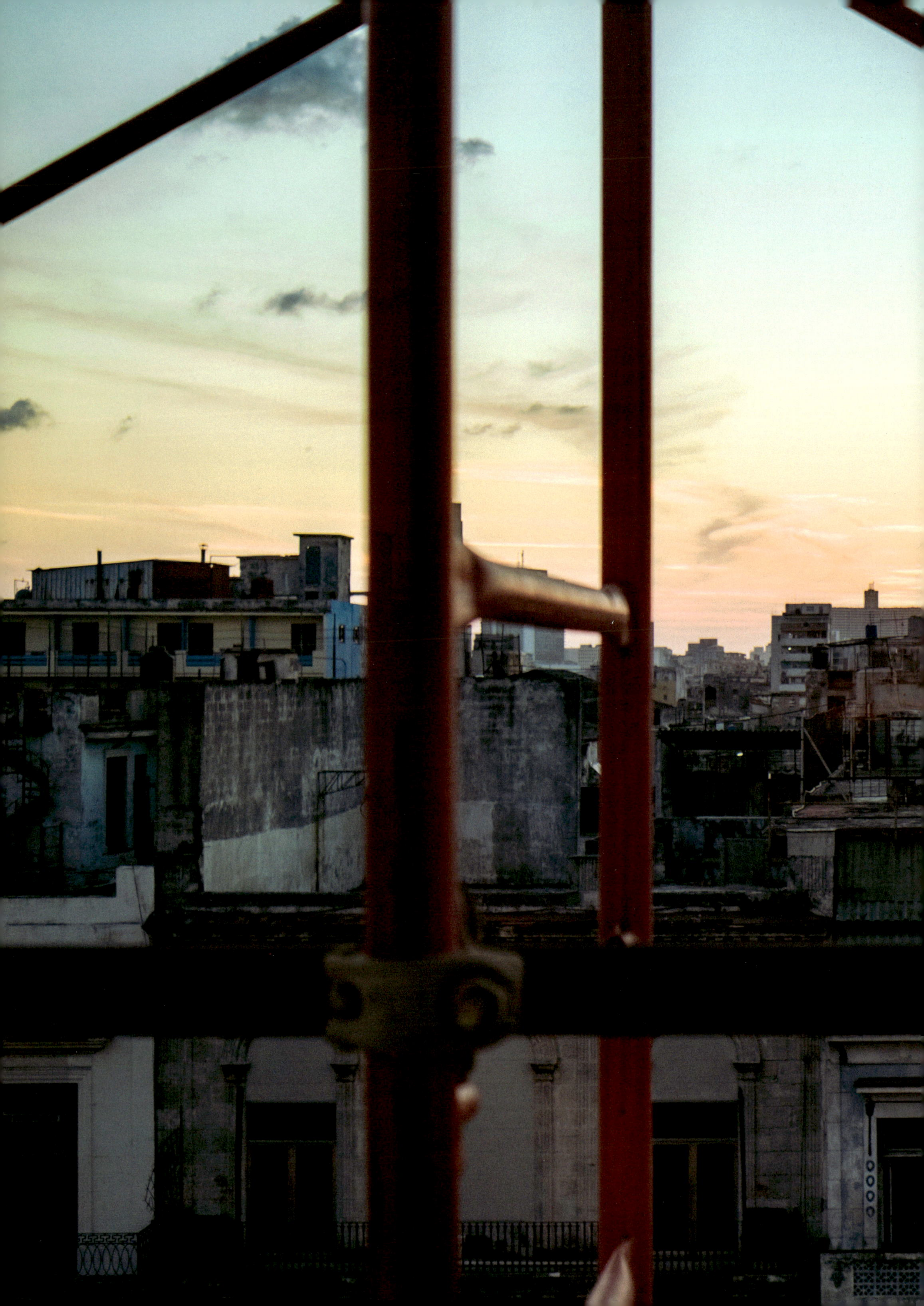

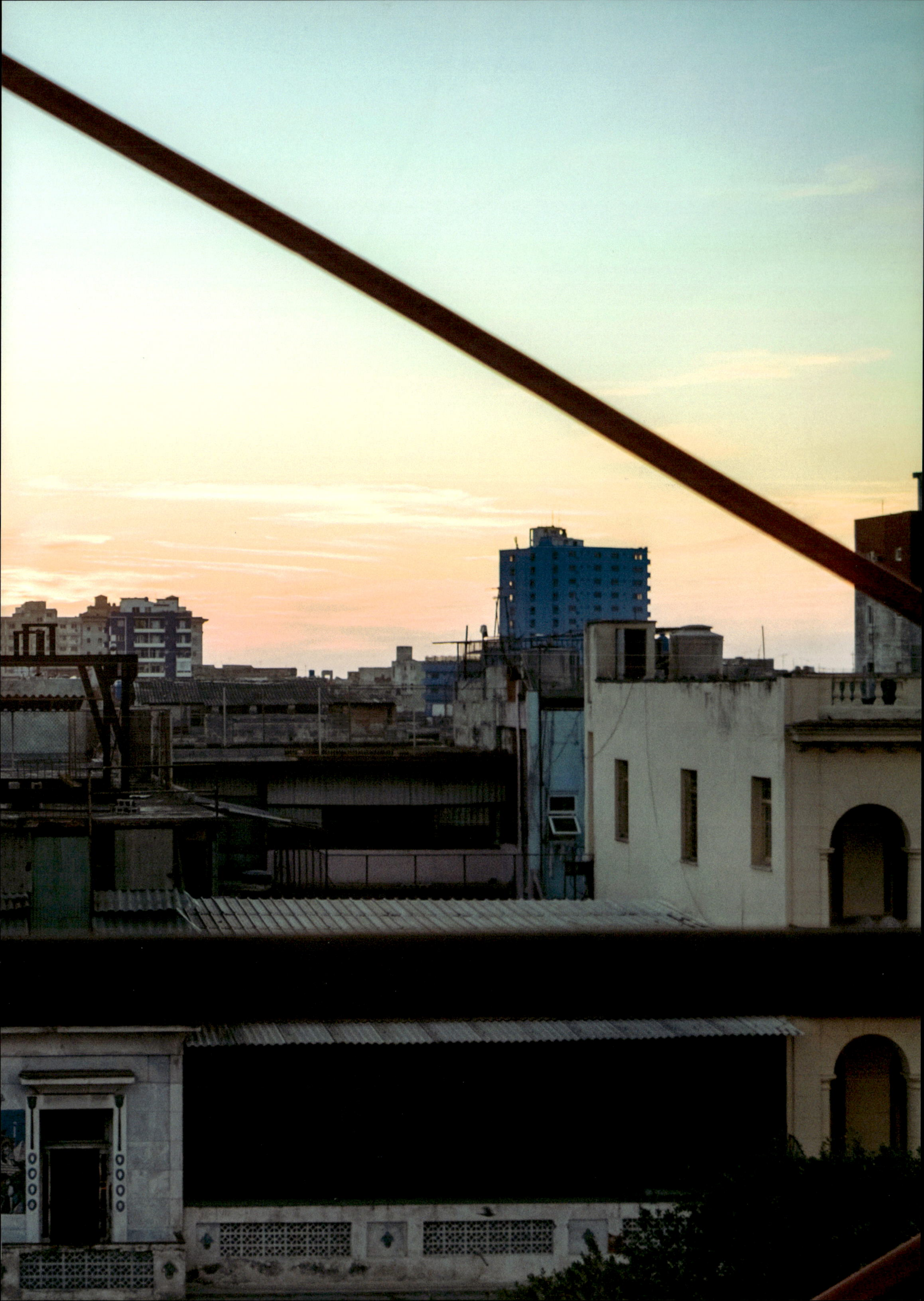

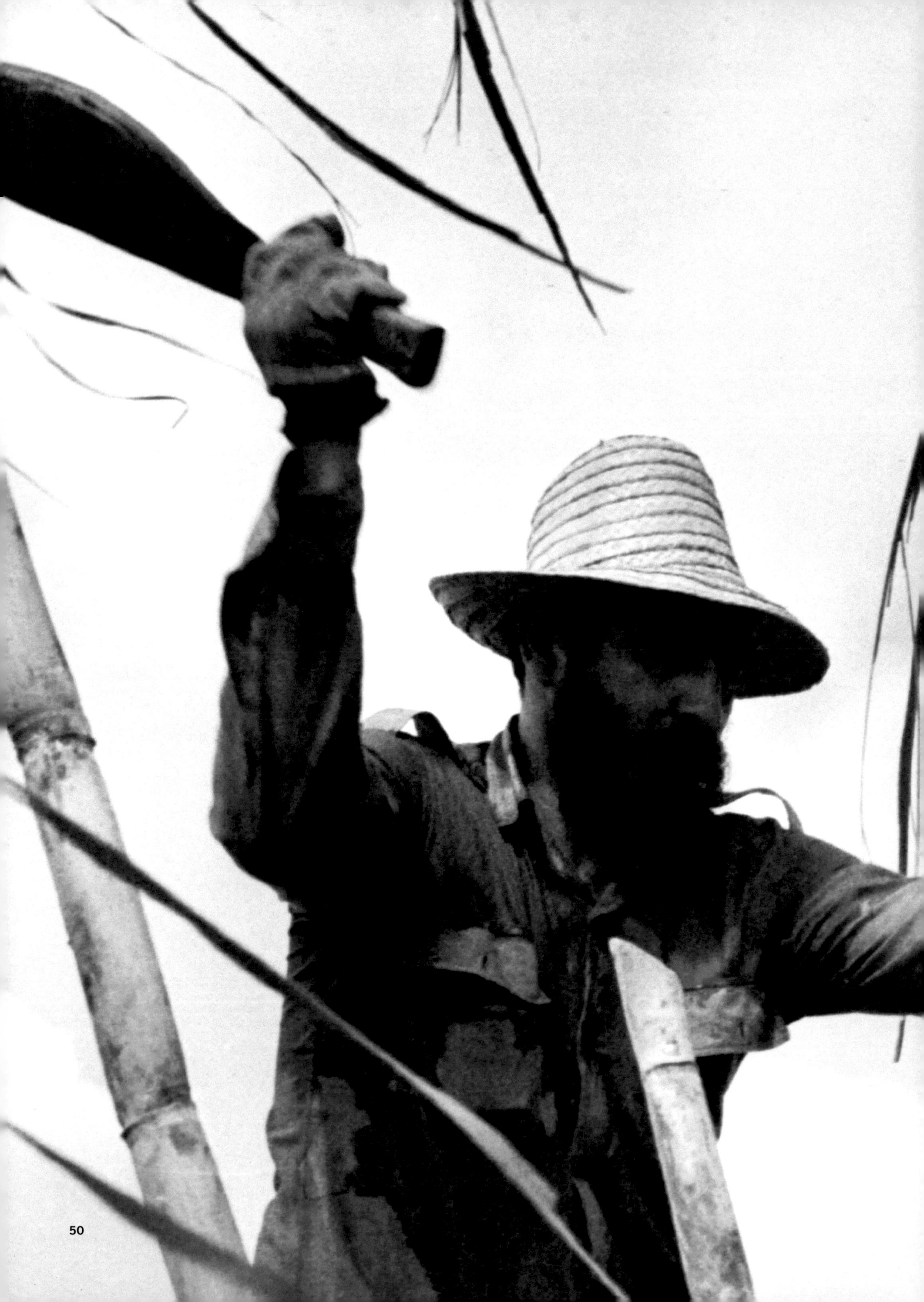

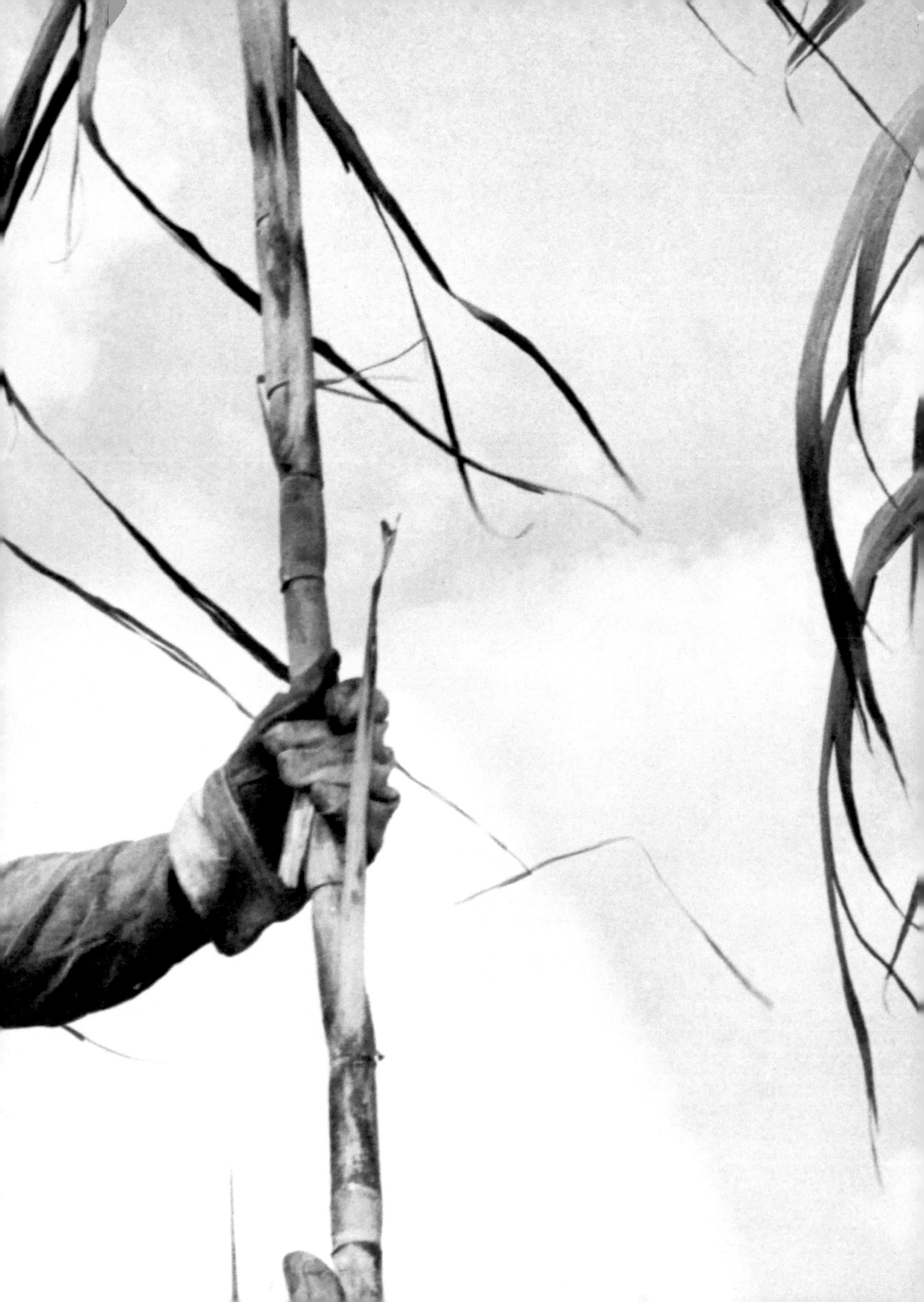

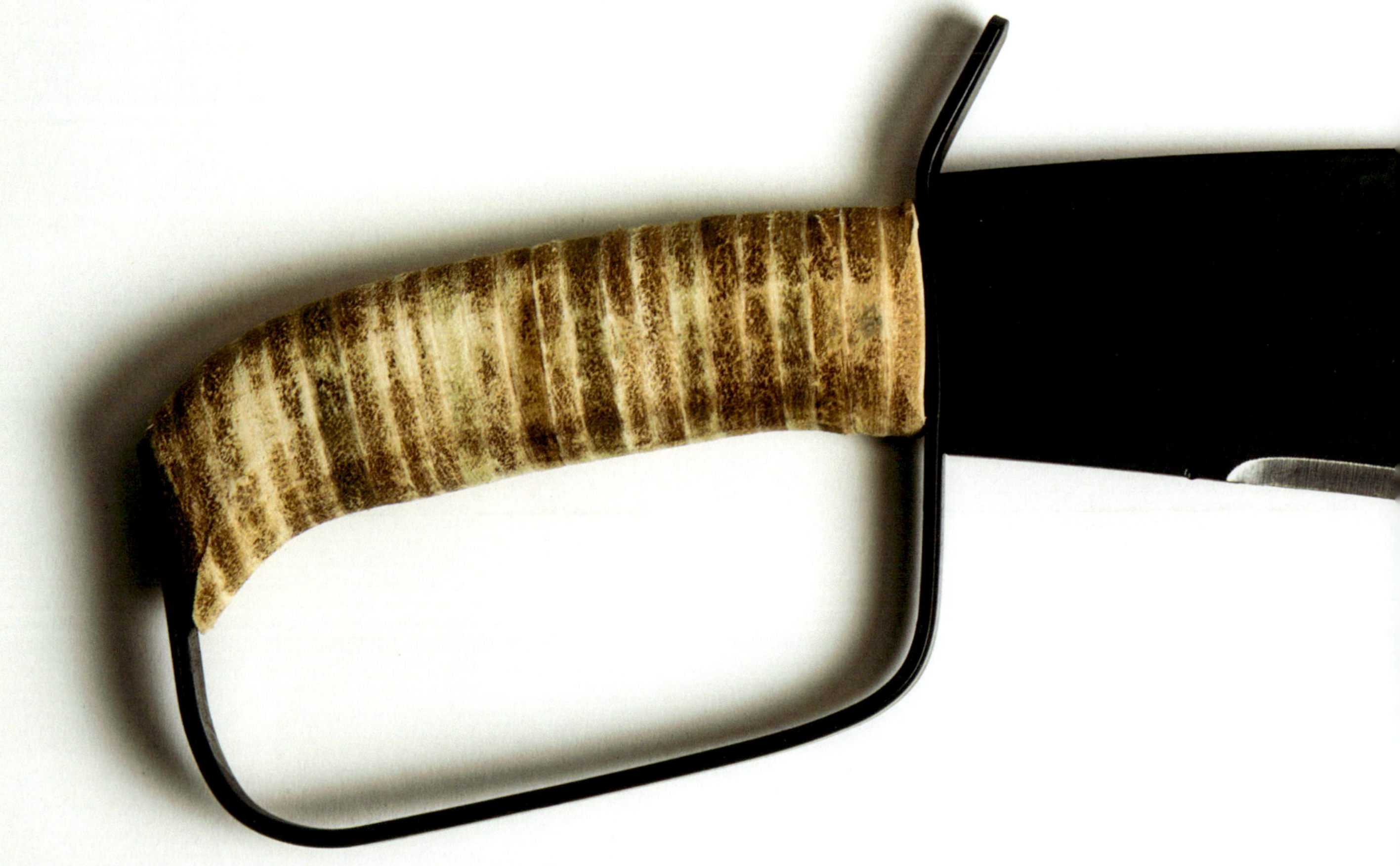

SOG SOGFARI MACHETE / $20

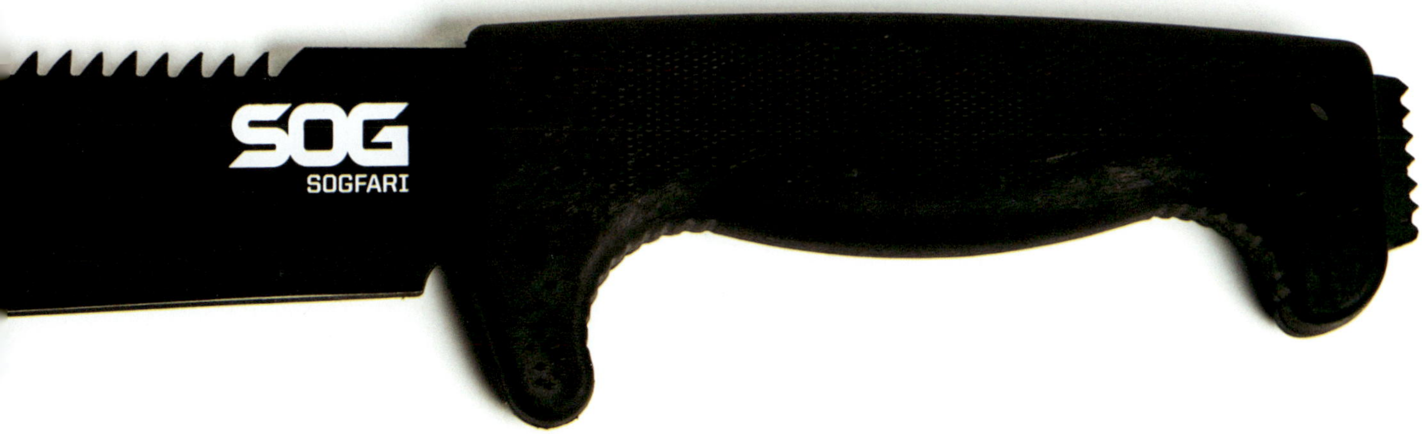

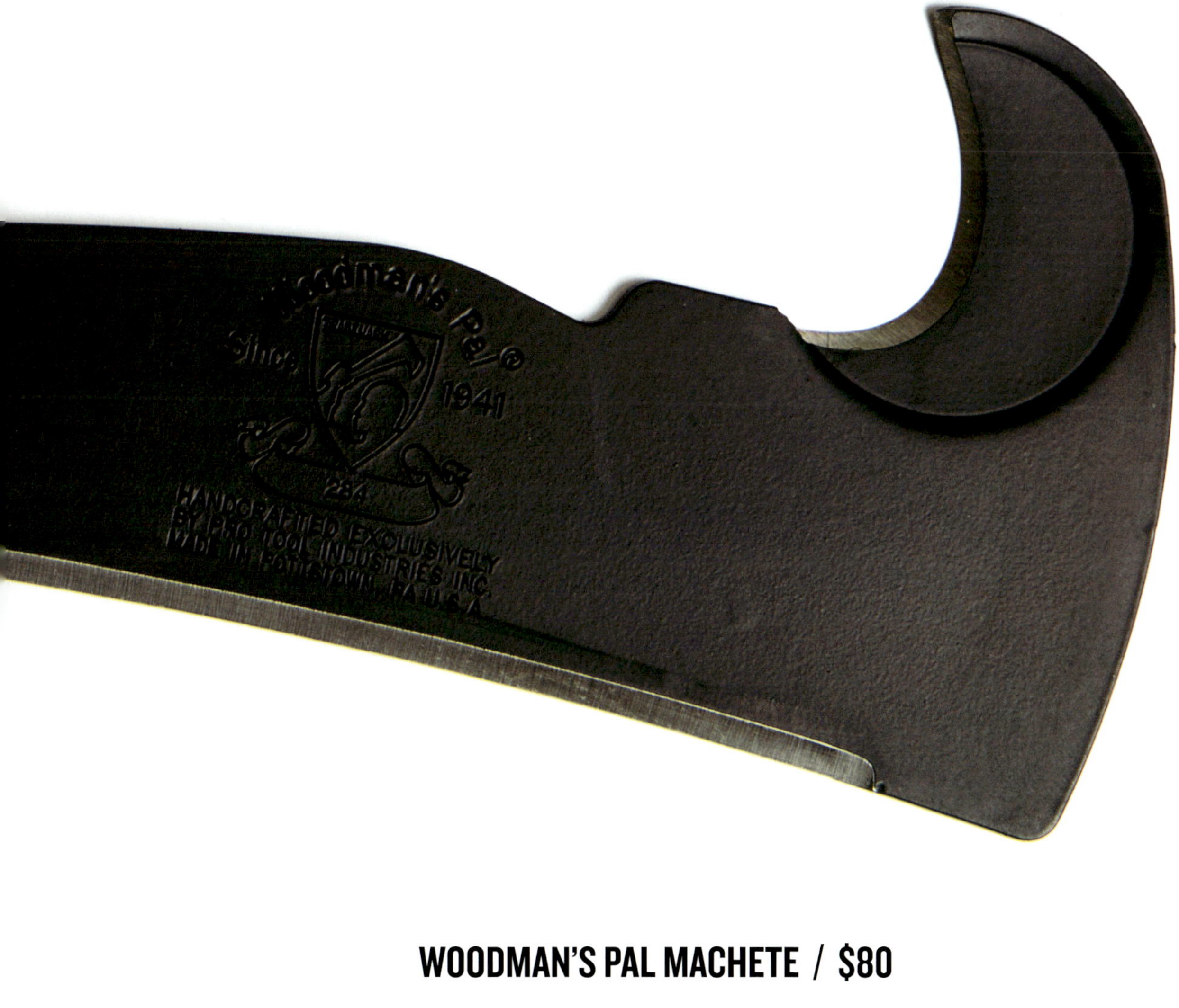

WOODMAN'S PAL MACHETE / $80

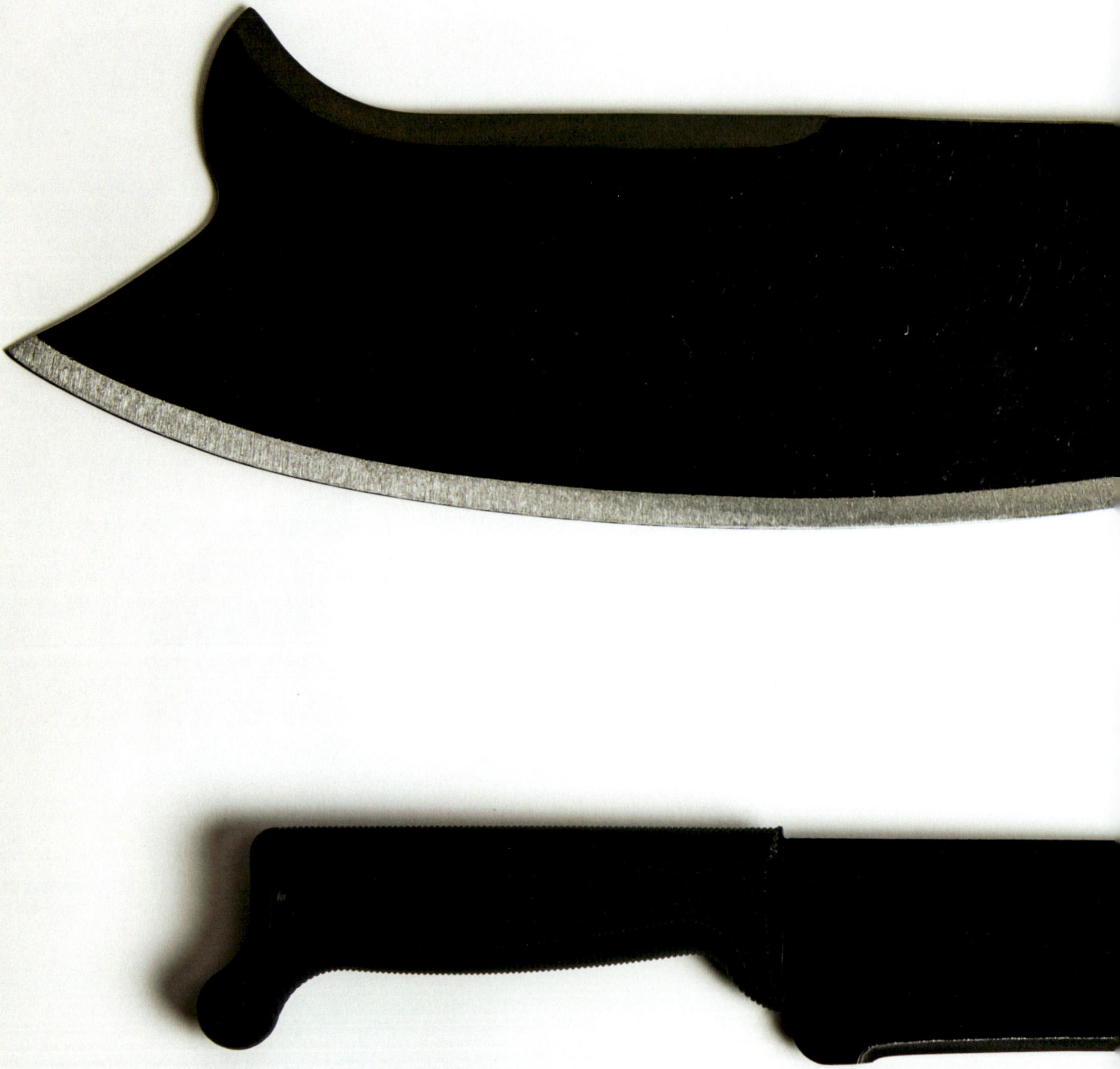

BARYONYX MACHETE / $70

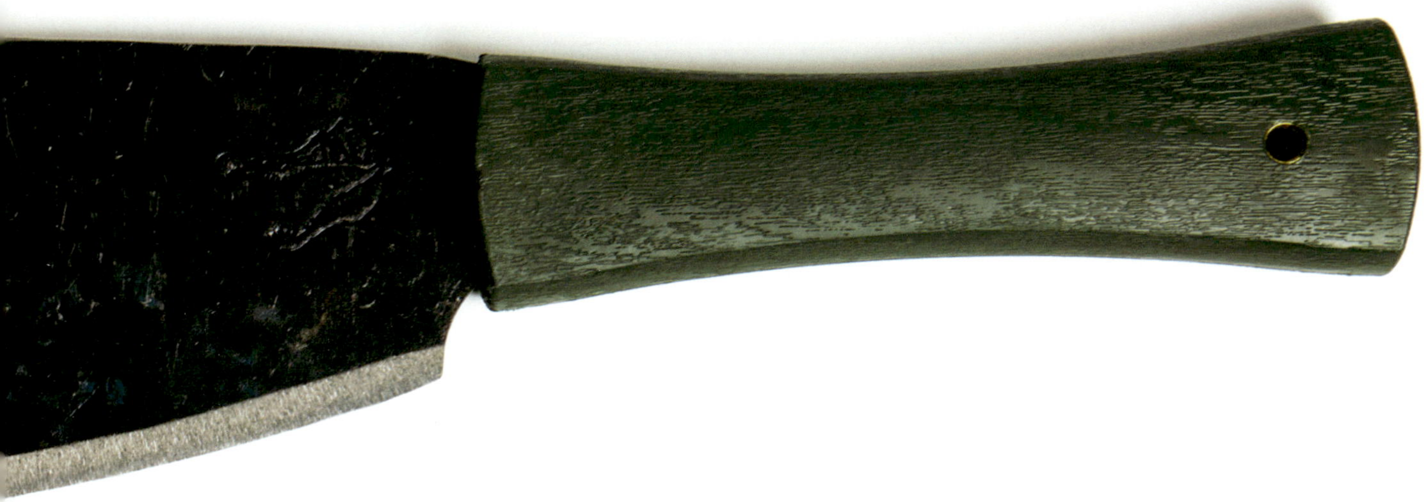

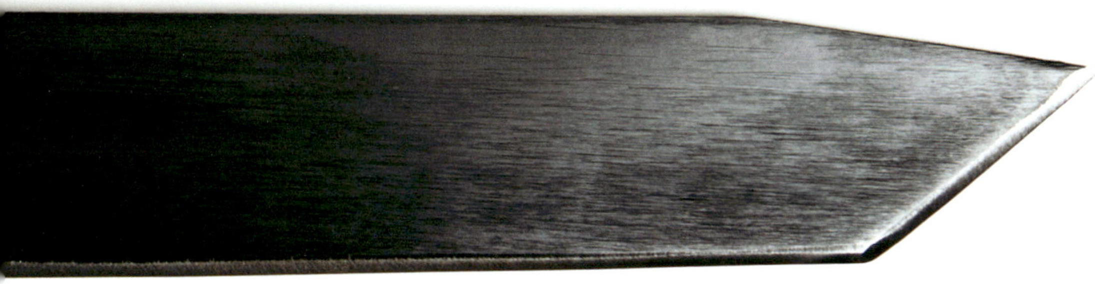

COLD STEEL TANTO MACHETE / $20

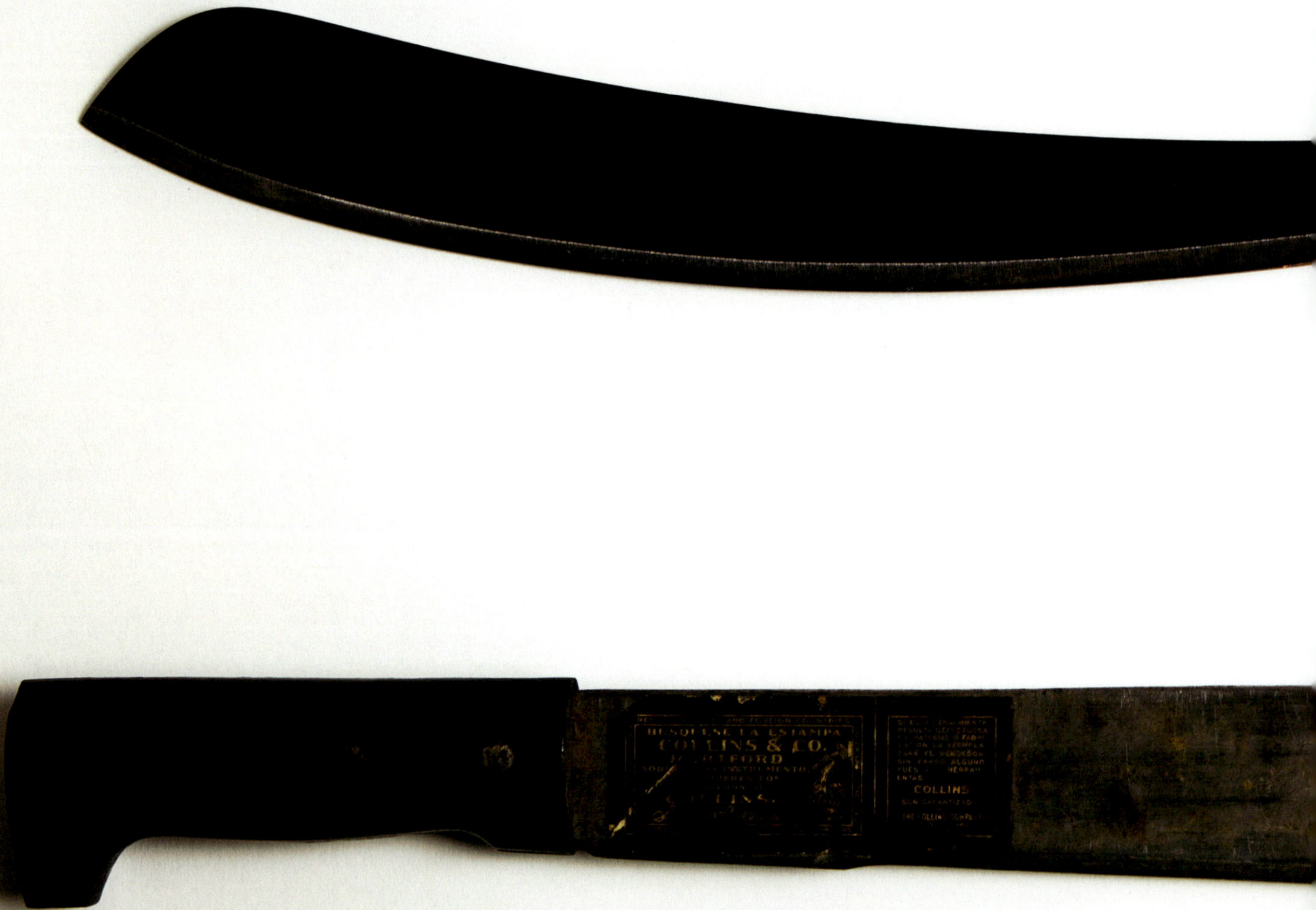
COLLINS & CO.
COLLINS

CONDOR PARANG MACHETE / $55

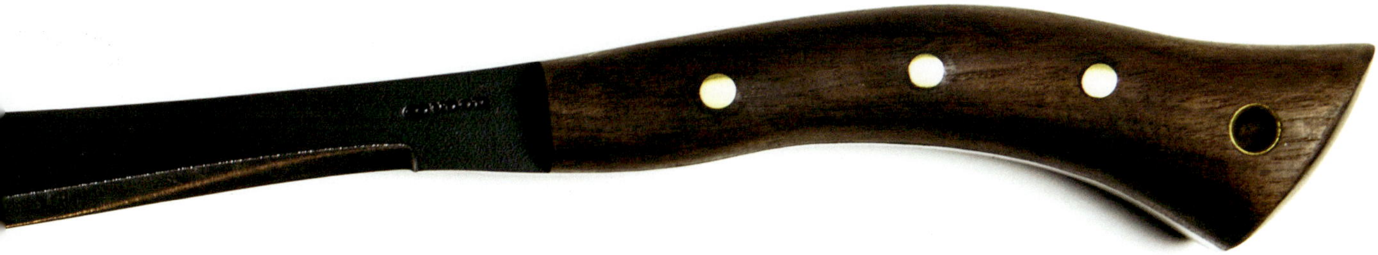

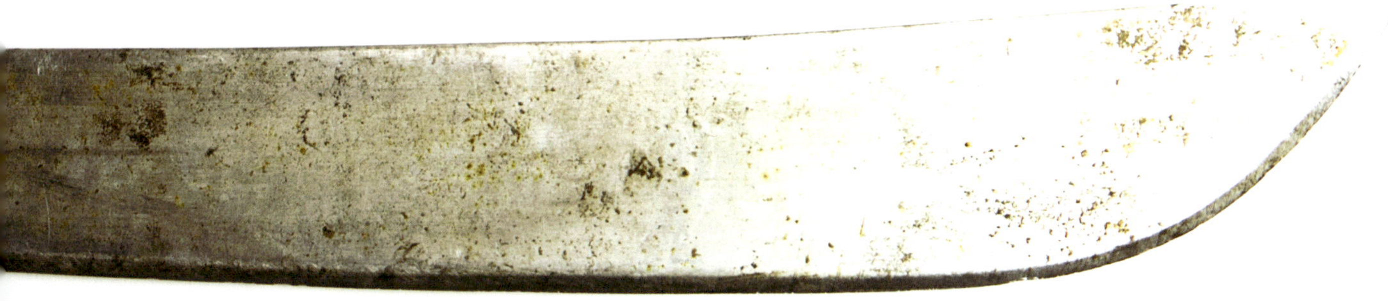

ARTIFACT

COLLINS & CO. HARTFORD MACHETE / $175

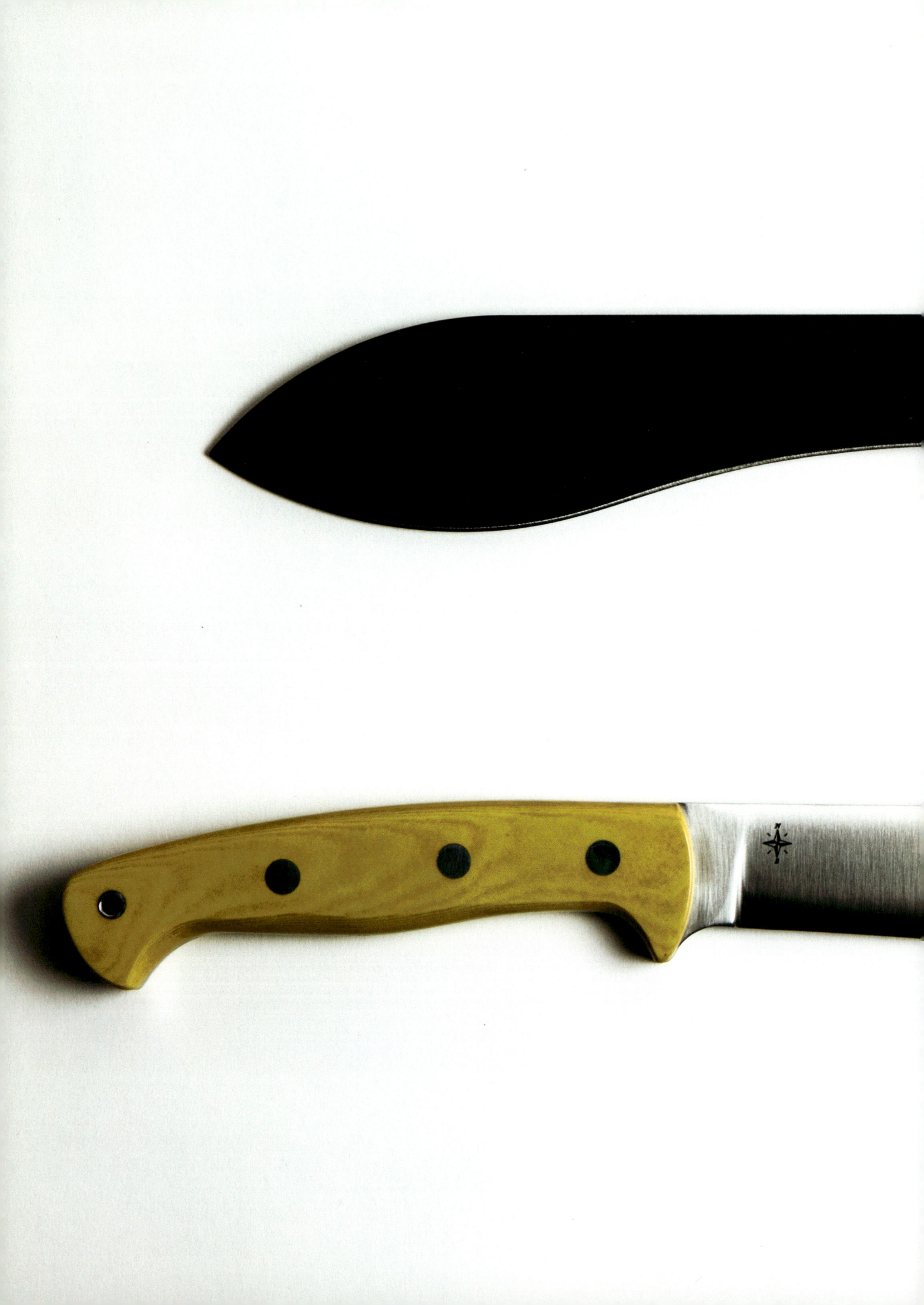

BENCHMADE JUNGLE BOLO / $110

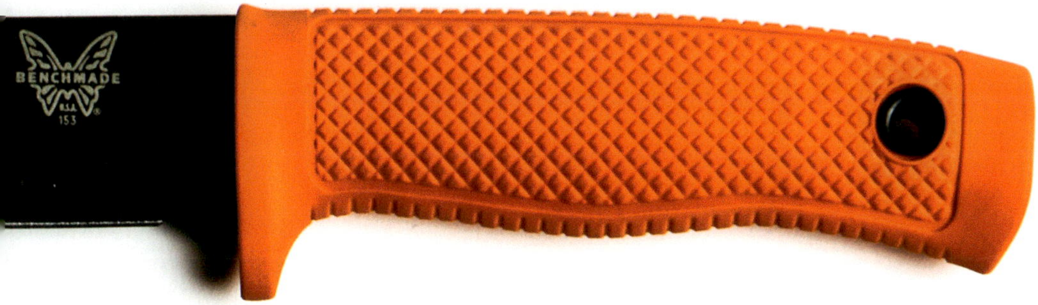

BARK RIVER GOLOK / $270

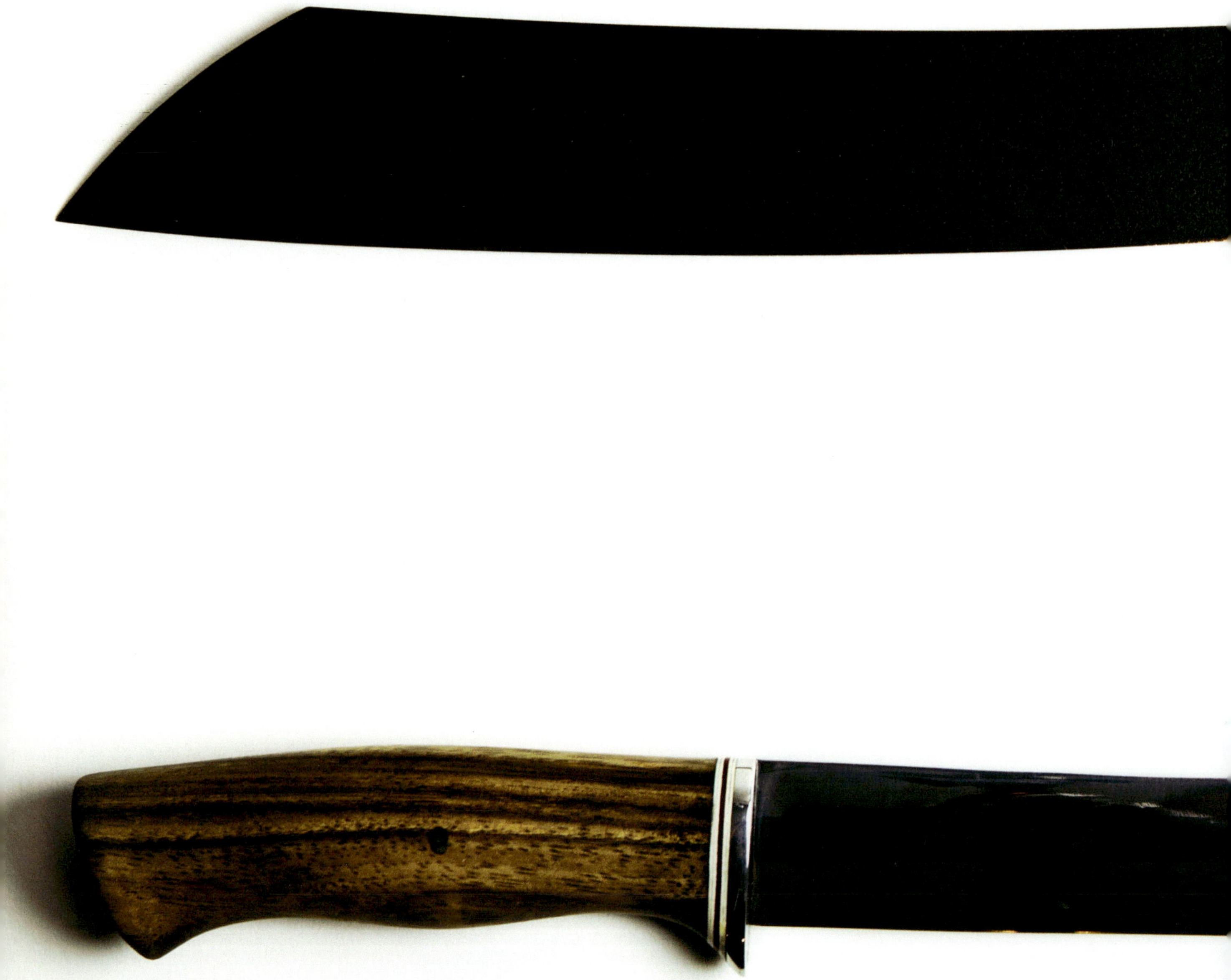

KA-BAR ADVENTURE PARANGATANG MACHETE / $70

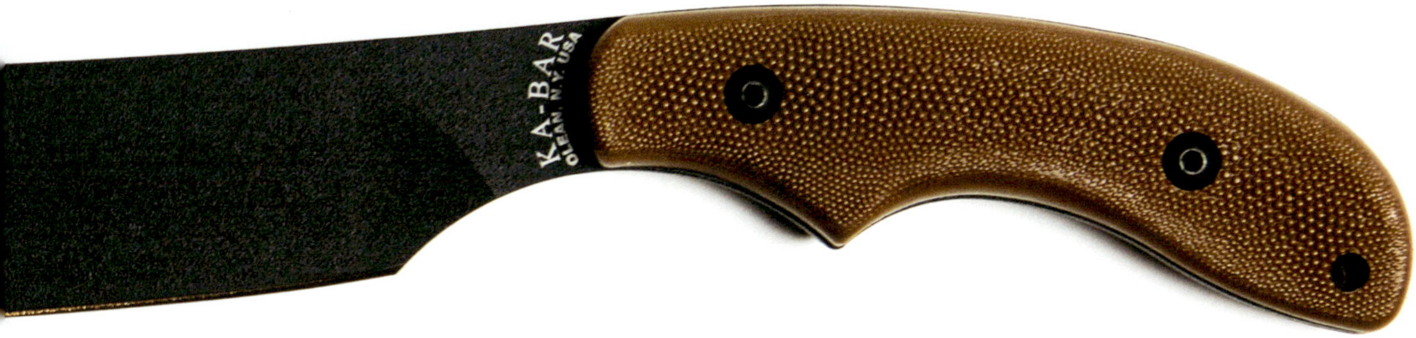

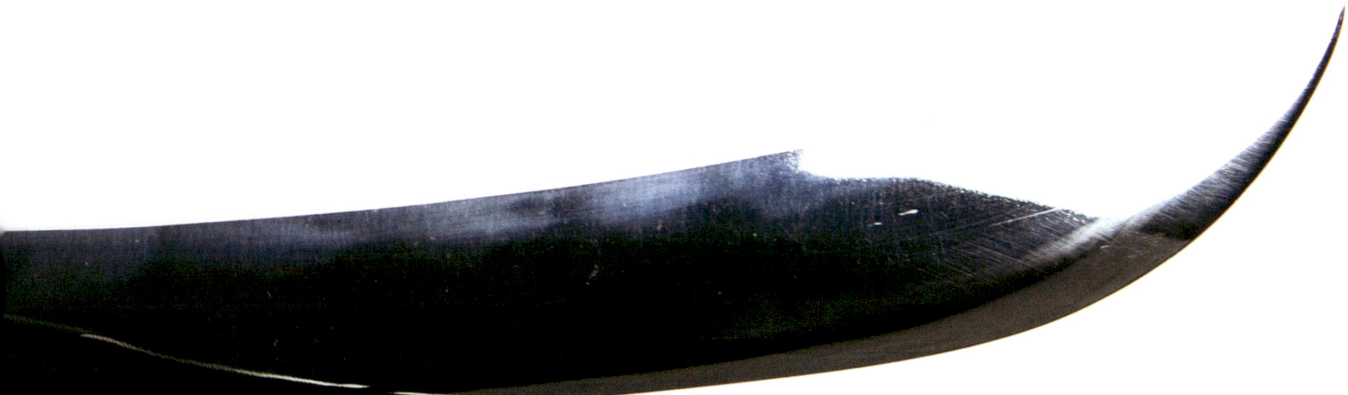

CITADEL PARANG JUNGLE MACHETE / $425

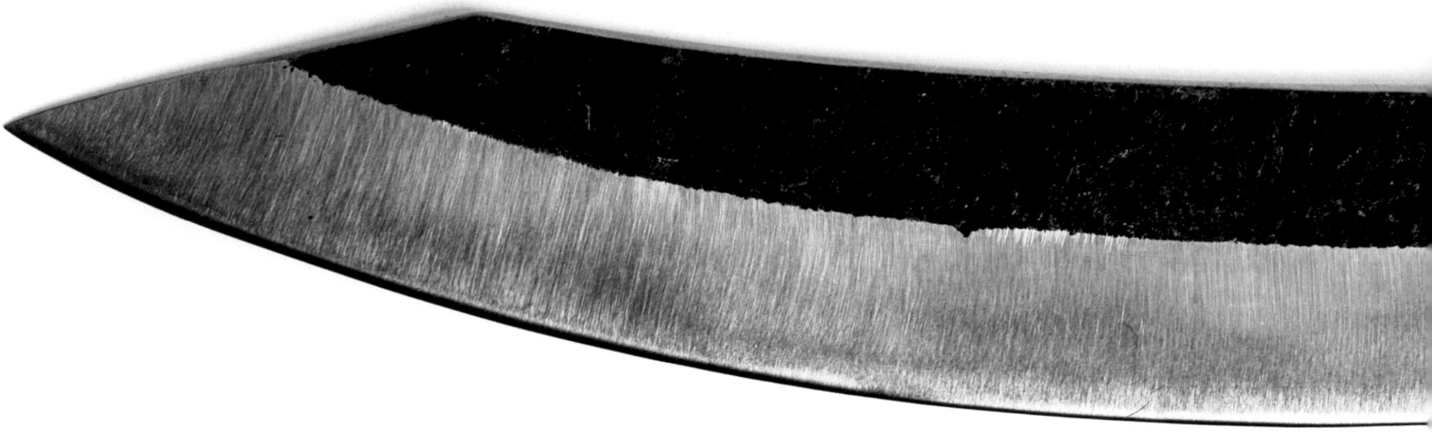

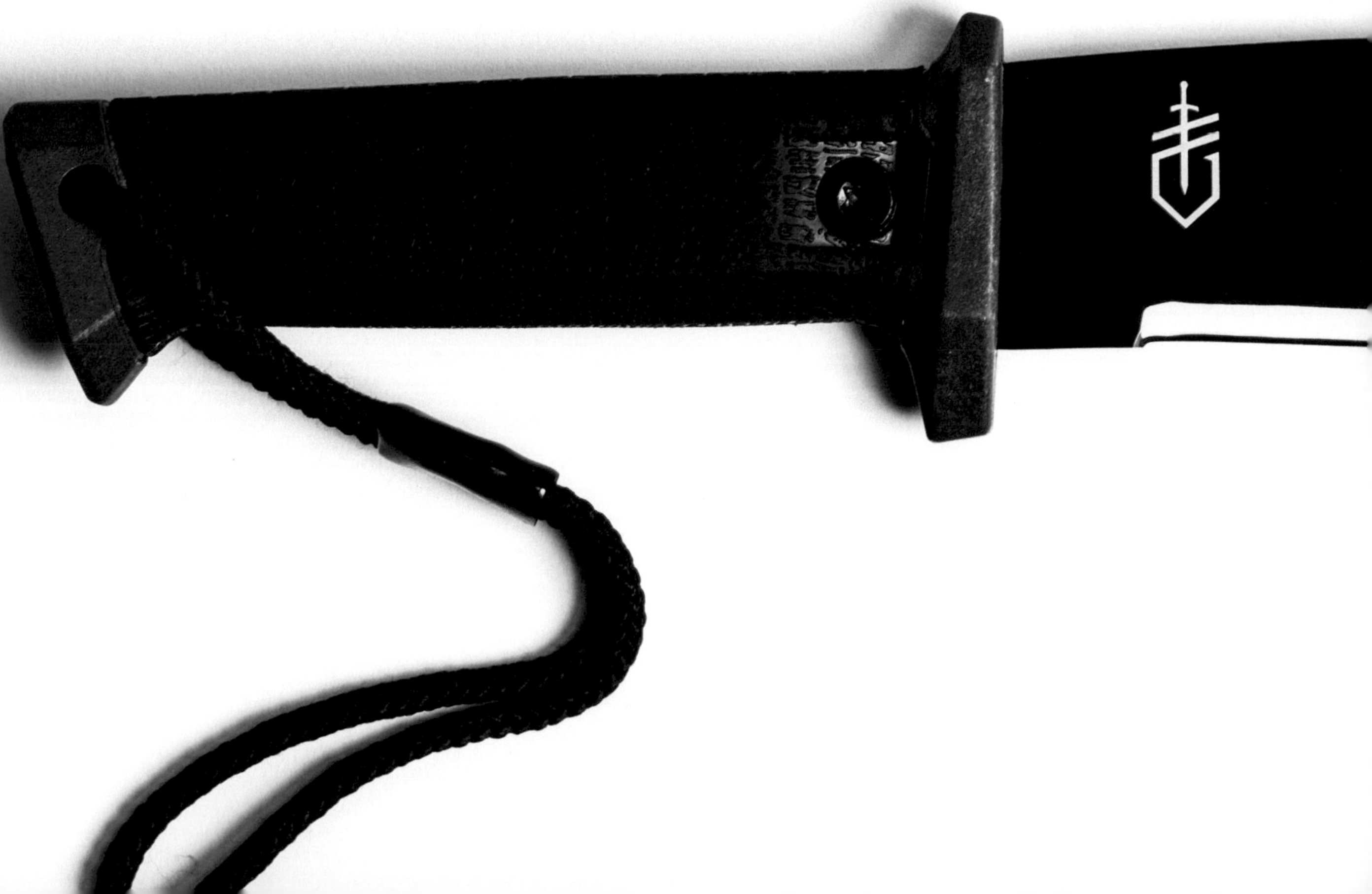

CRKT MAH-CHETE KNIFE / $60

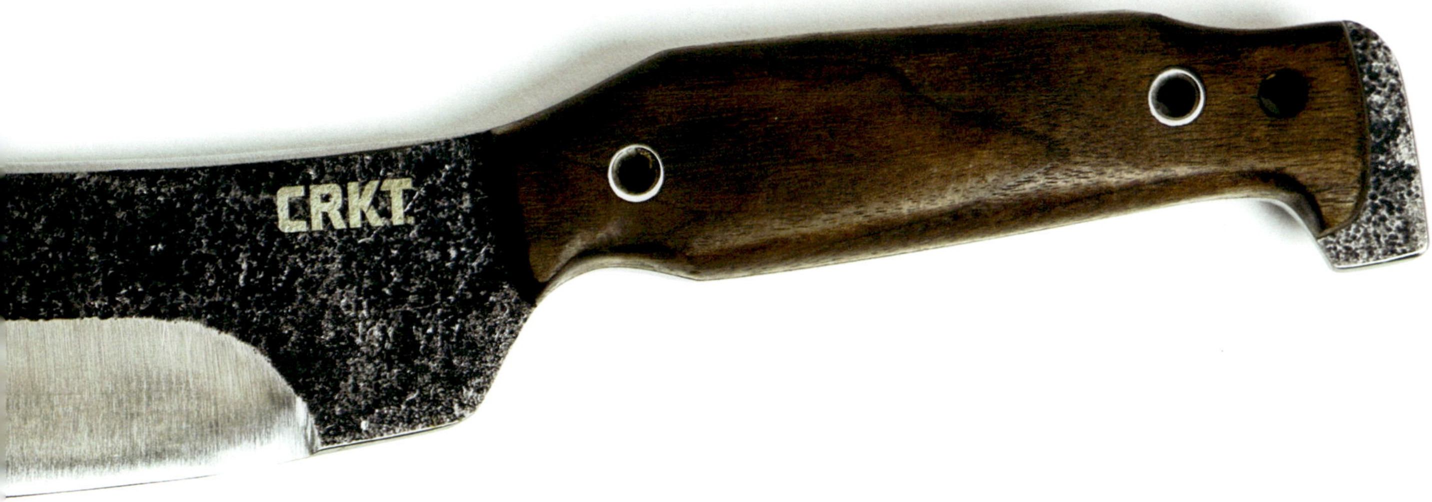

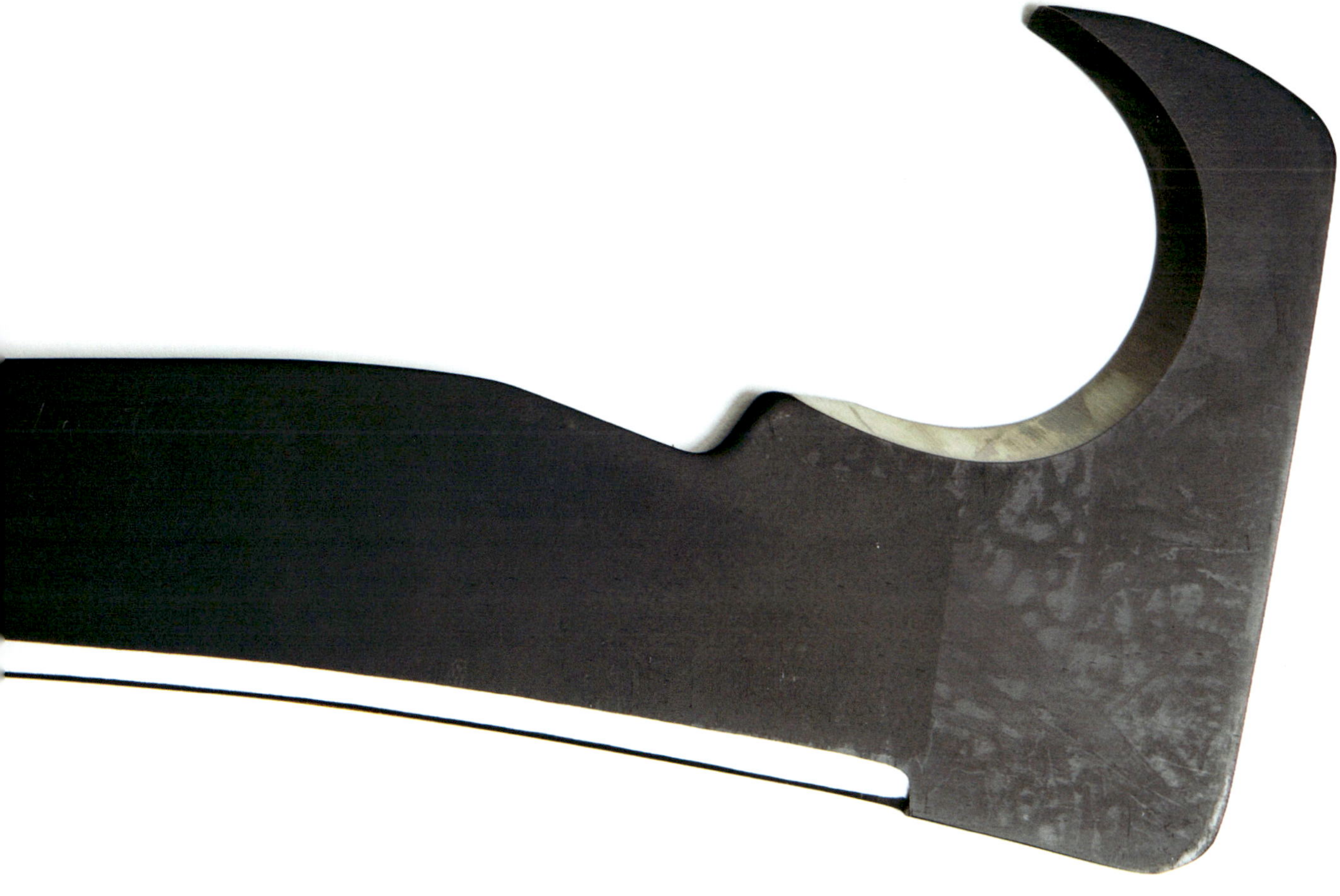

GERBER GATOR MACHETE PRO / $60

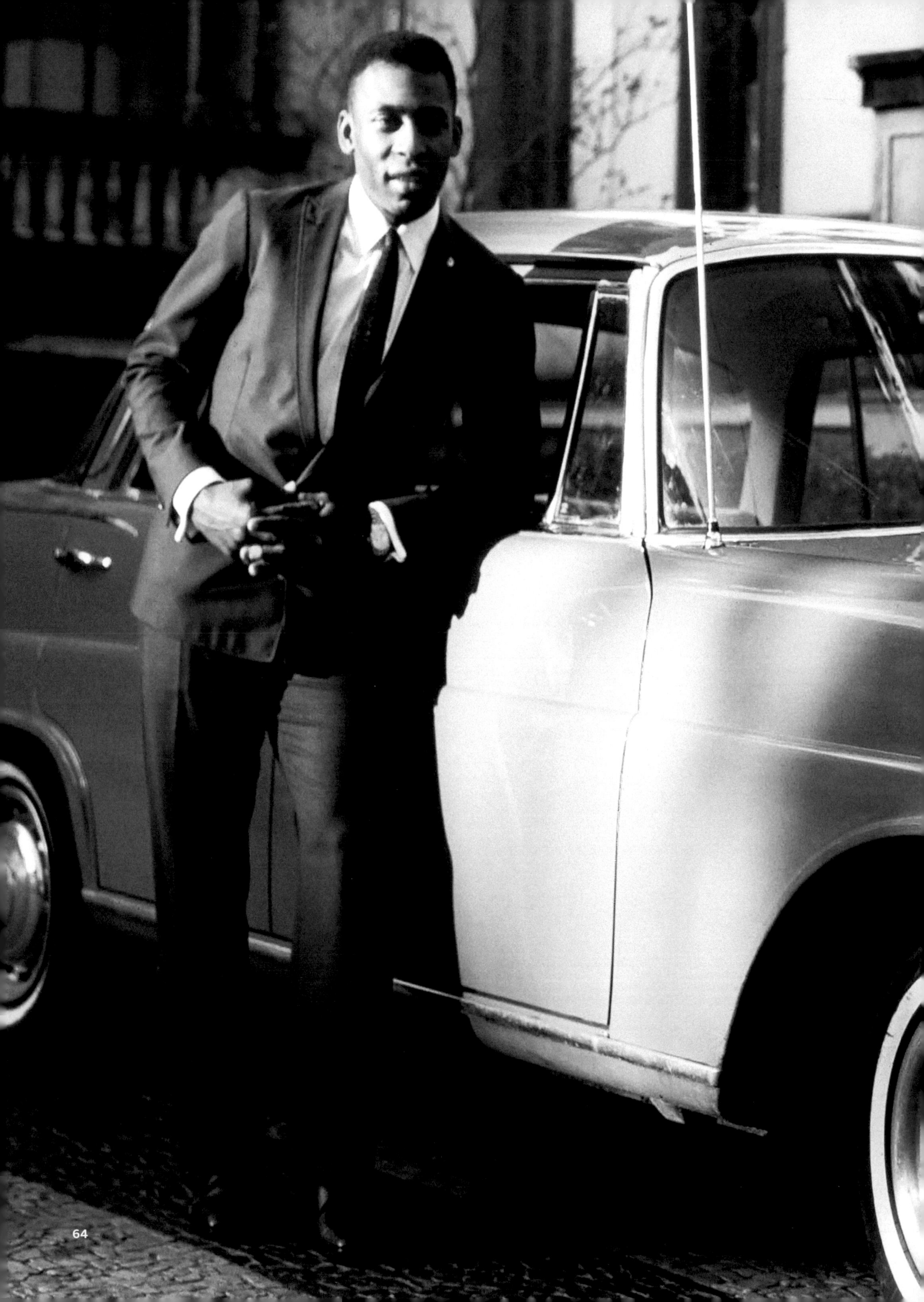

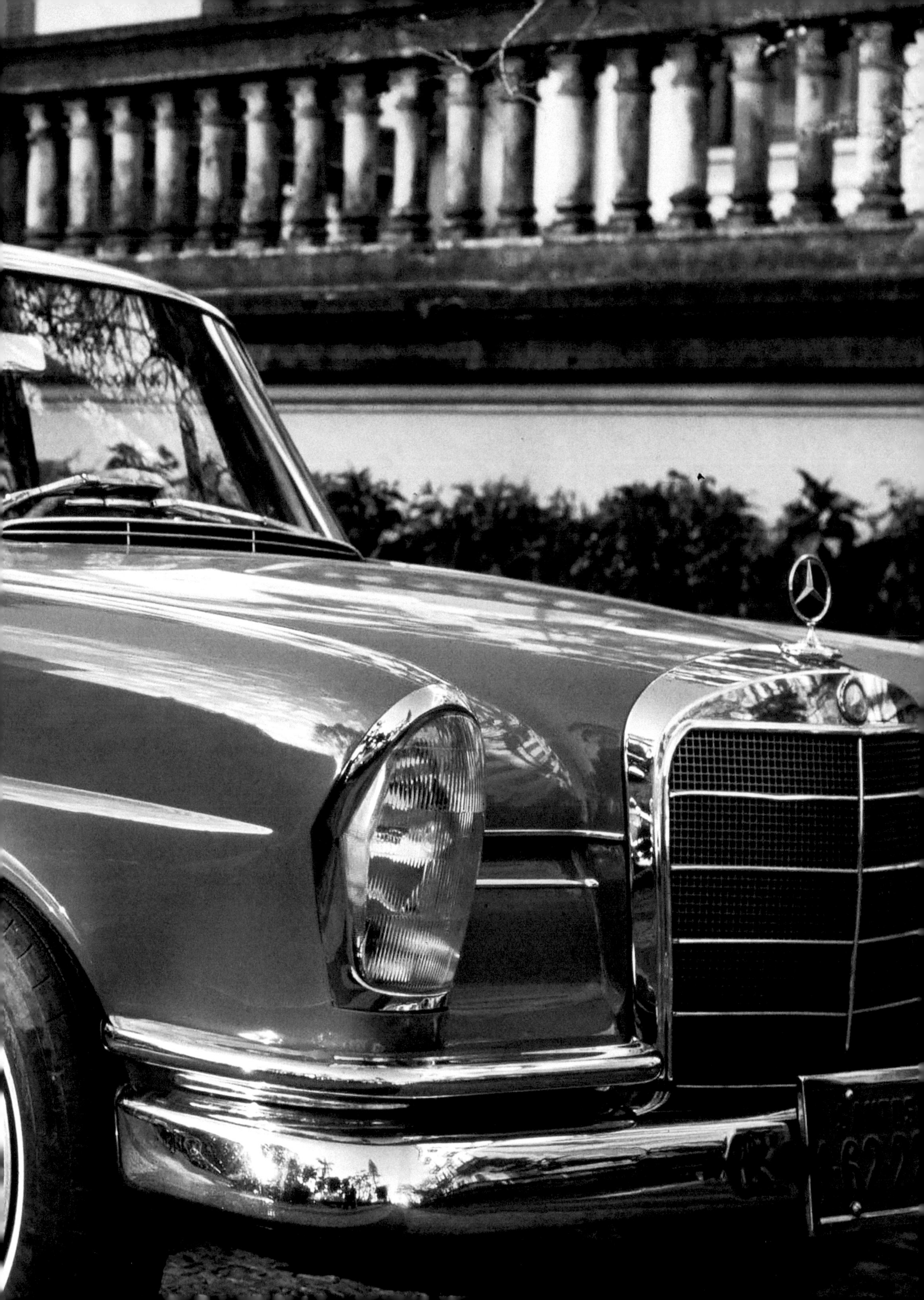

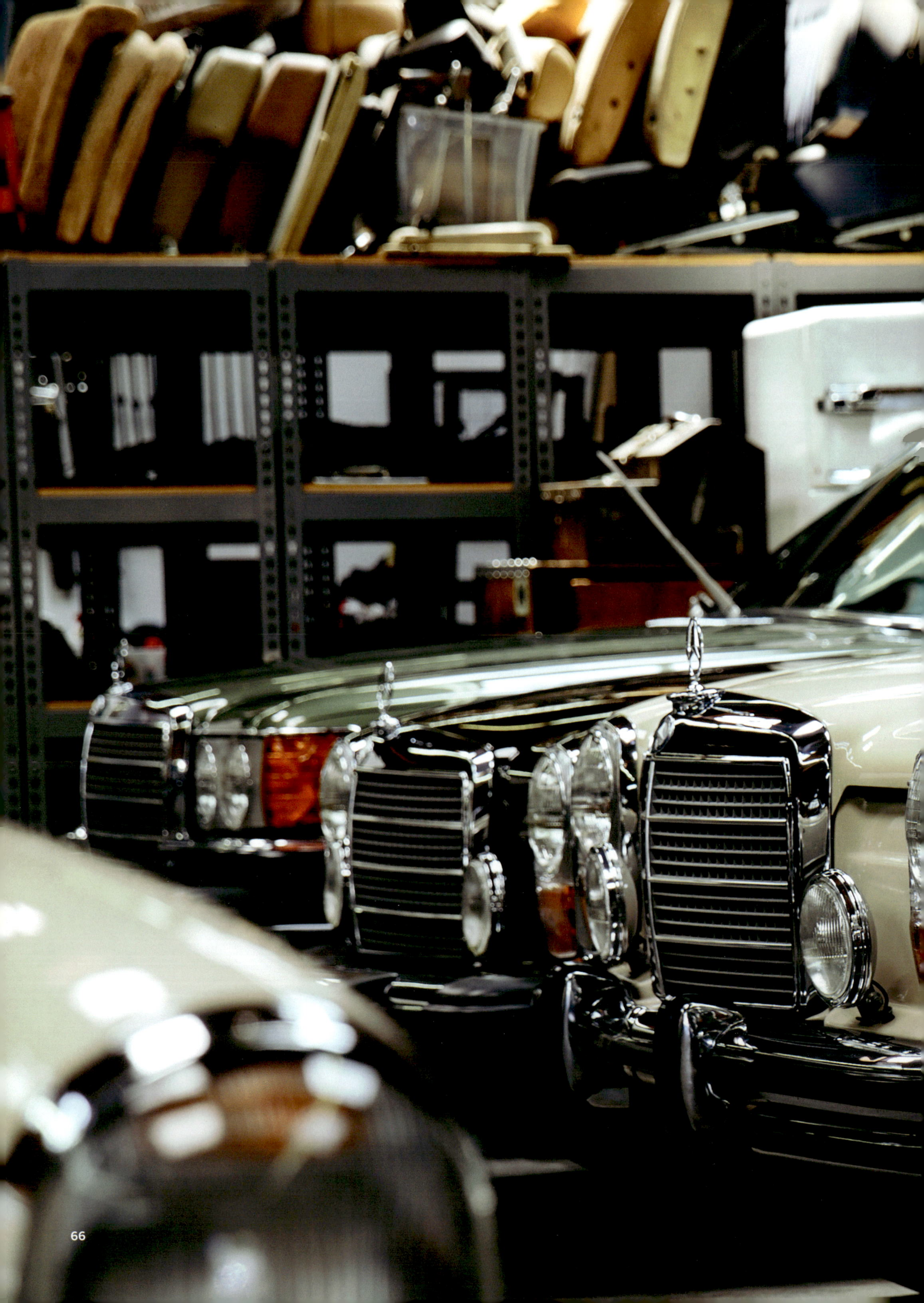

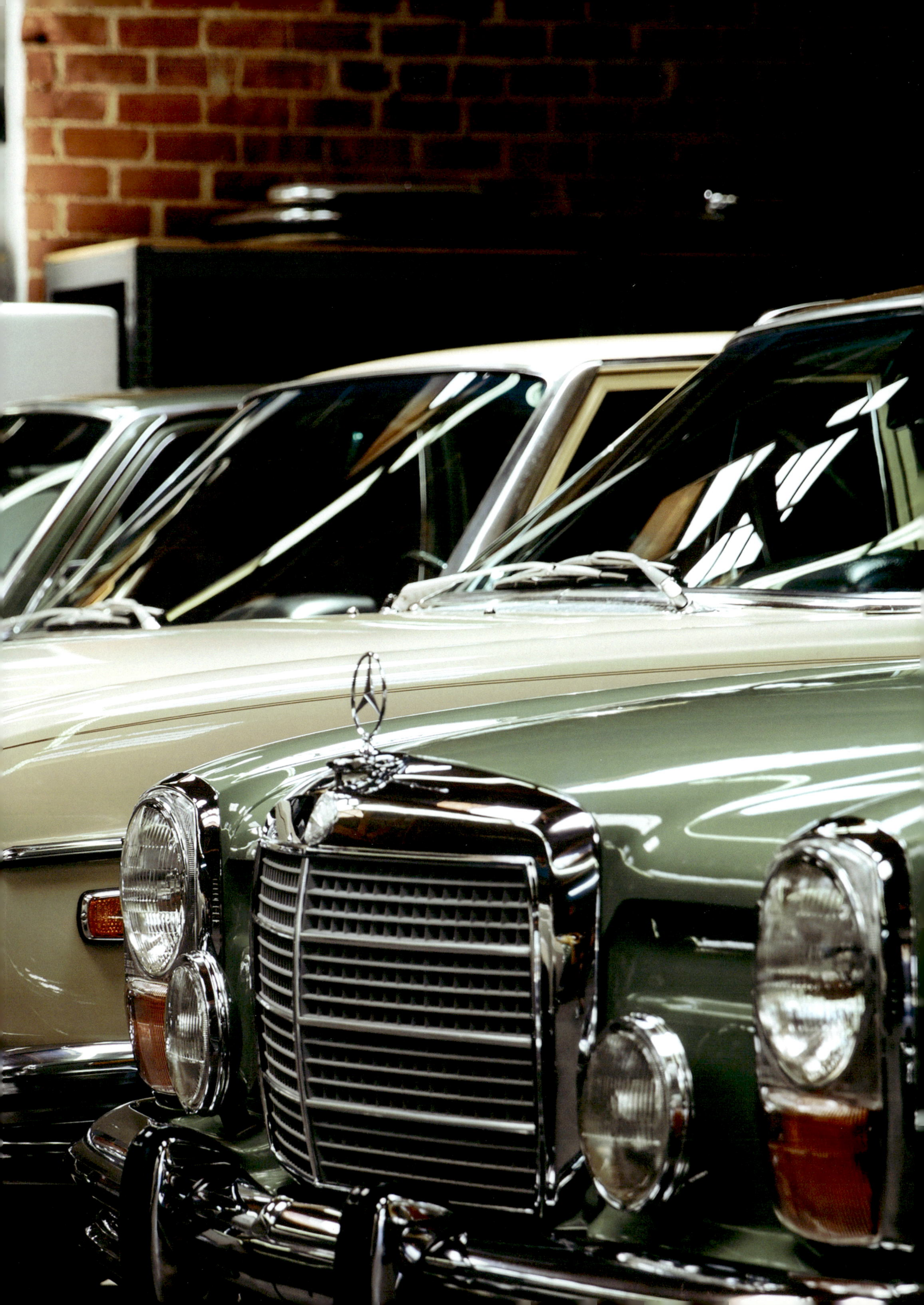

MERCEDES MOTORING

The Mercedes-Benz diesels made from the late 1960s to the mid 1980s might be the best mass-produced cars in history. At least that's the opinion of J.G. Francis, owner of Mercedes Motoring in Glendale, California. And after you've seen or driven a car he's restored, it's hard to disagree.

Francis grew up in Nevada watching a local mechanic work on and extoll the virtues of the Mercedes-Benz diesel. But despite always being obsessed with moving parts and having a penchant for working with his hands, the bug to restore cars didn't really bite J.G. until 2003. That's when he came across a maple yellow 300SD for $700 and dove in headfirst, starting Mercedes Motoring shortly after. It remains a small operation more than a decade later.

When hunting for fresh candidates, J.G. says you should be on the lookout for examples with low mileage, no rust or accidents, original paint, only a few owners, and, if possible, a garage they called home. Maintenance records are also vital, mostly because they validate the actual mileage of the vehicle. A pre-purchase inspection is another valuable exercise he recommends, calling it "the best $200 bucks you'll ever spend in the classic car world."

Once a prospective vehicle gets the green light for restoration, it's taken apart, piece by piece, polished and perfected in hopes of returning it to that just-rolled-off-the-assembly-line condition. The shop focuses mostly on lighter, semi-restorations, targeting cars that keep as much factory originality as possible. According to J.G., light restorations have about 200 points of interest to restore. And on full restorations, you can expect the crew to be hard at work for more than a year.

The Mercedes Motoring HQ can hold up to 20 cars, with overflow ending up at J.G.'s home, ensuring he has the option to tinker 24/7. Amongst the head turners you'll find at the shop is a 1973 250C that is J.G.'s personal vehicle. The unique color combo, in conjunction with the lack of options, make it more attractive to Francis, who says the no-frills model is "pretty much just four wheels and a motor." It's also lowered and outfitted with 15-inch wheels with matching hubcaps — an ultra-rare modification.

Like all of the rides he and his crew build, the green coupe is driven on a regular basis, from state-crossing road trips to daily commutes, which is how Francis recommends these restored beauties should be conserved. "I'm thinking of all the bat-shit crazy experiences I've had in my life and none of them revolve around staring at my car in the garage. What a dagger through the heart it would be had I let all those beautiful cars just sit there."

And so it's the open road for most of the Benzes that Mercedes Motoring brings back to life, as more and more automotive enthusiasts experience the passion and attention to detail this shop employs on a daily basis. In the meantime, J.G. will continue the hunt, taking the long way home.

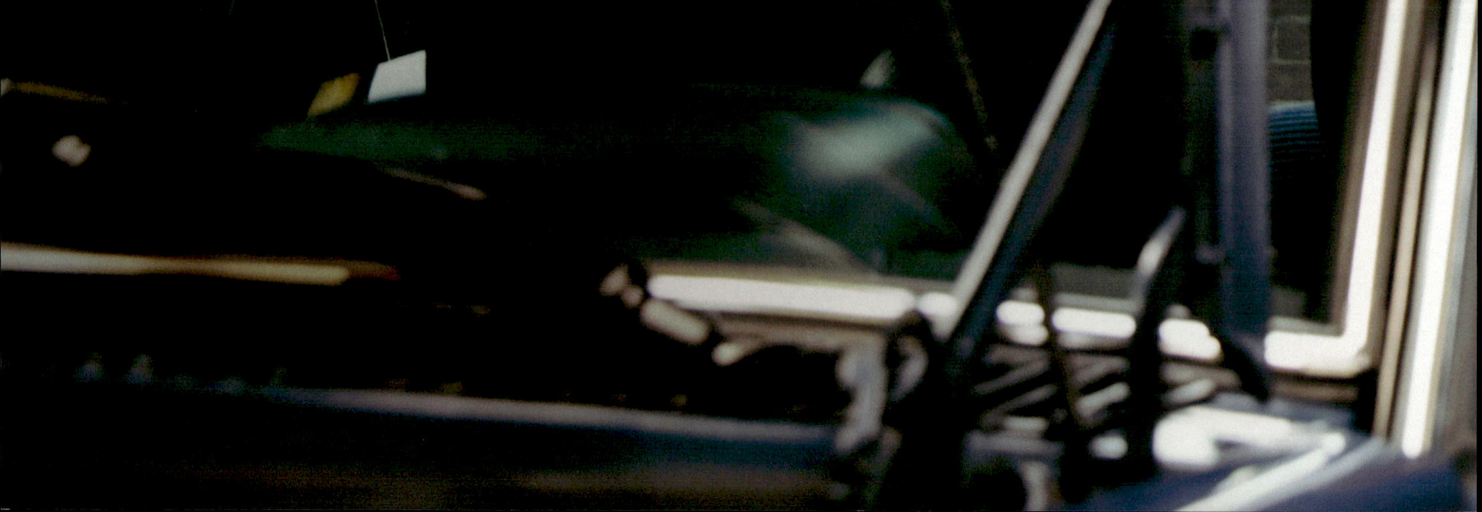

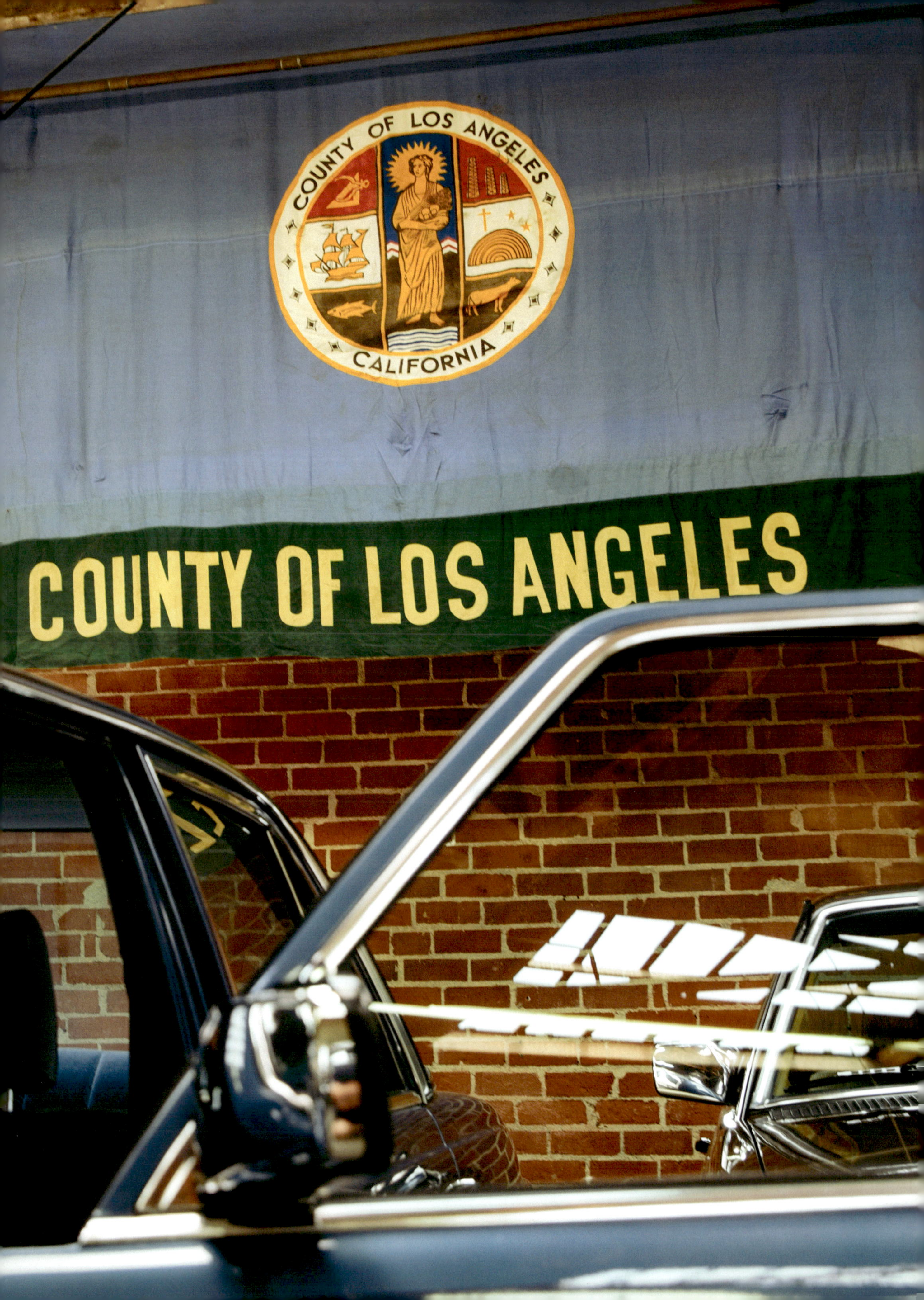
COUNTY OF LOS ANGELES
CALIFORNIA
COUNTY OF LOS ANGELES

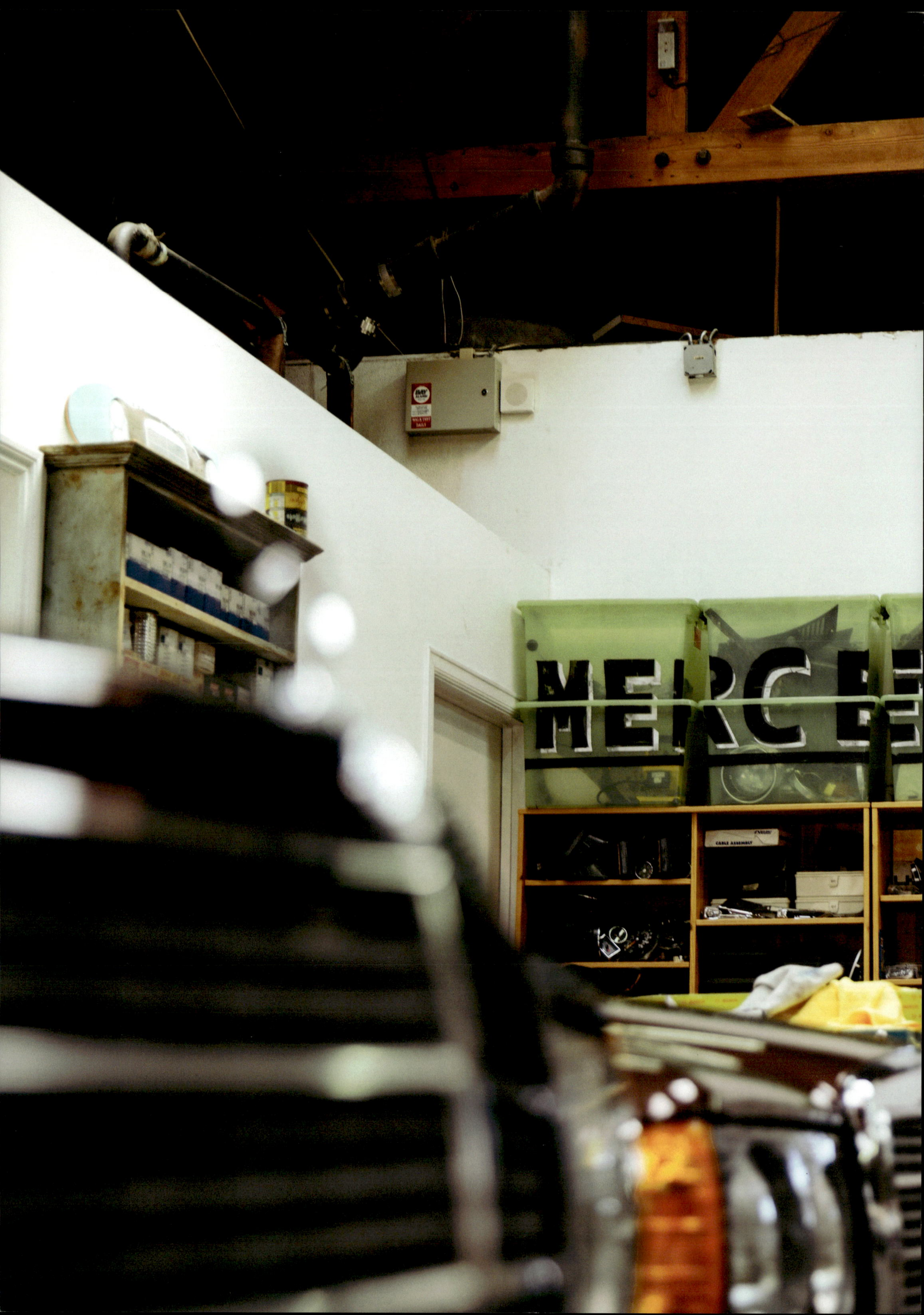

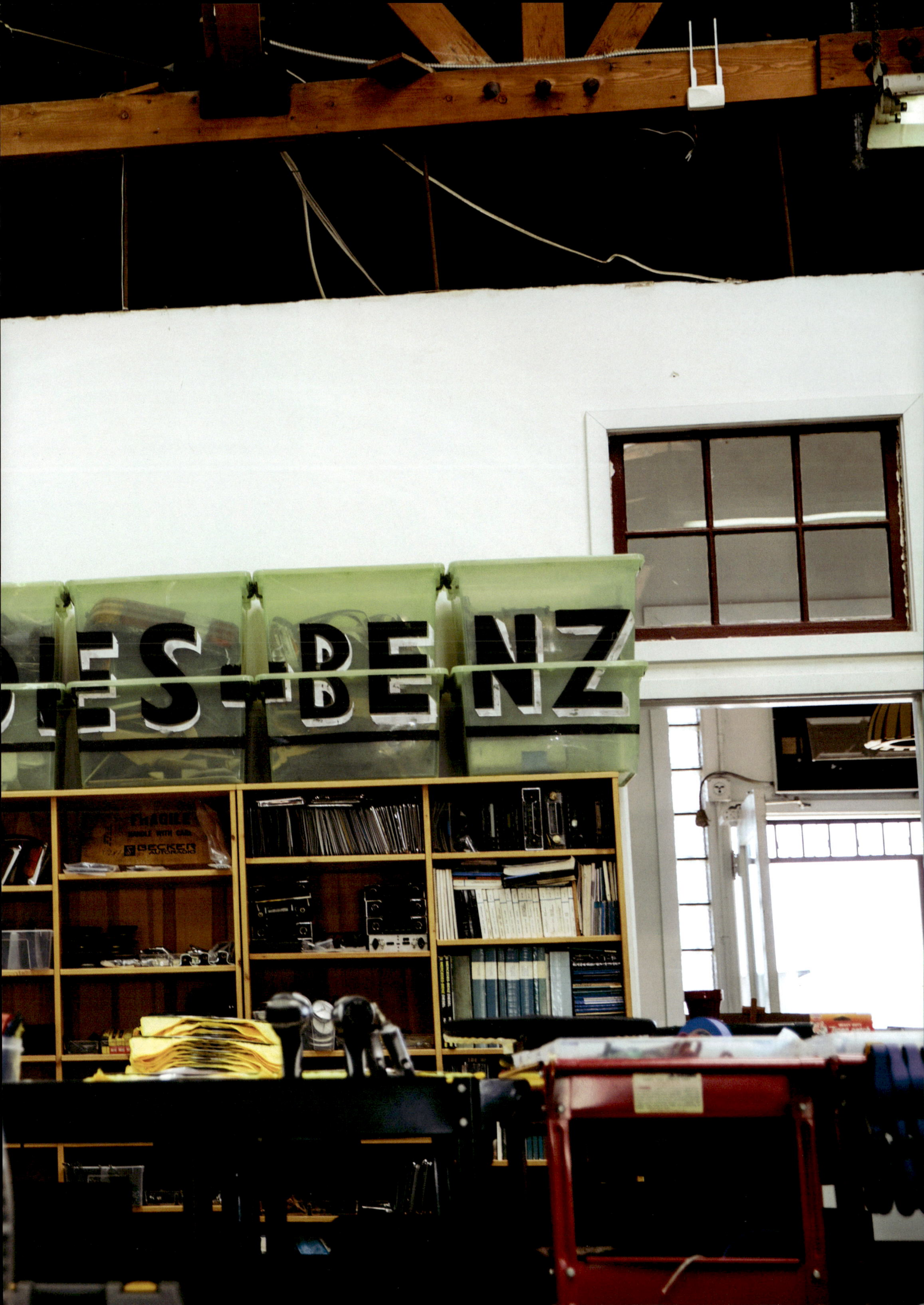
ES-BENZ
FRAGILE

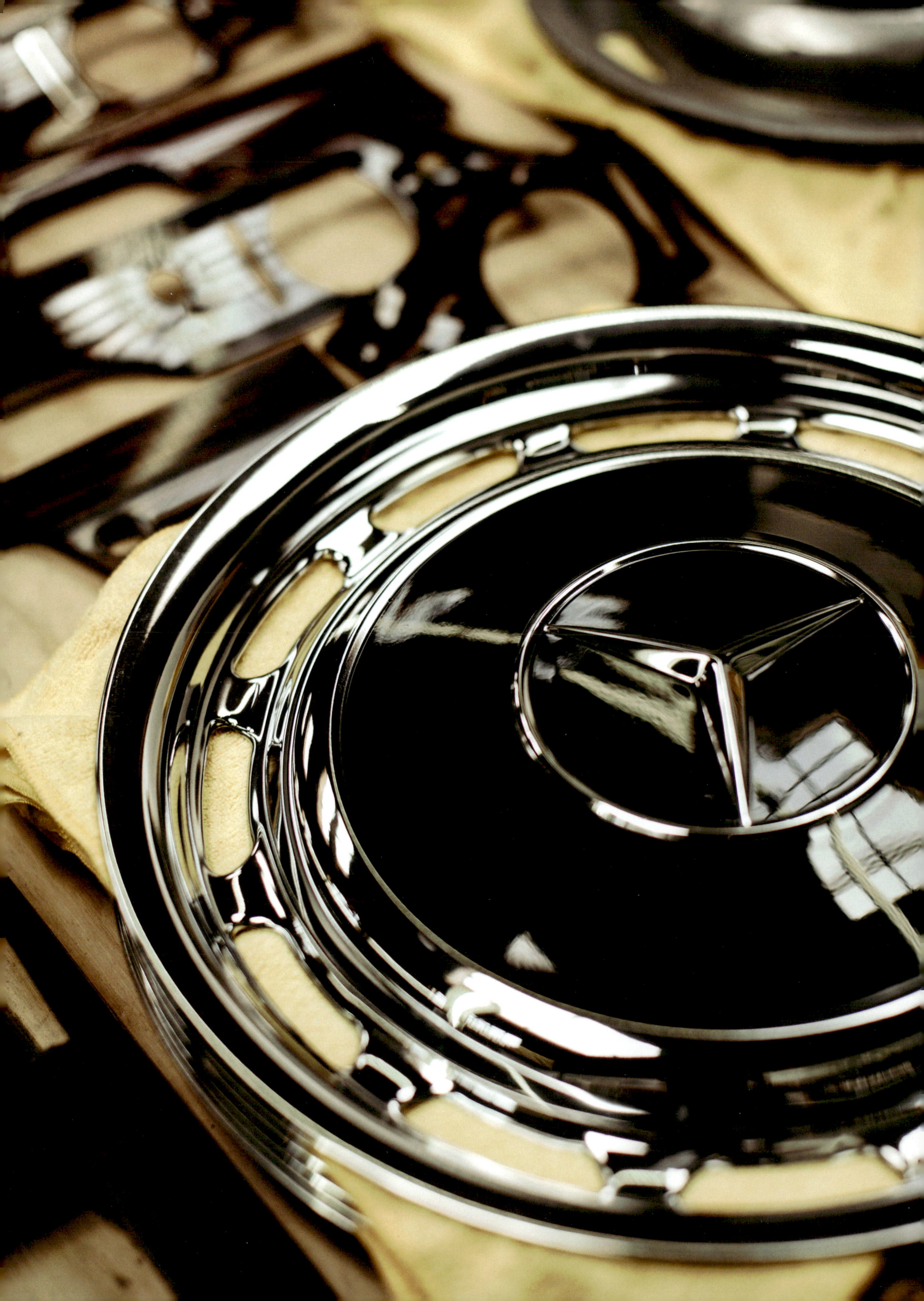

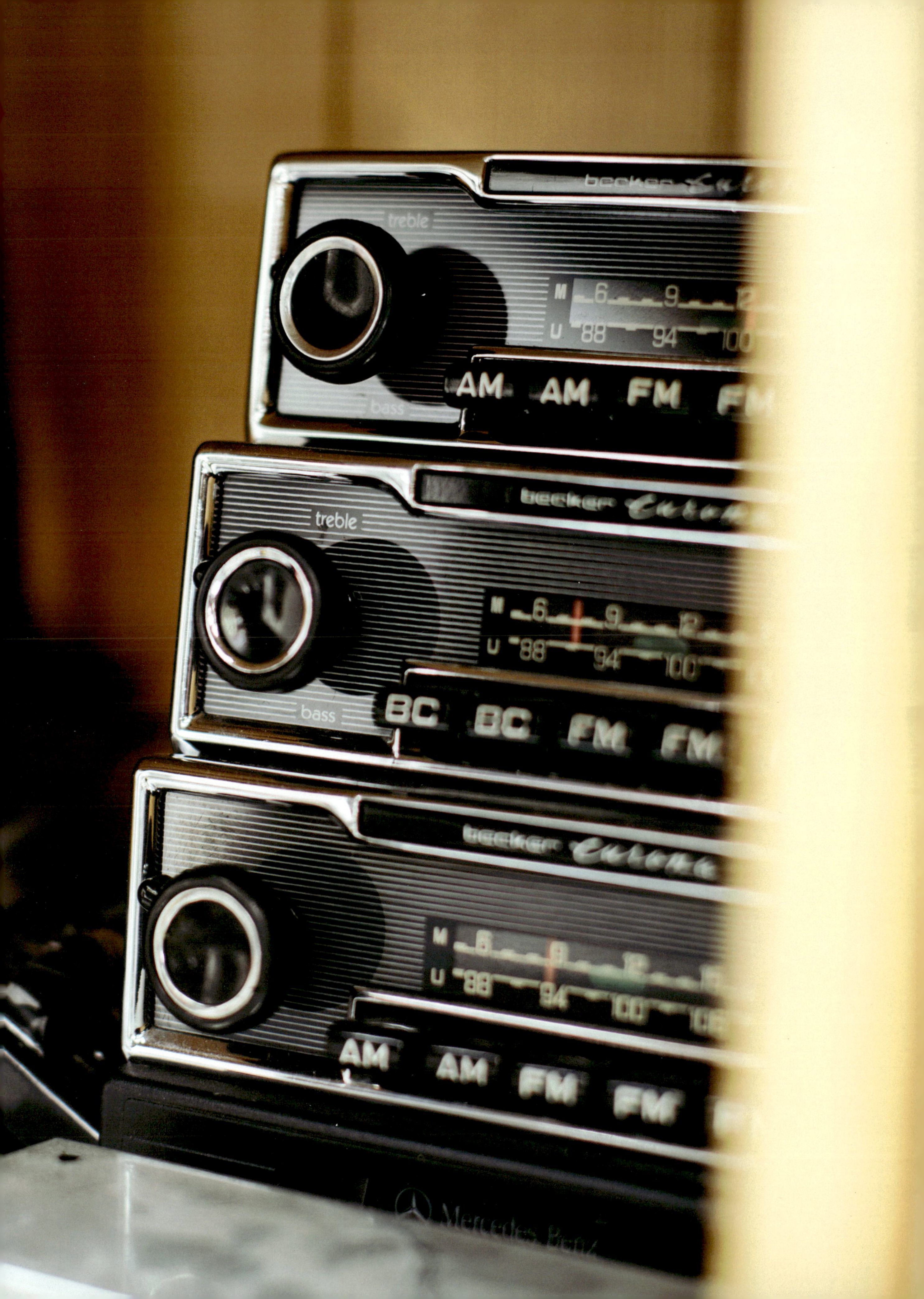
becker
treble
bass
AM AM FM
becker
treble
bass
BC BC FM FM
AM AM FM FM

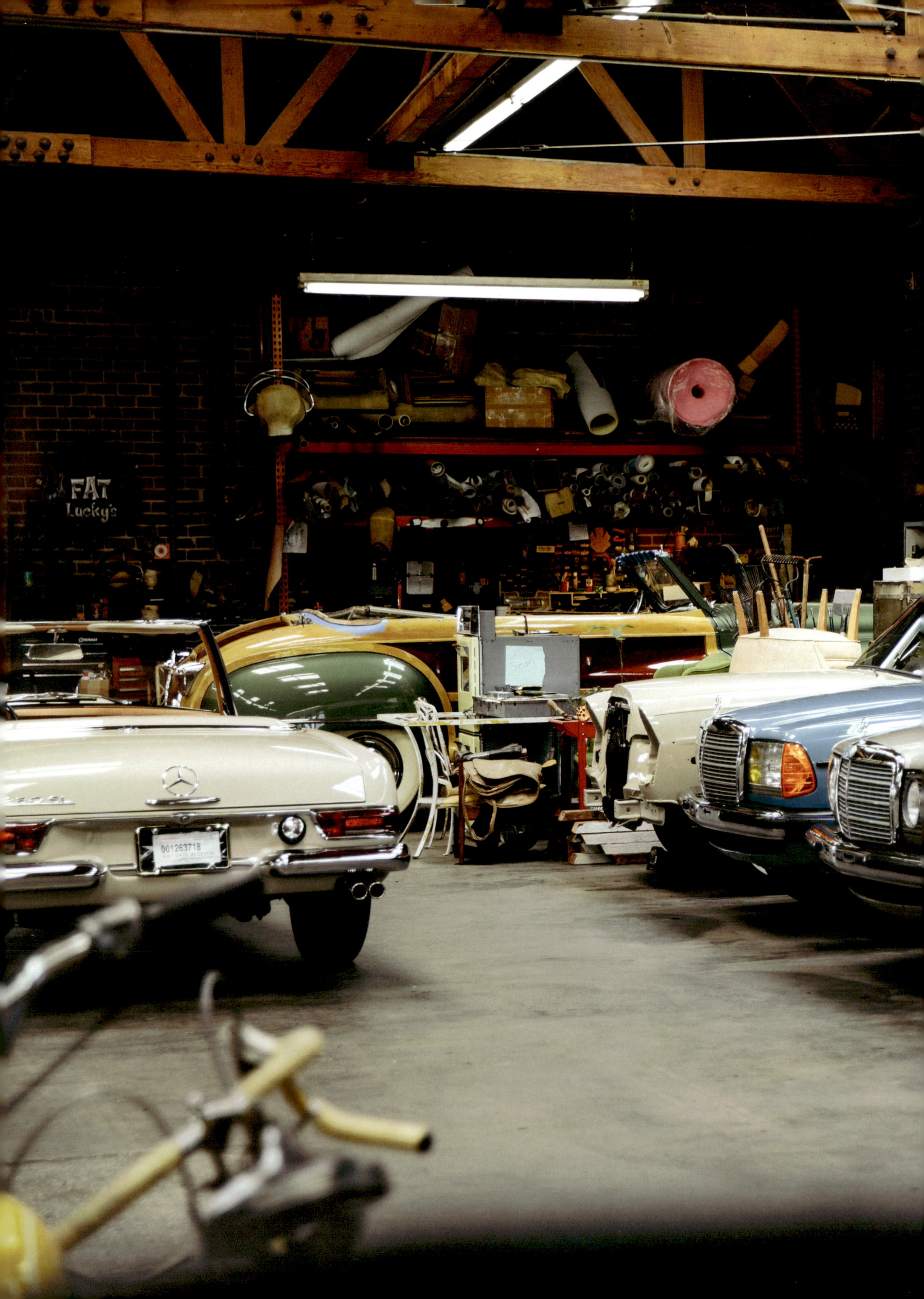
FAT
Lucky's
Sean
001263718

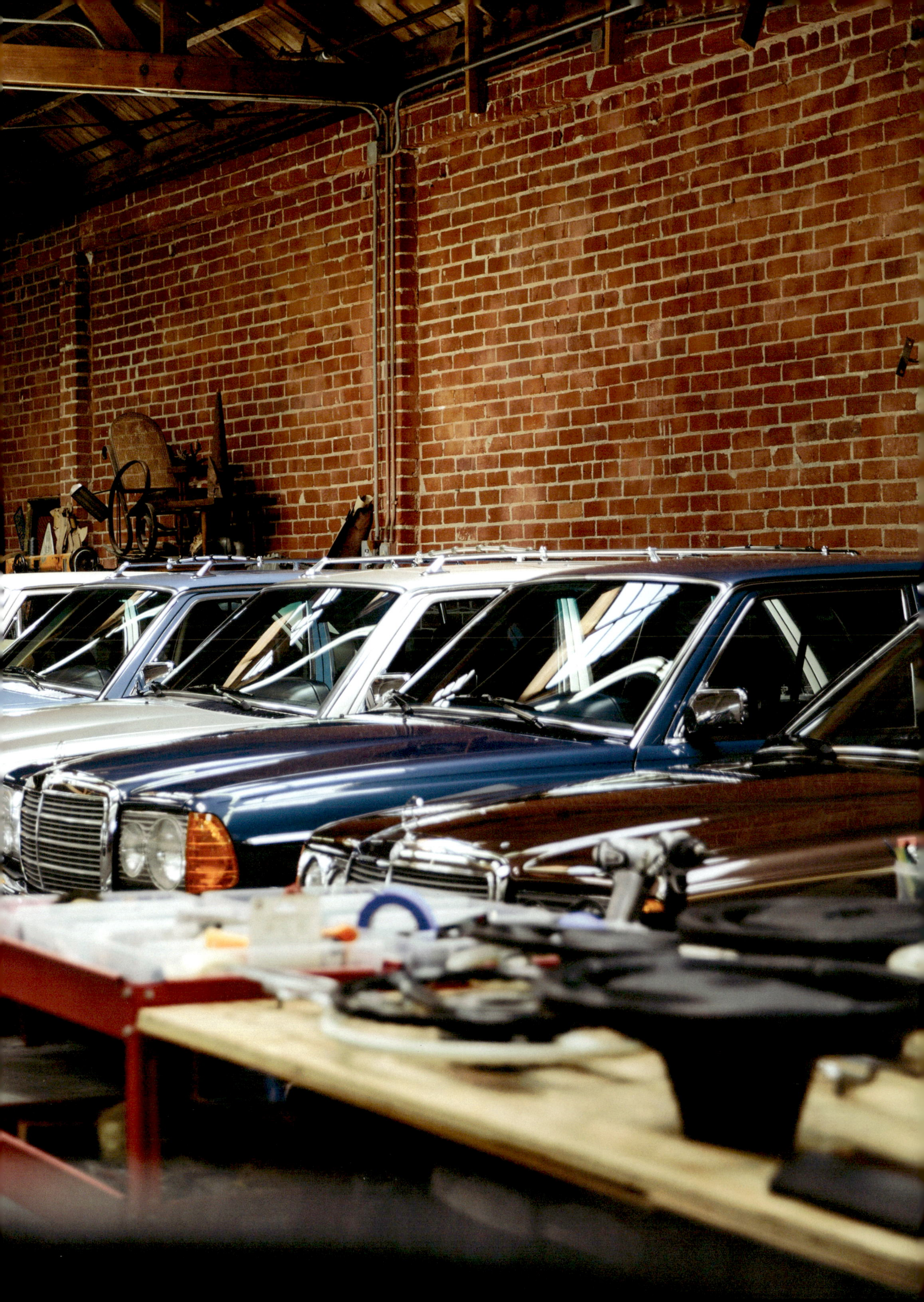

CASTAIC, CALIFORNIA

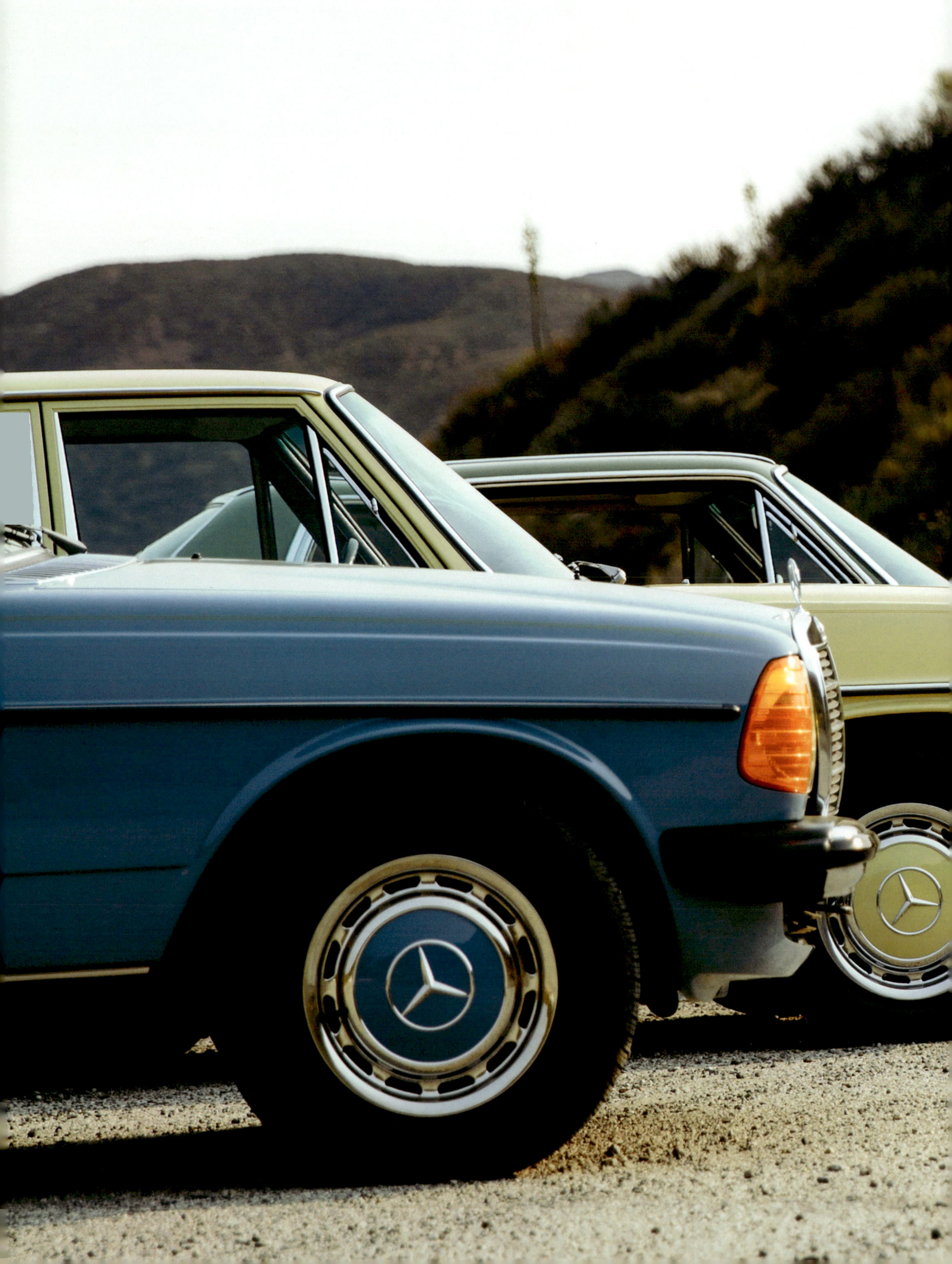

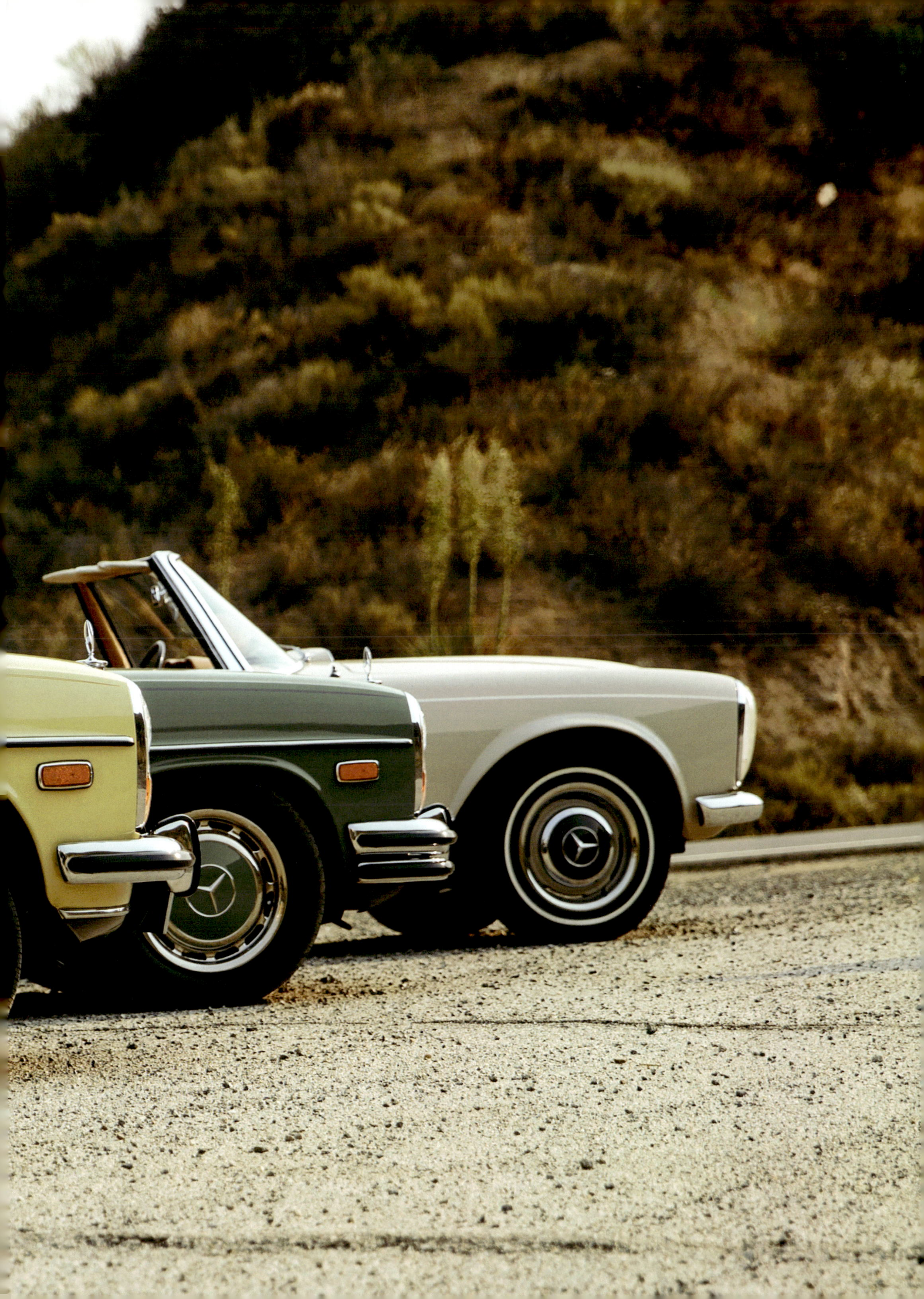

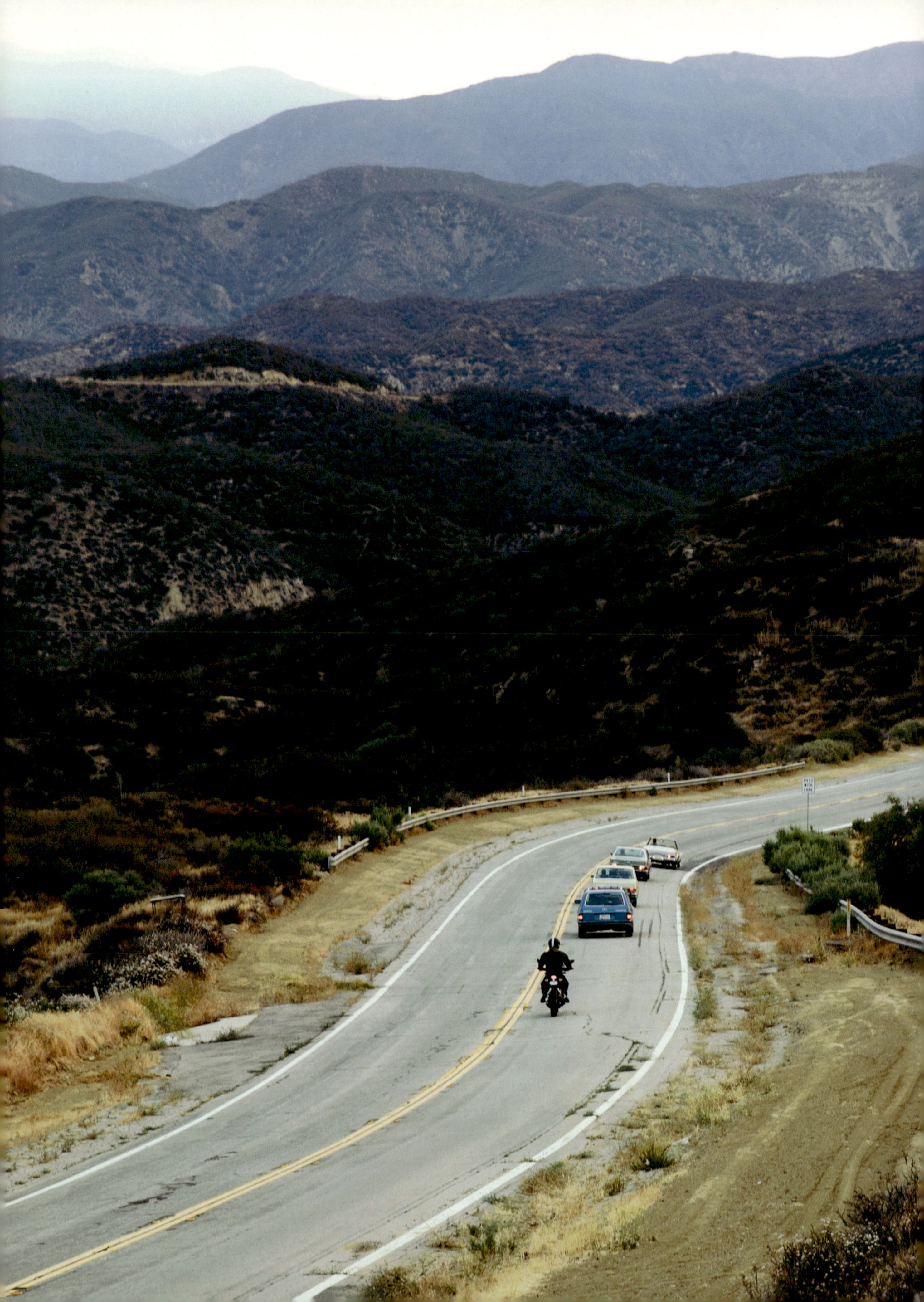

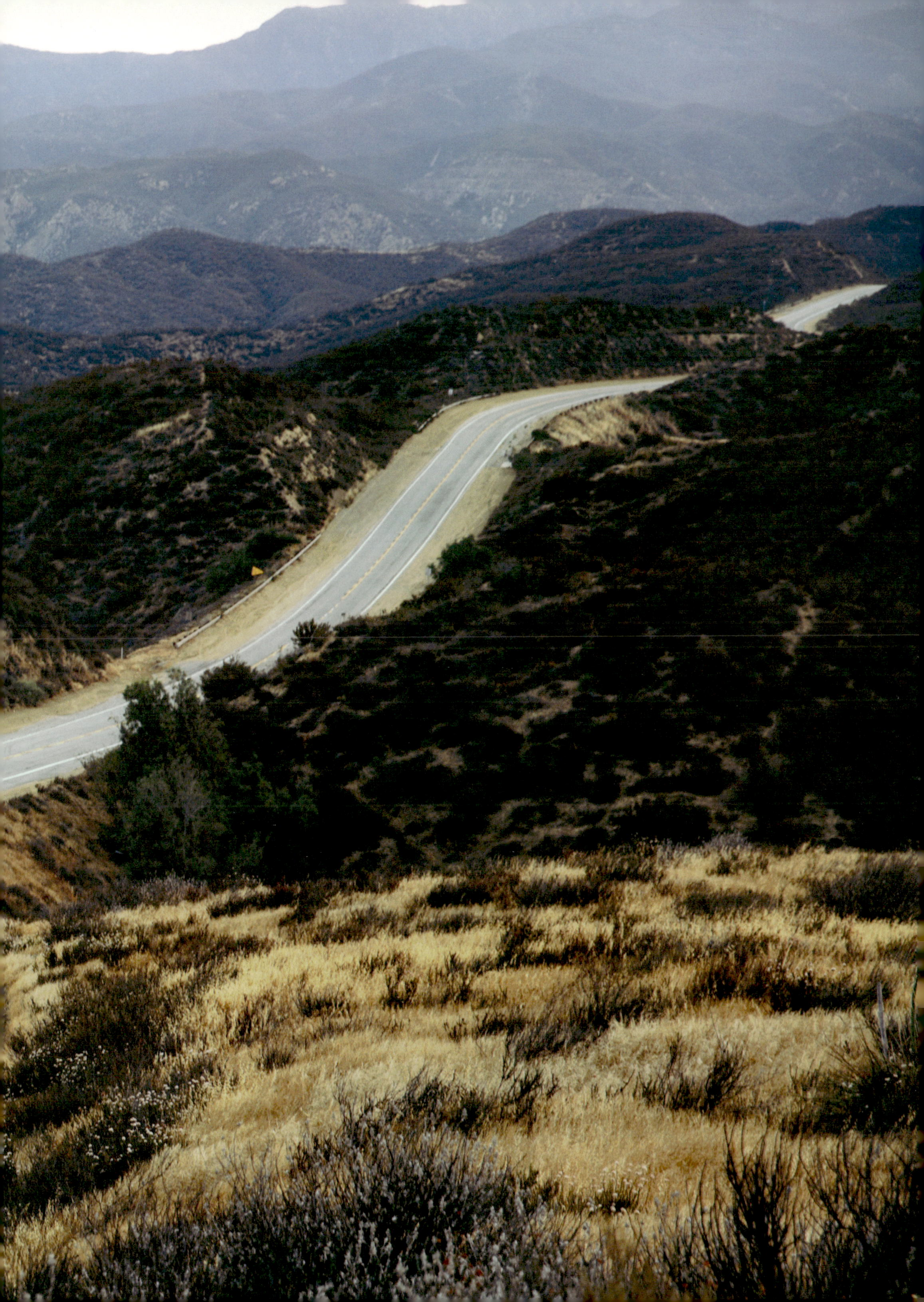

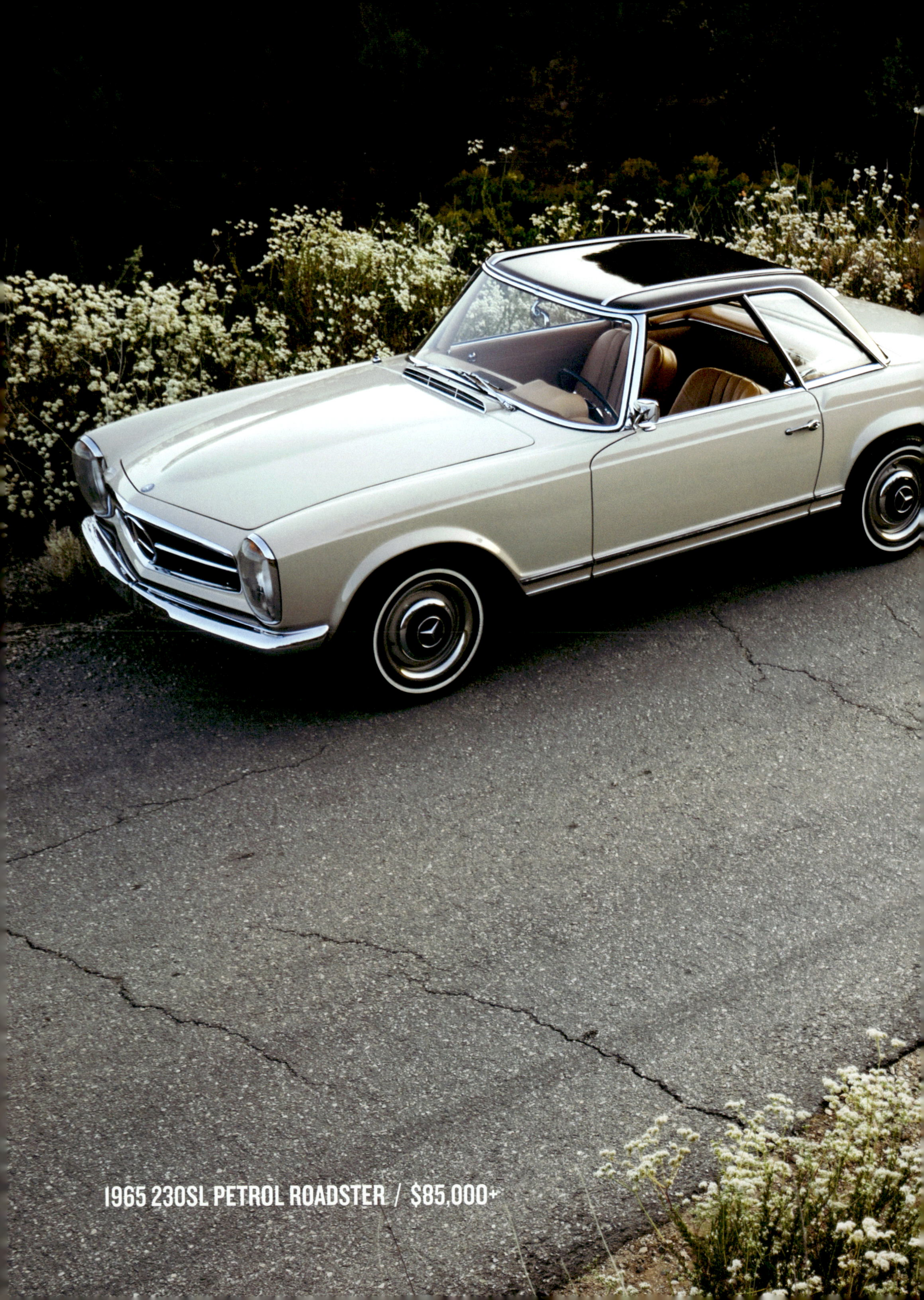

1965 230SL PETROL ROADSTER / $85,000+

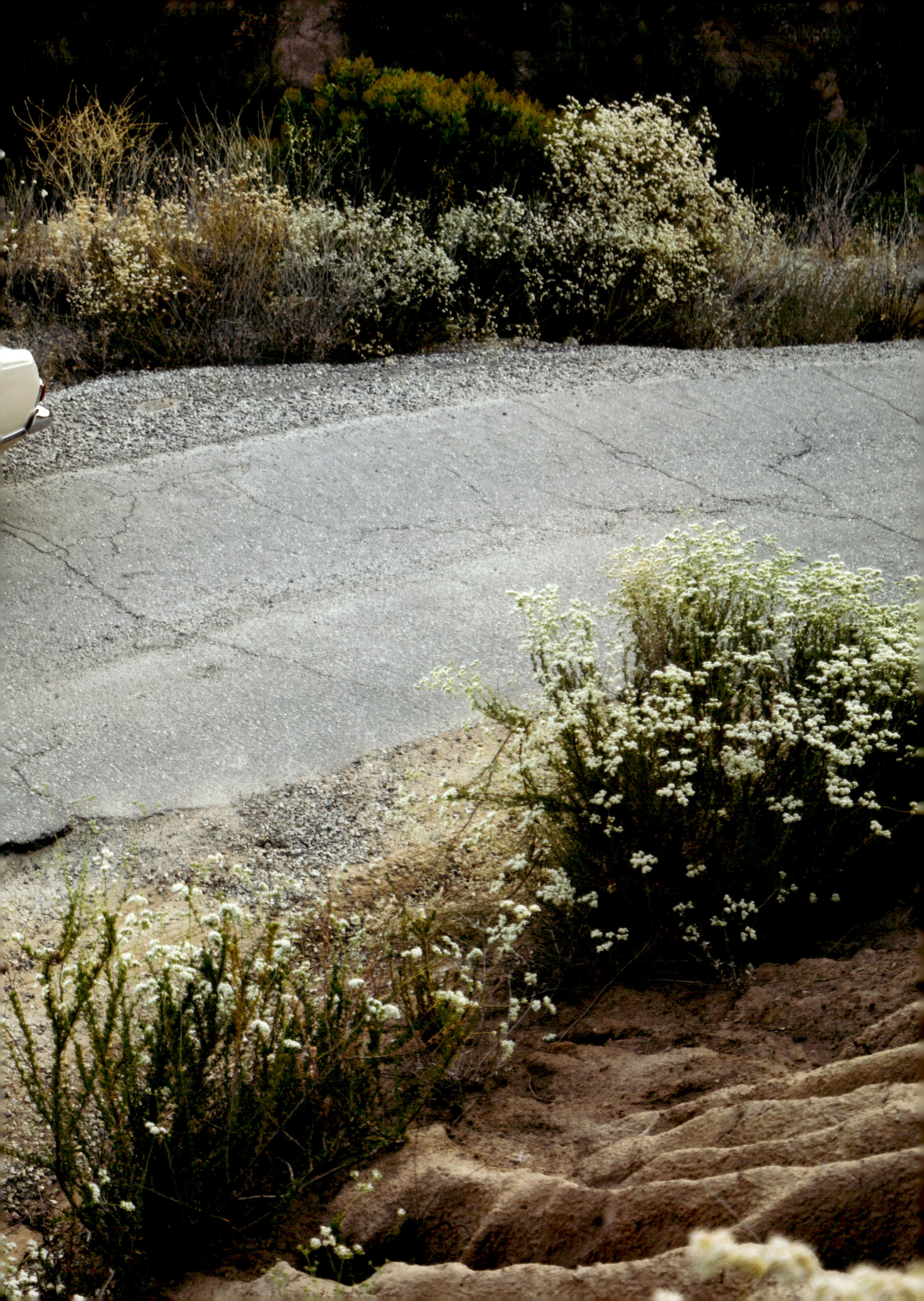

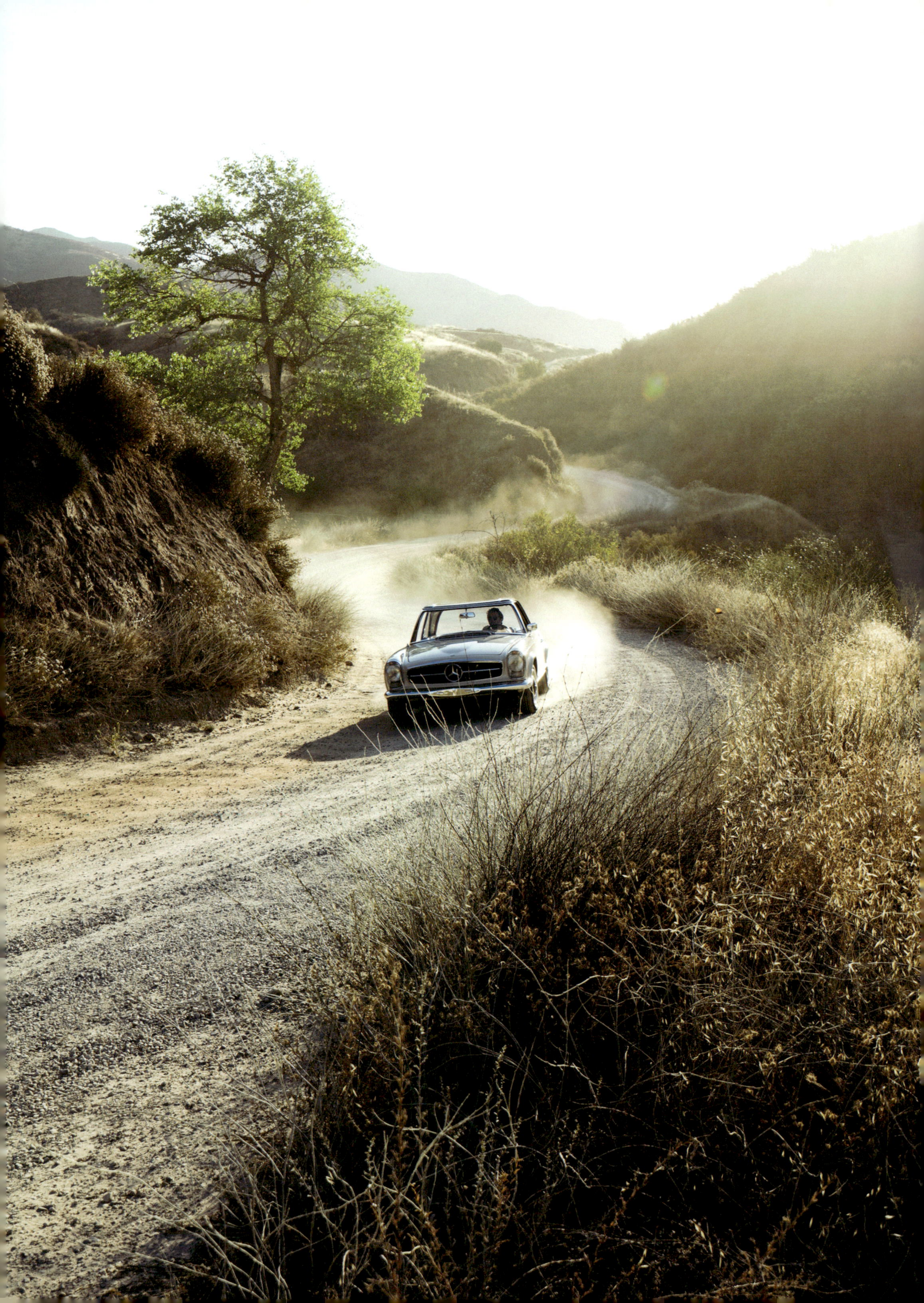

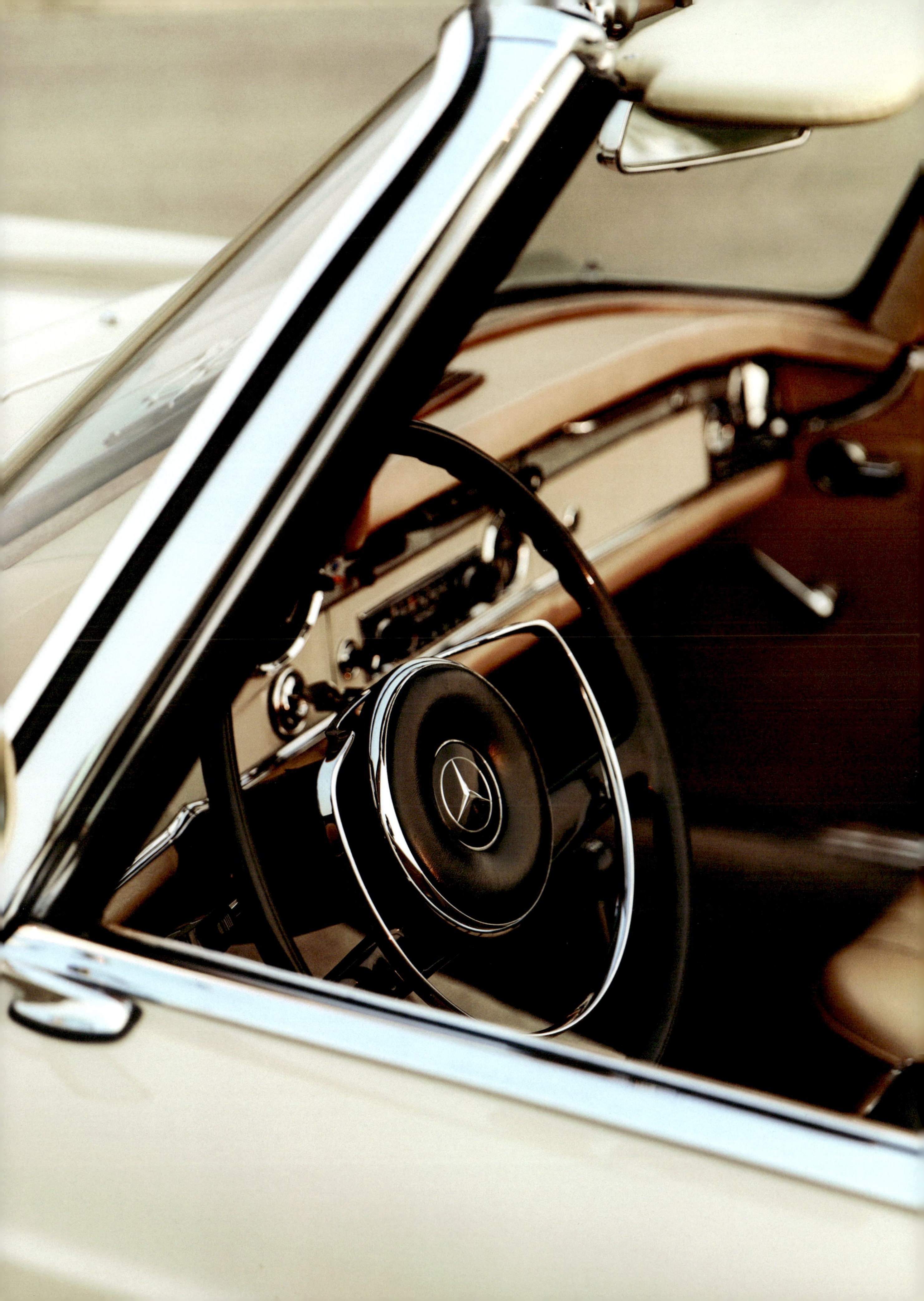

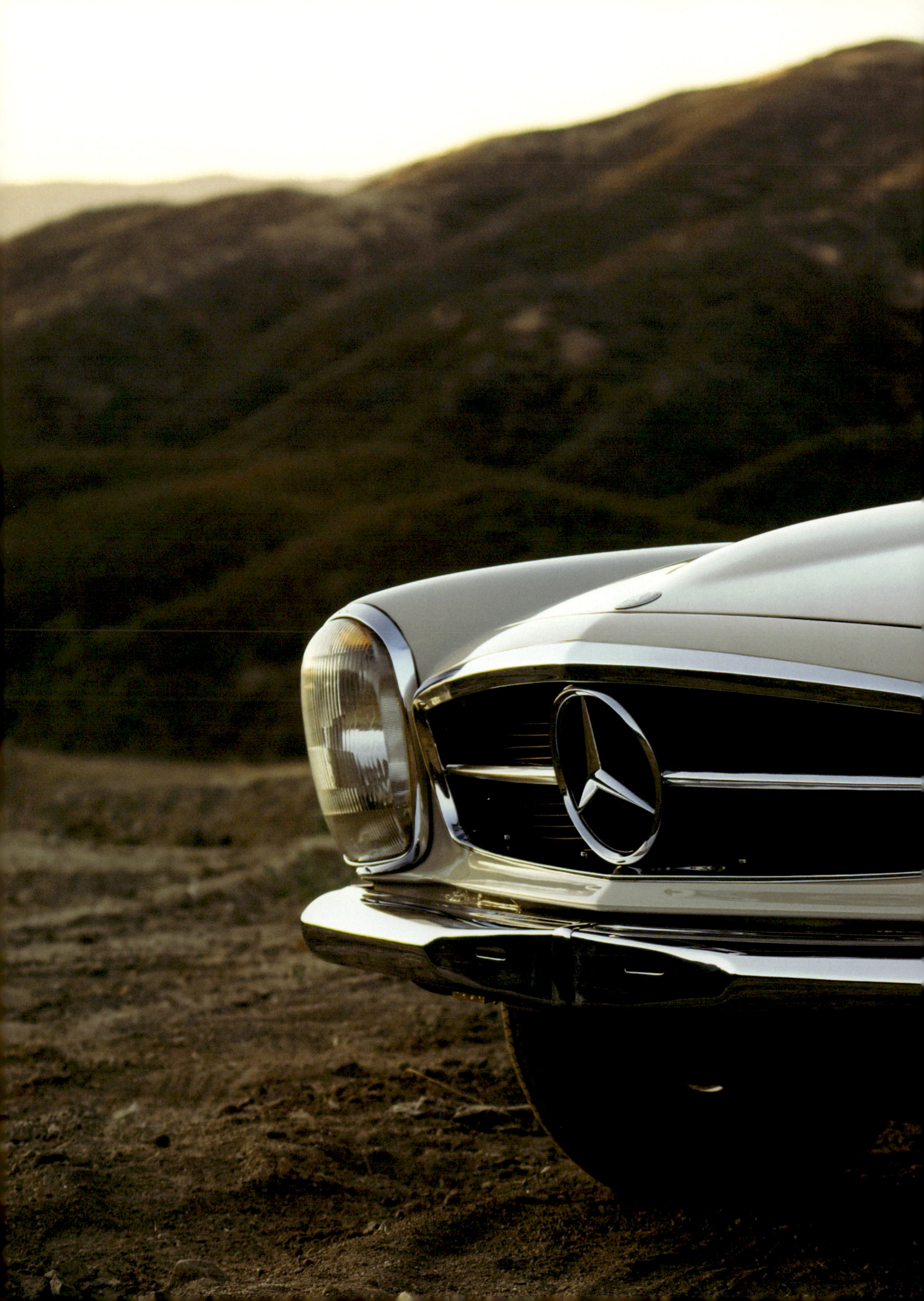

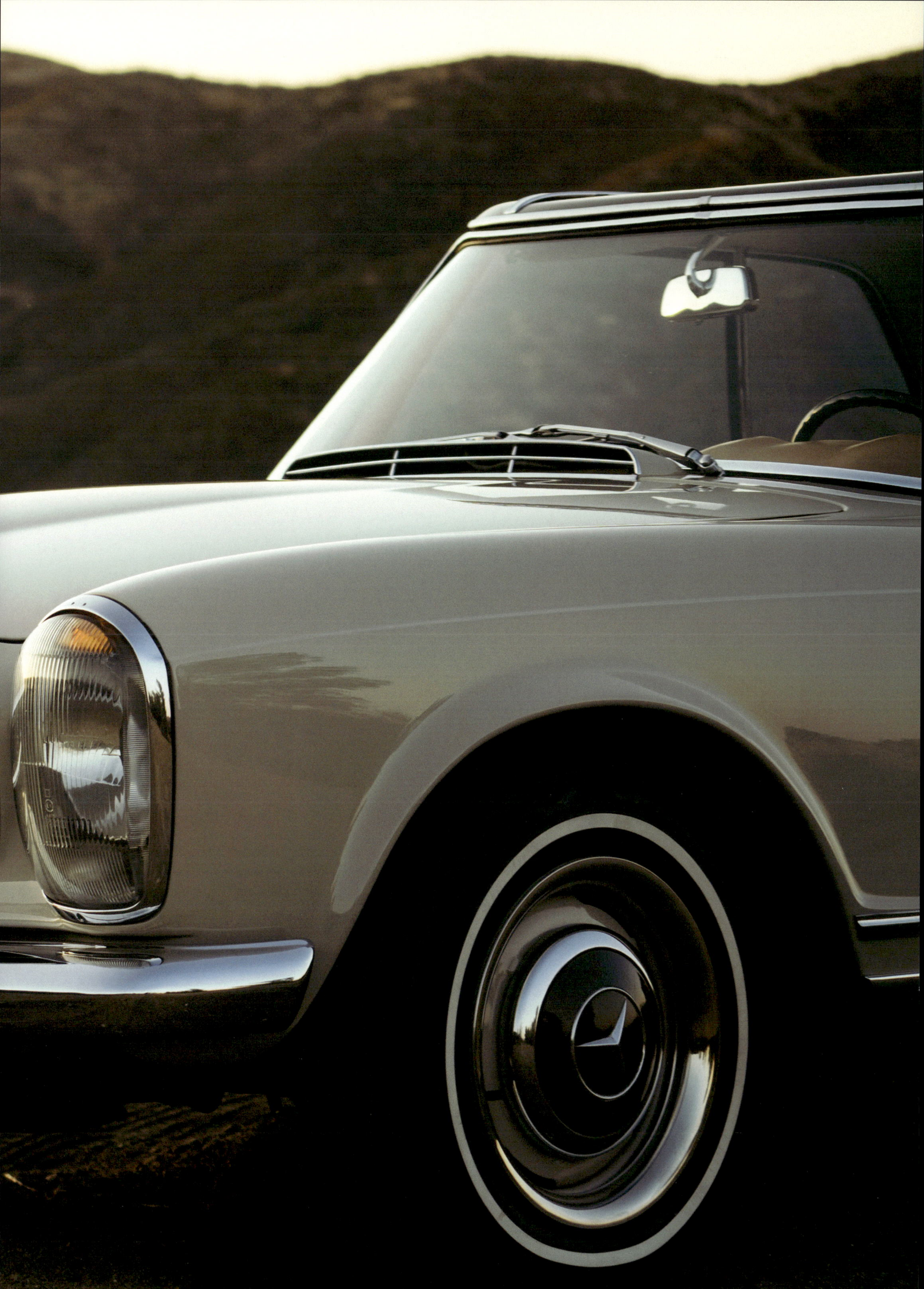

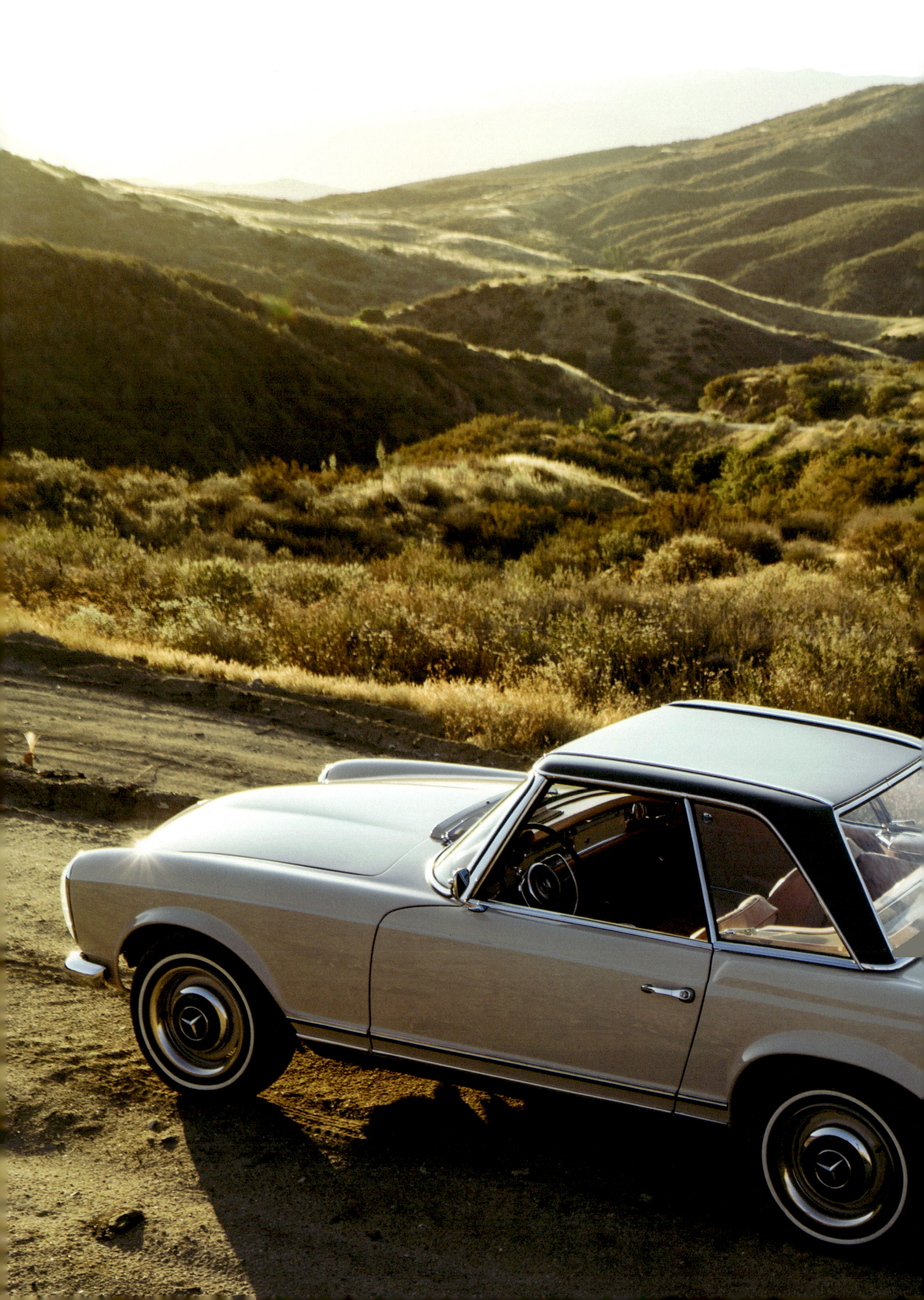

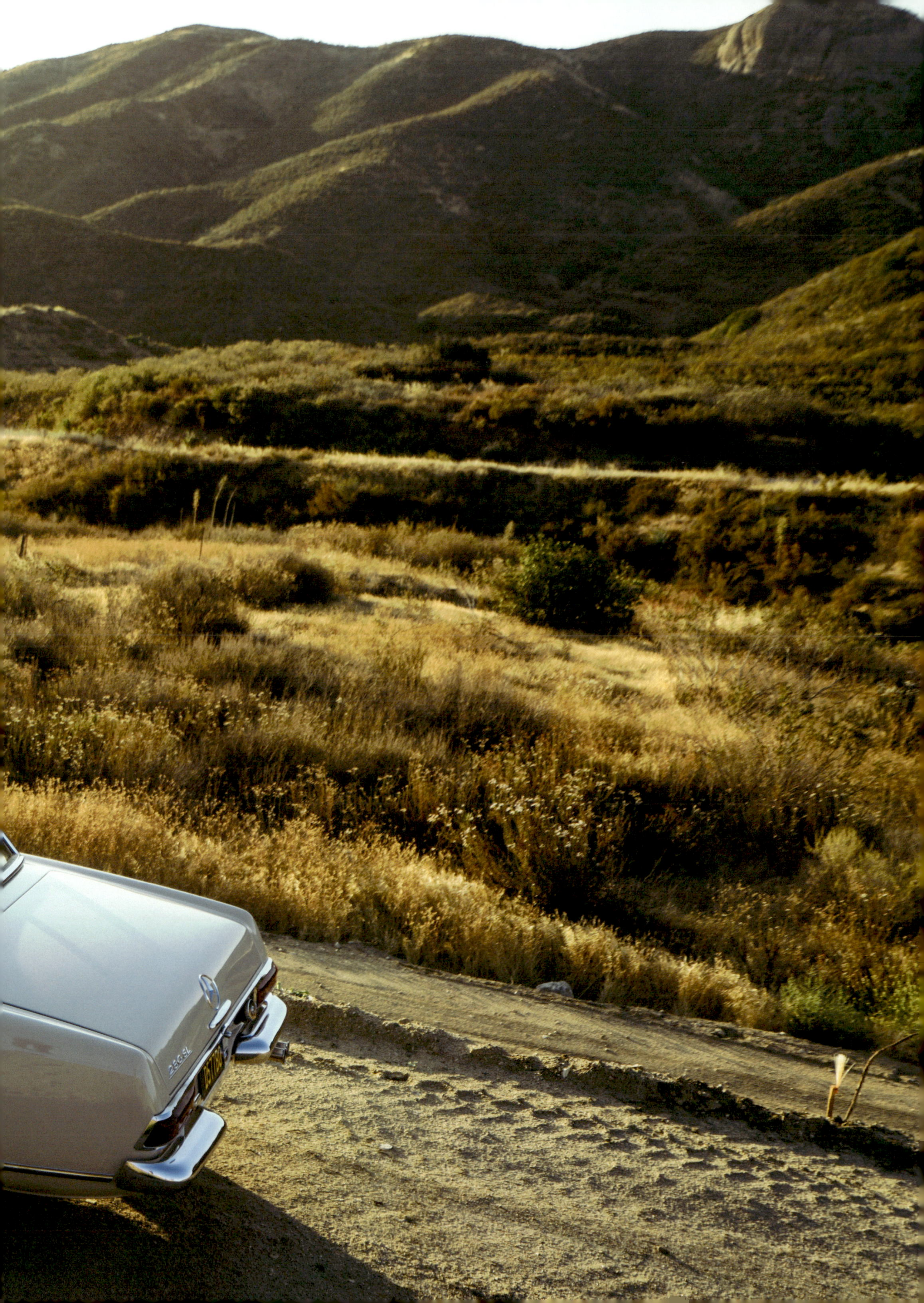

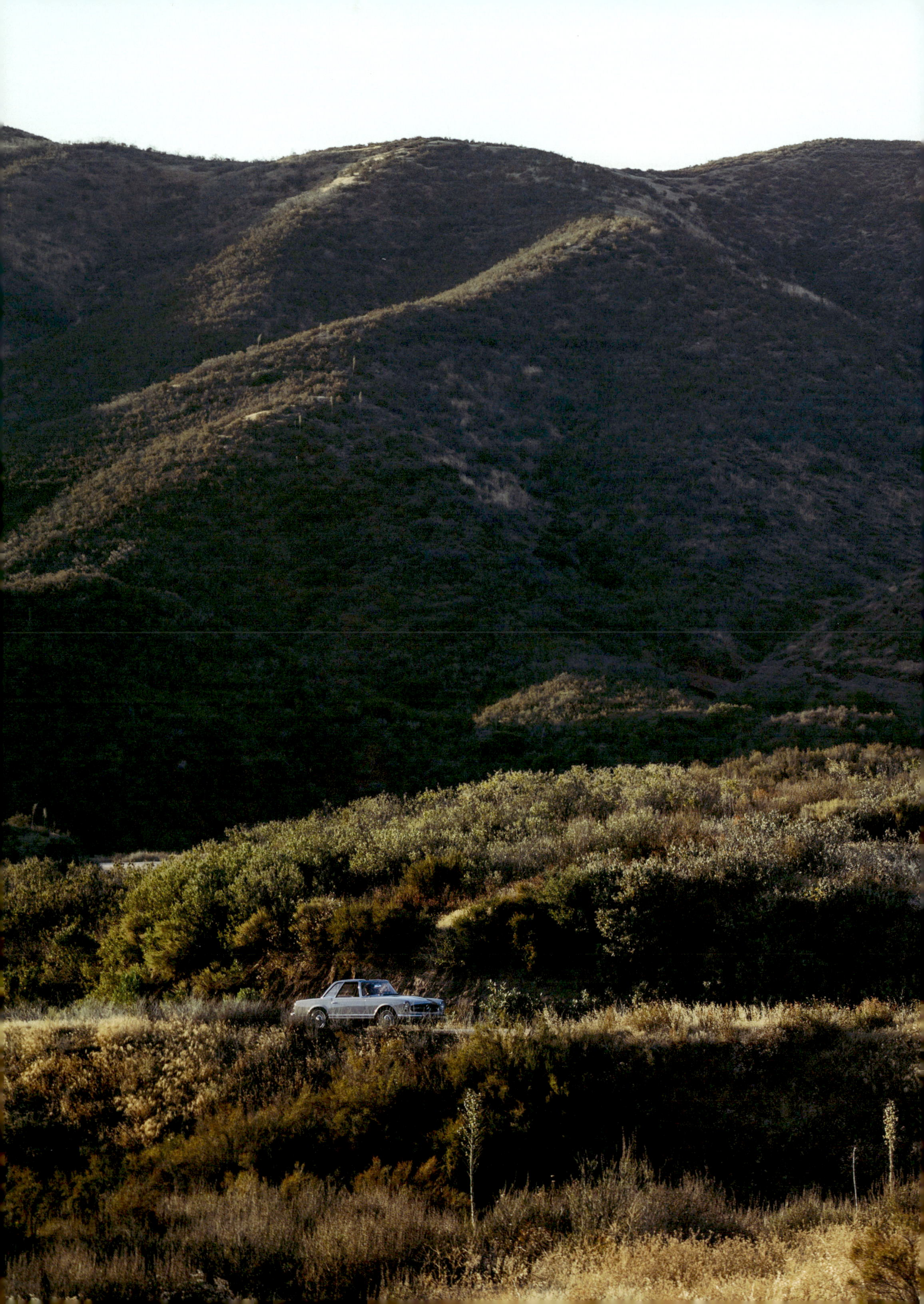

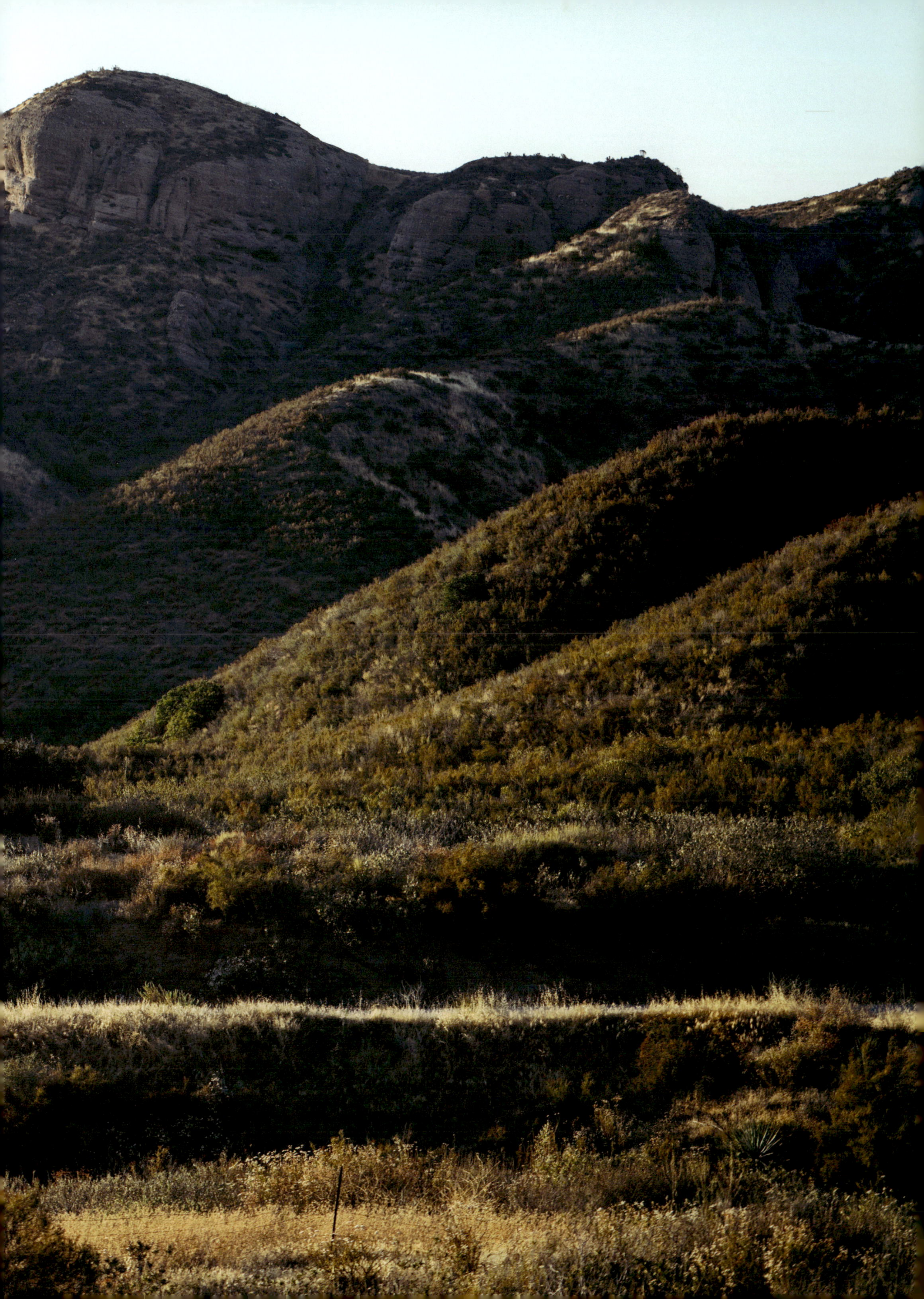

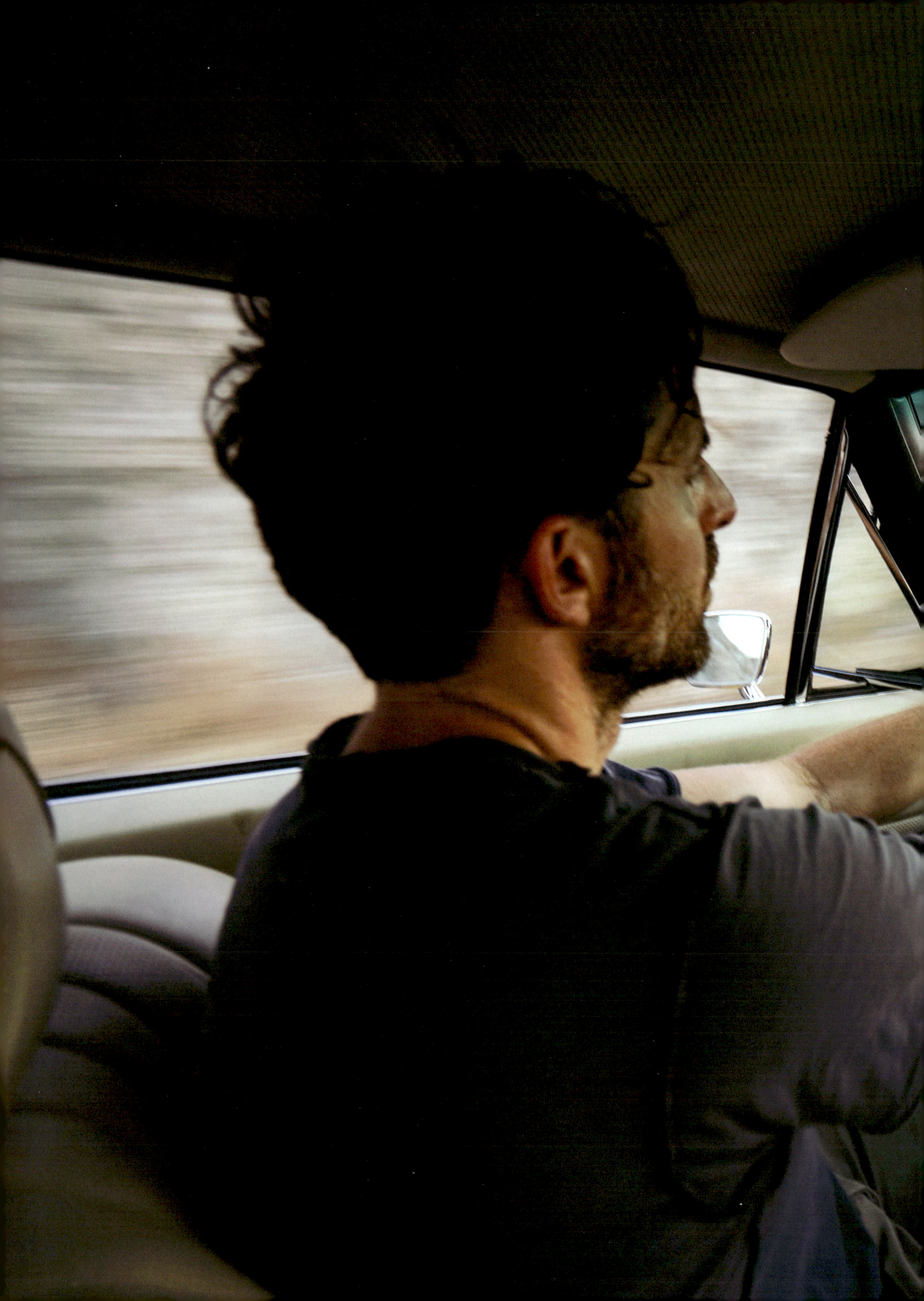

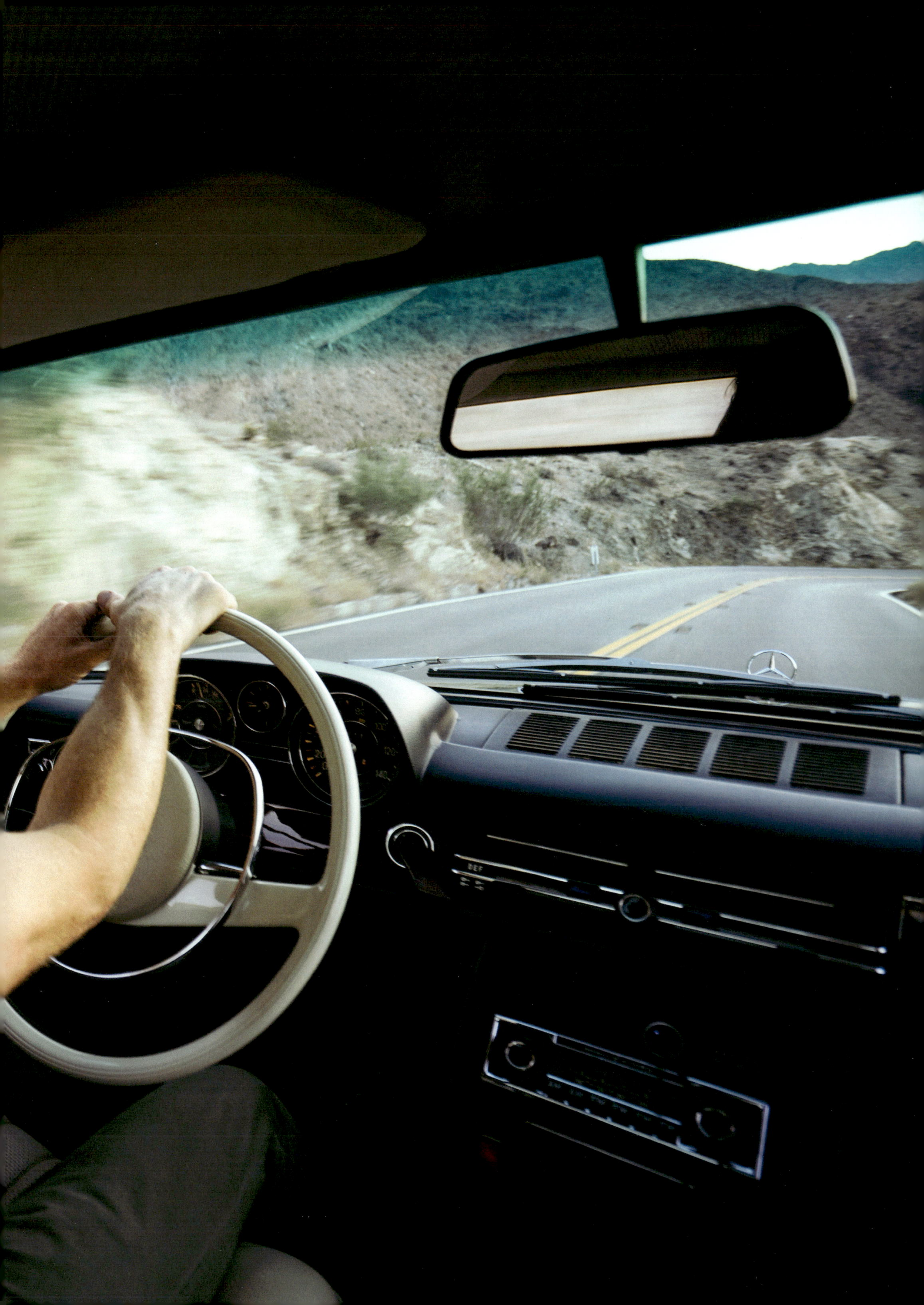

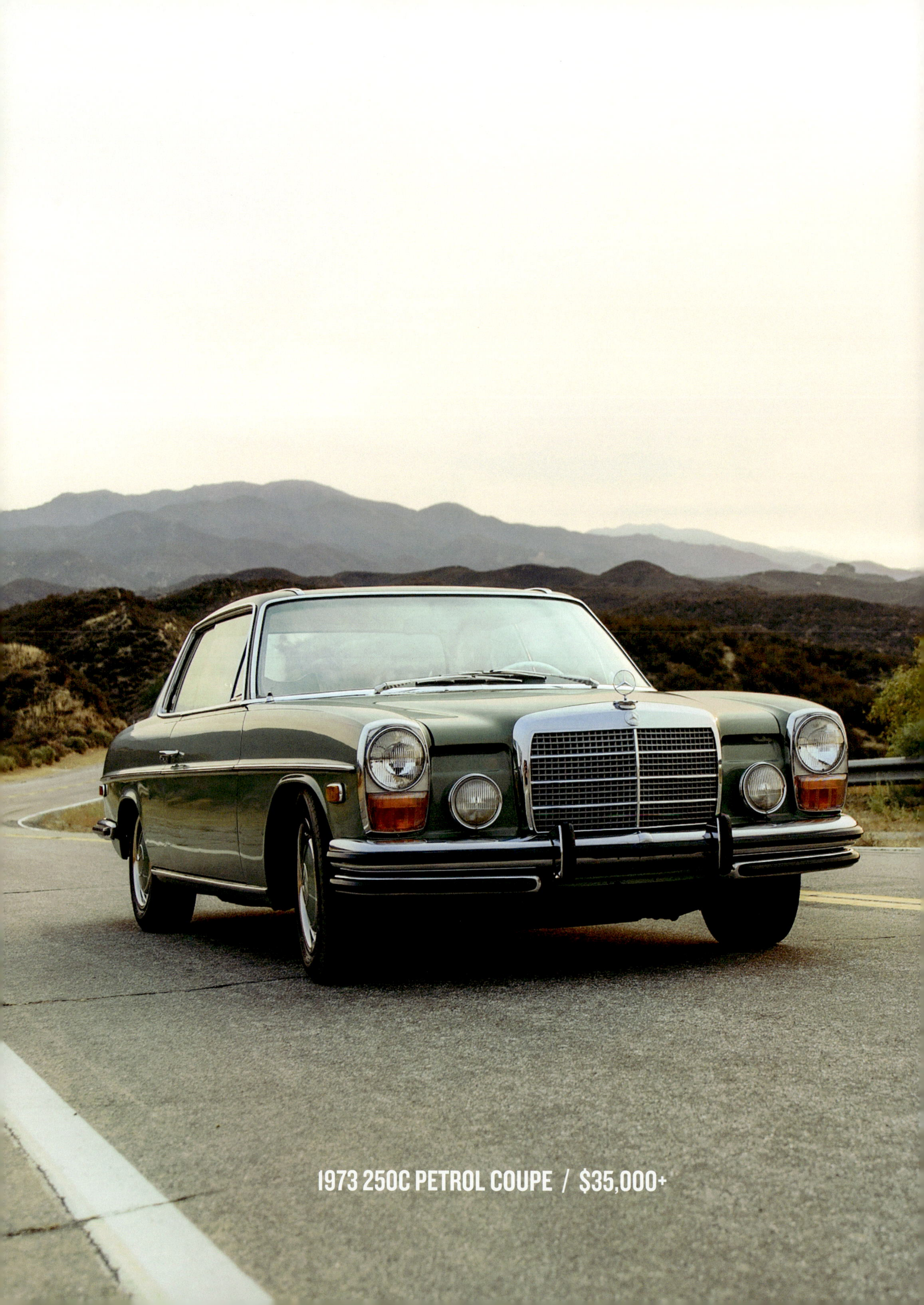

1973 250C PETROL COUPE / $35,000+

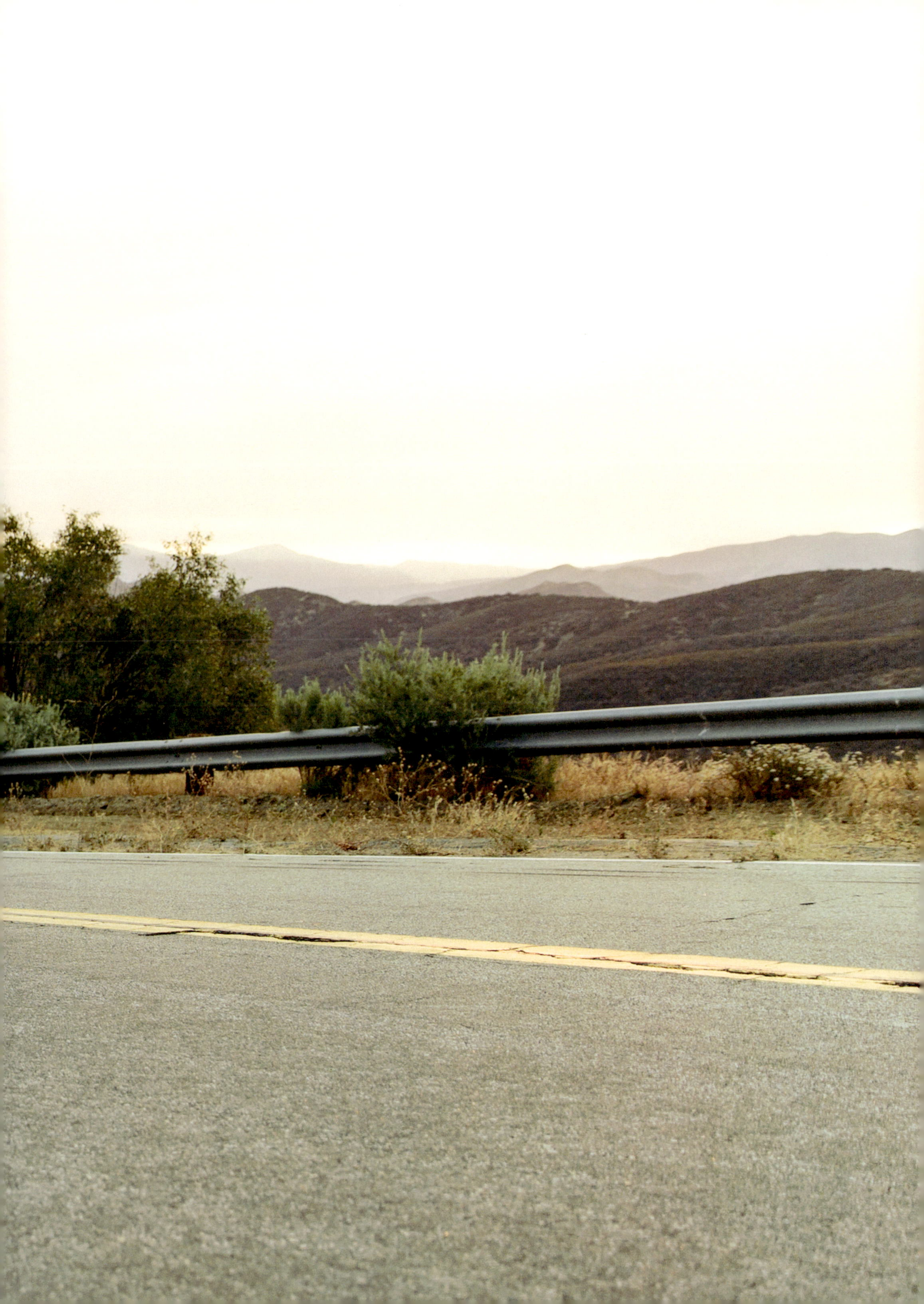

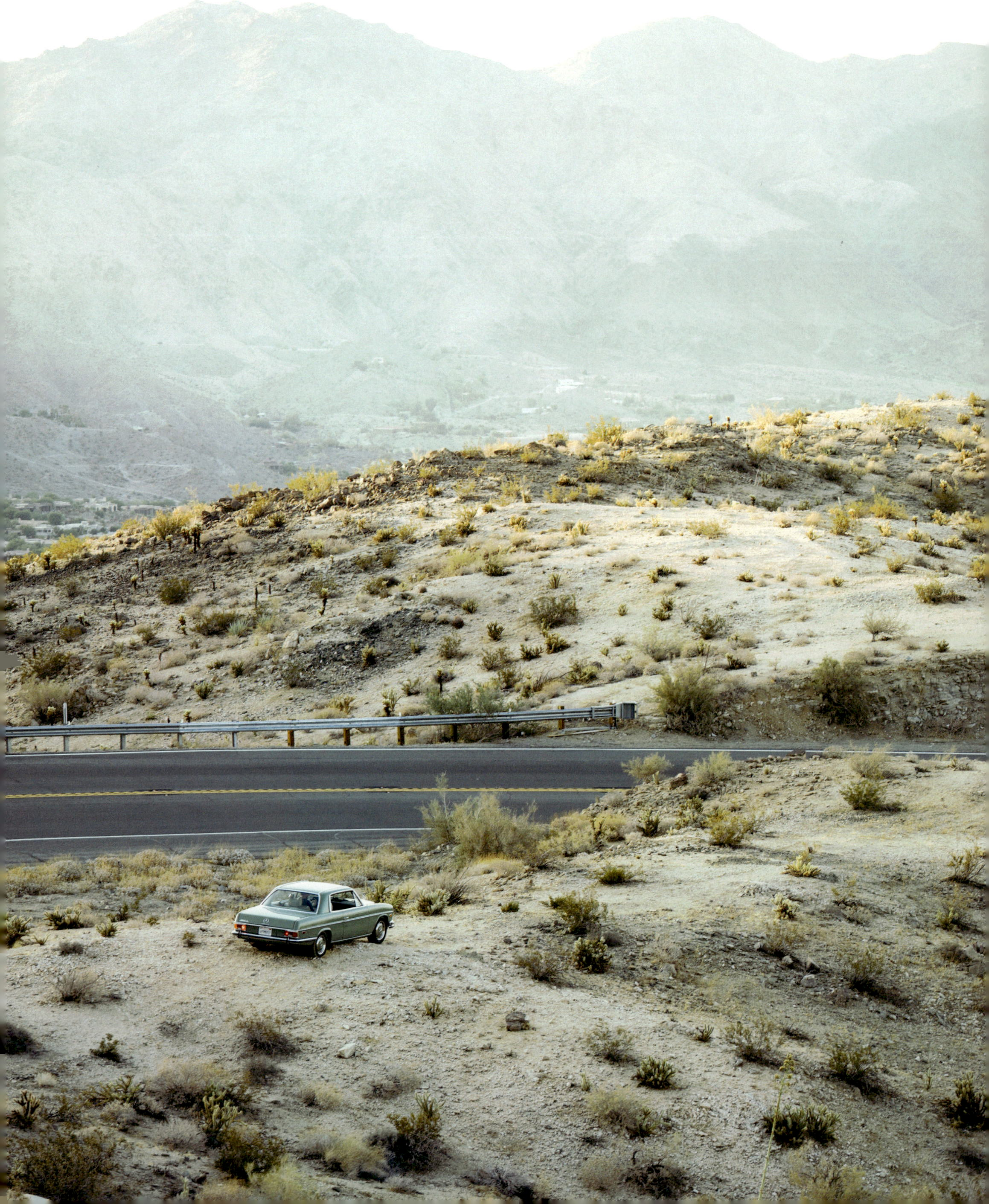

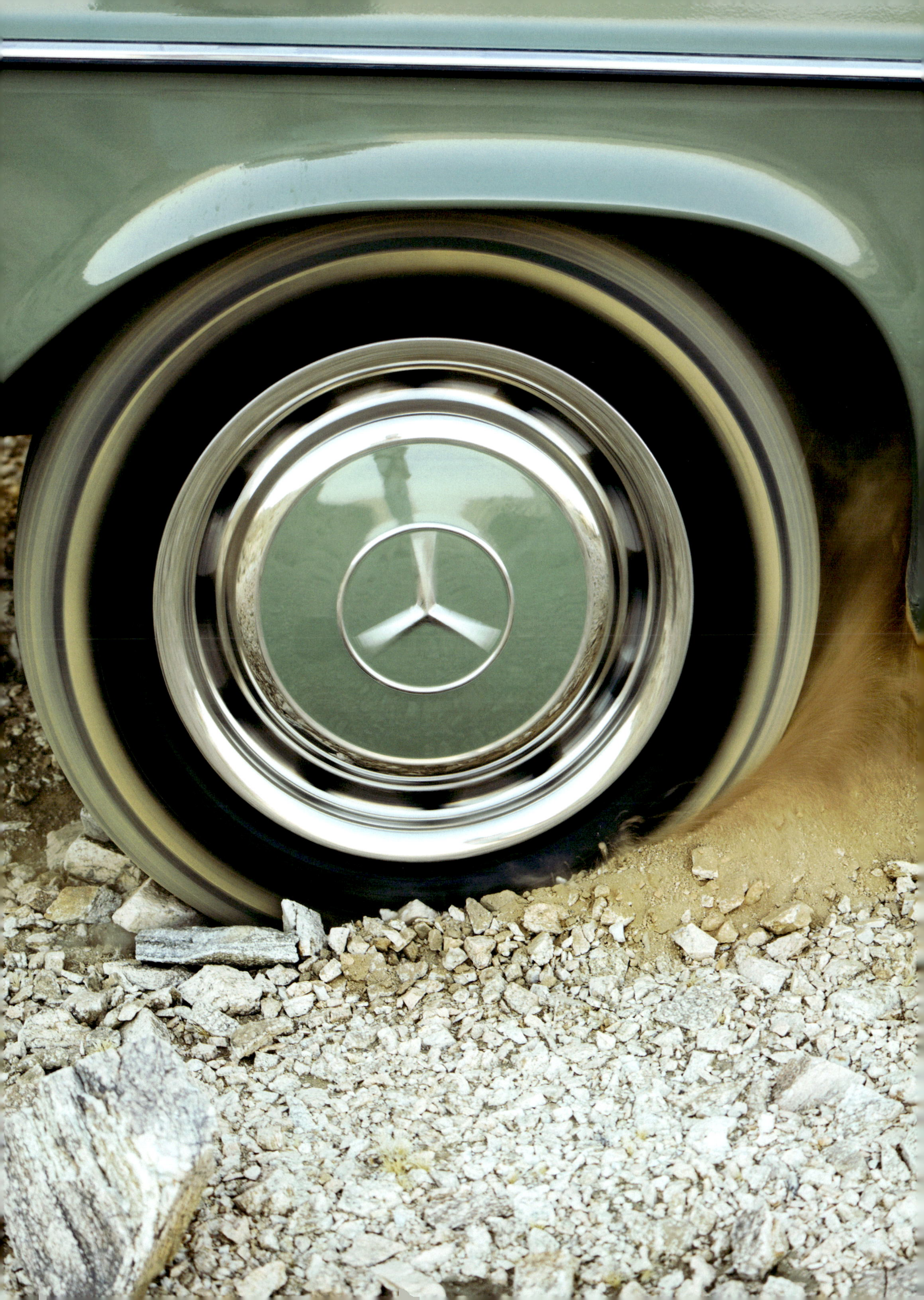

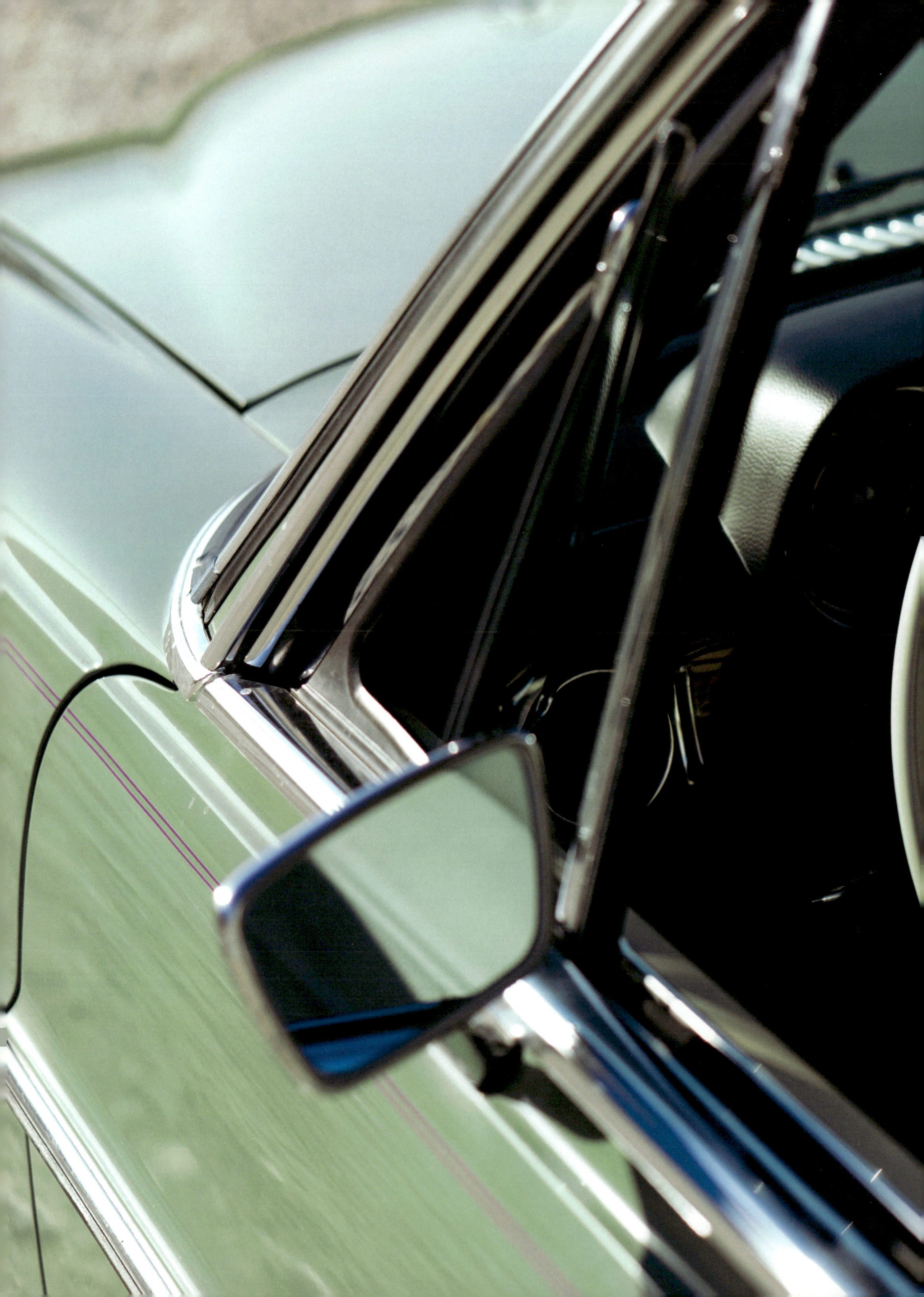

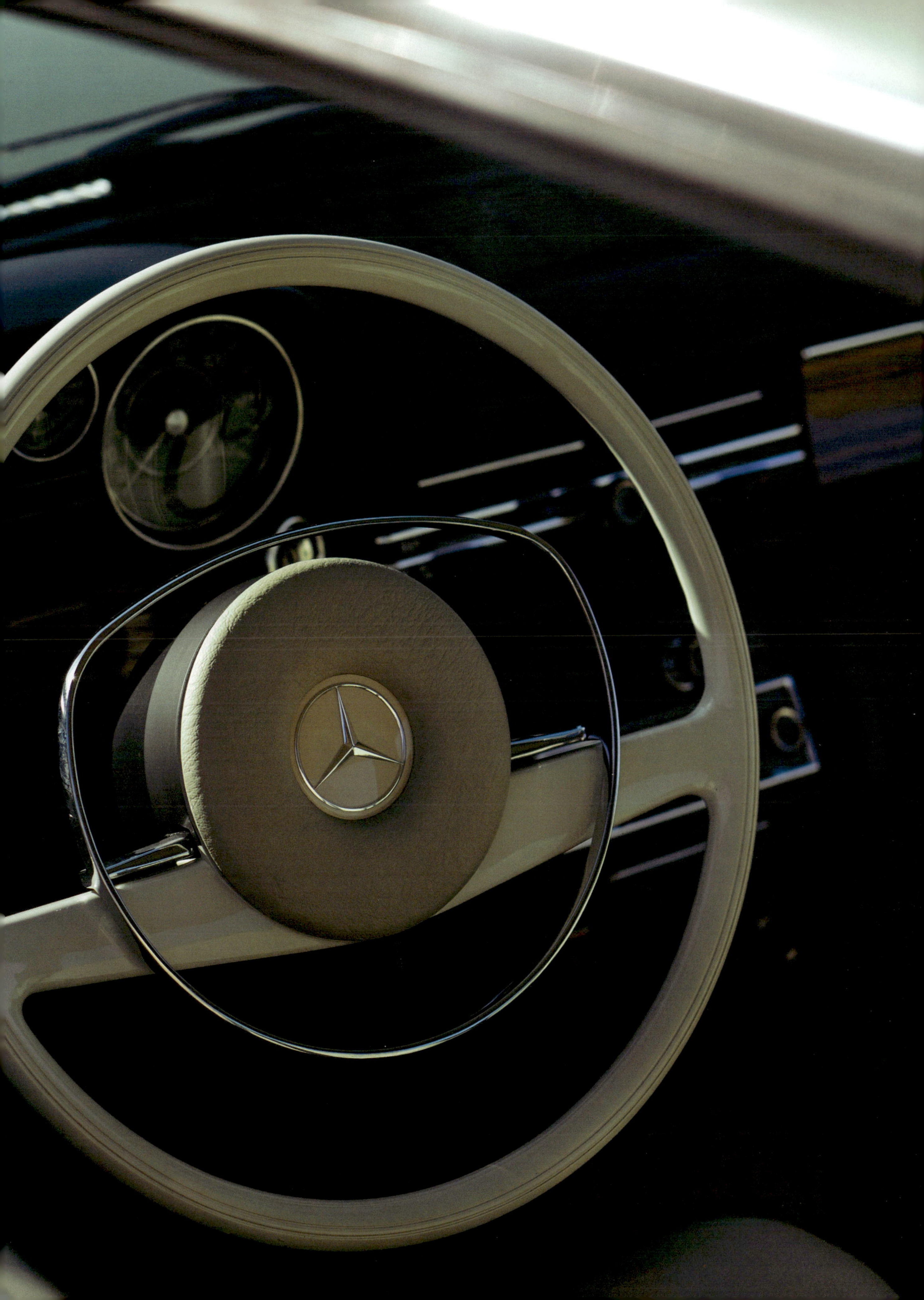

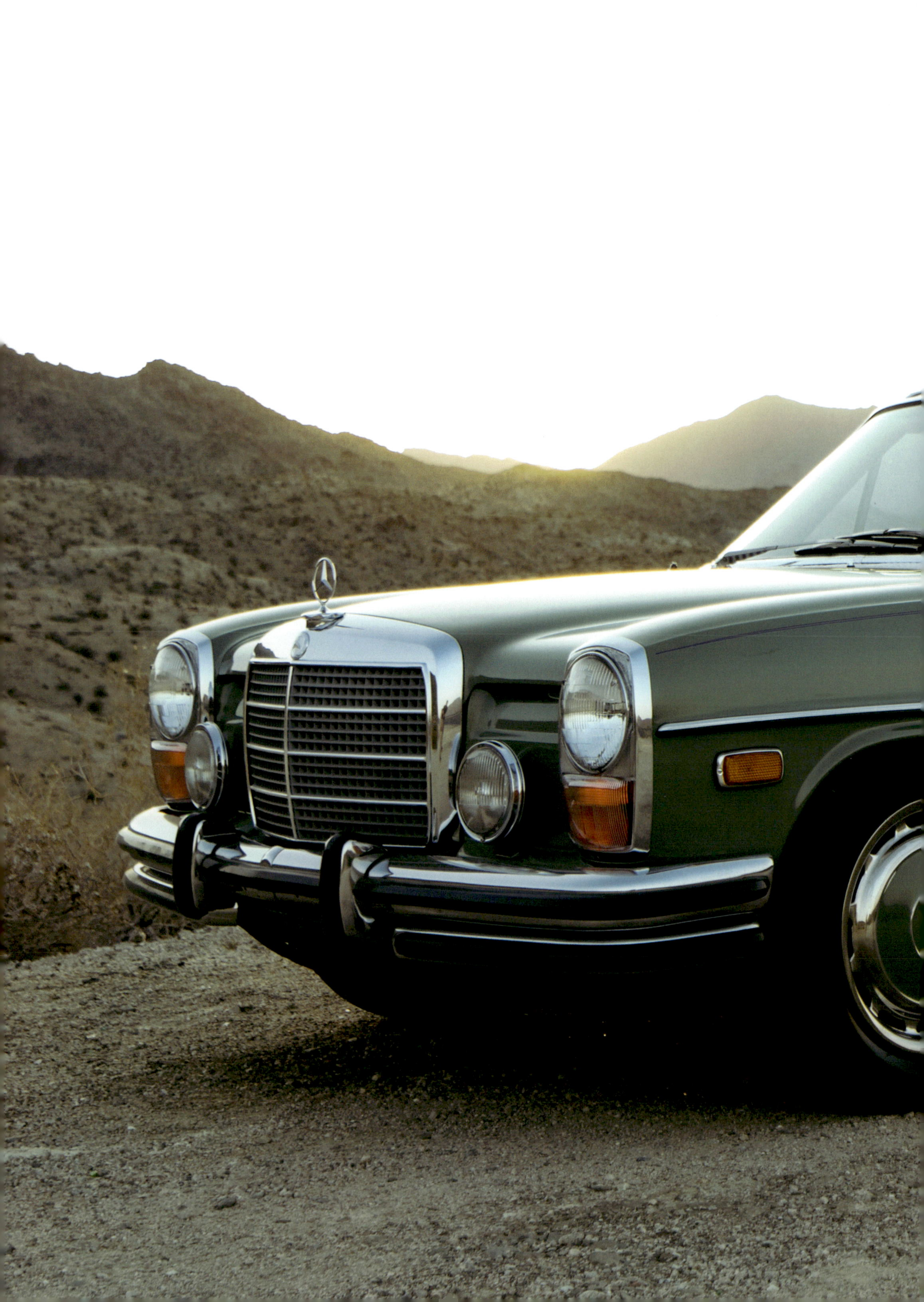

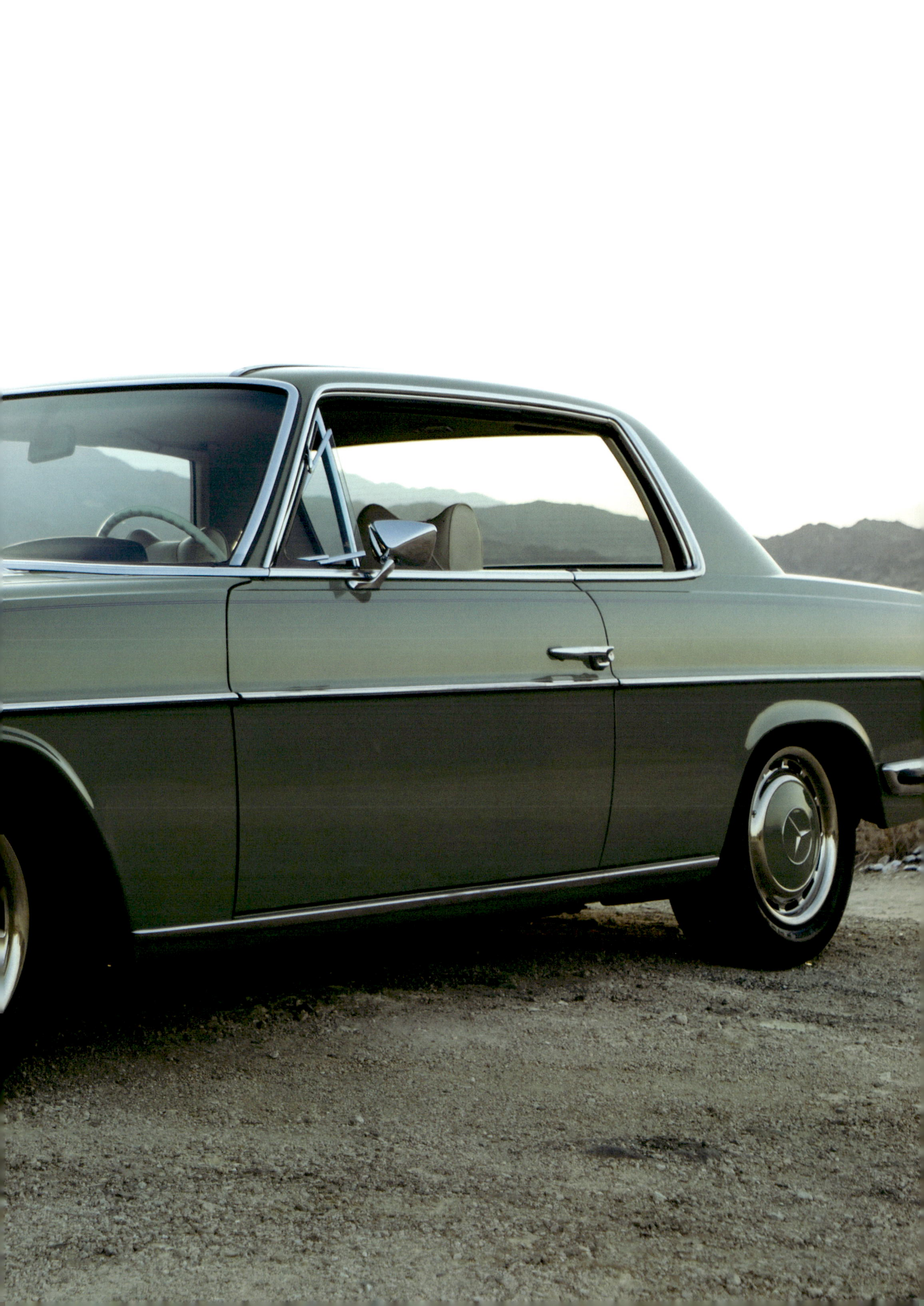

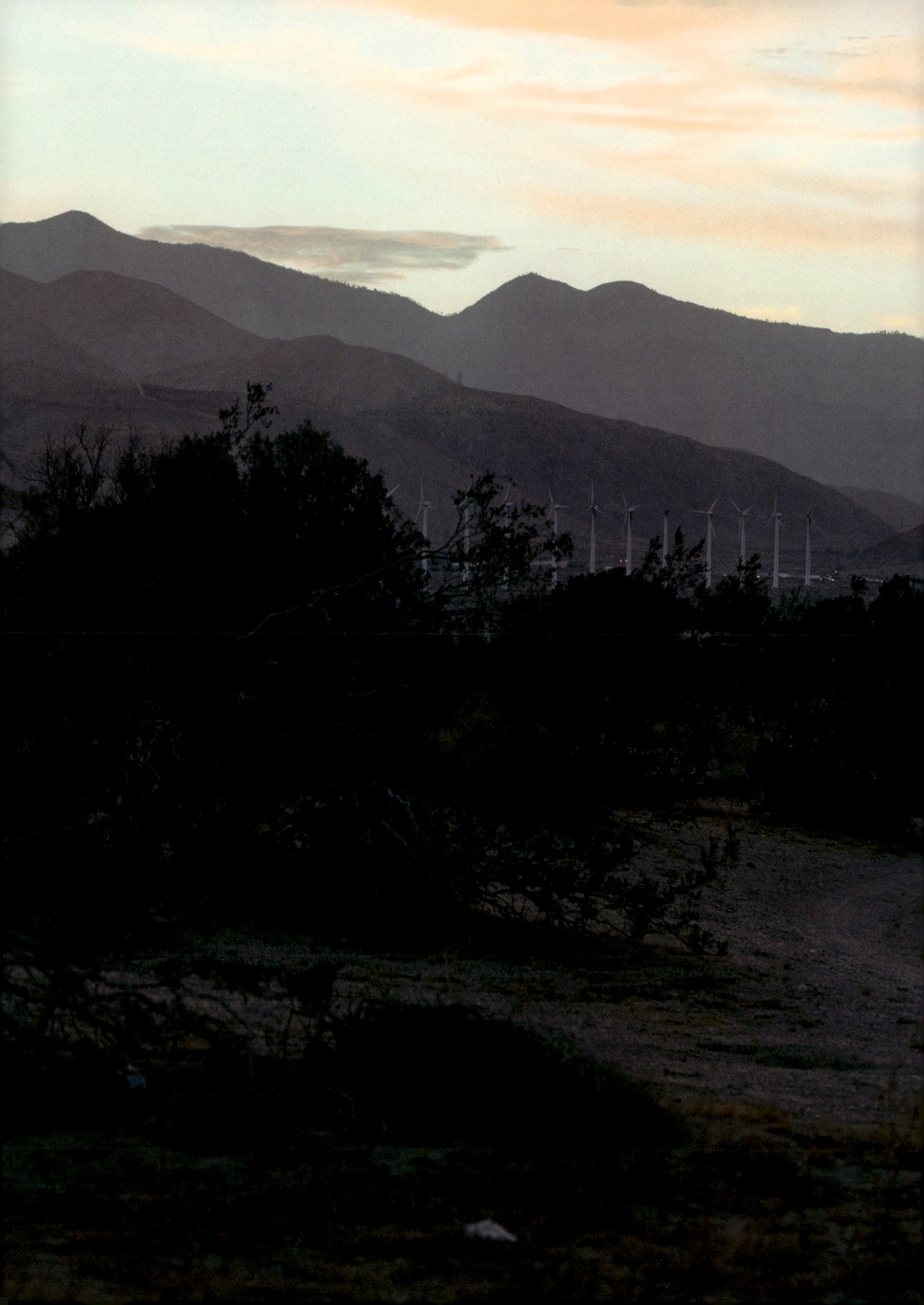

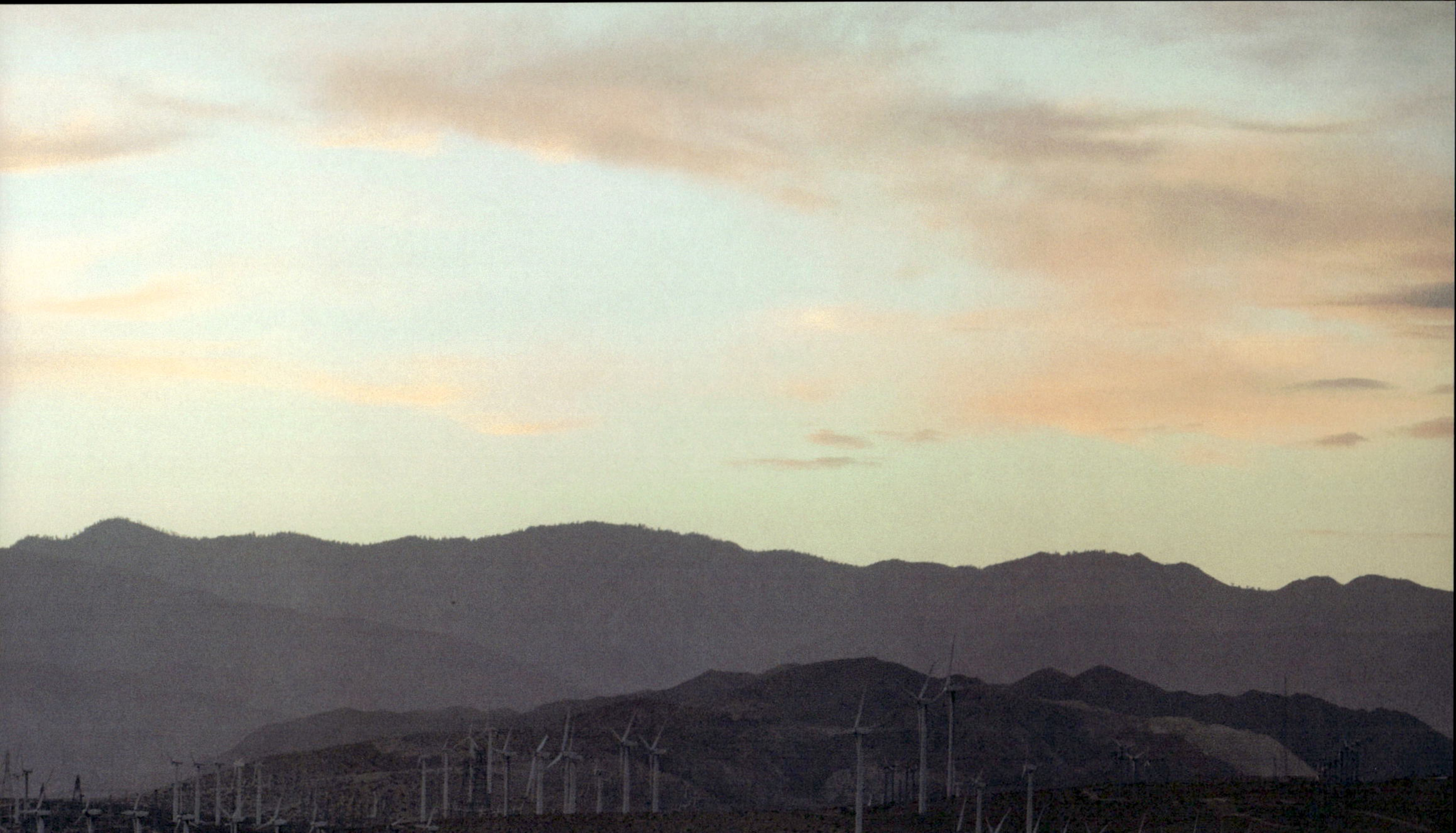

MALIBU, CALIFORNIA

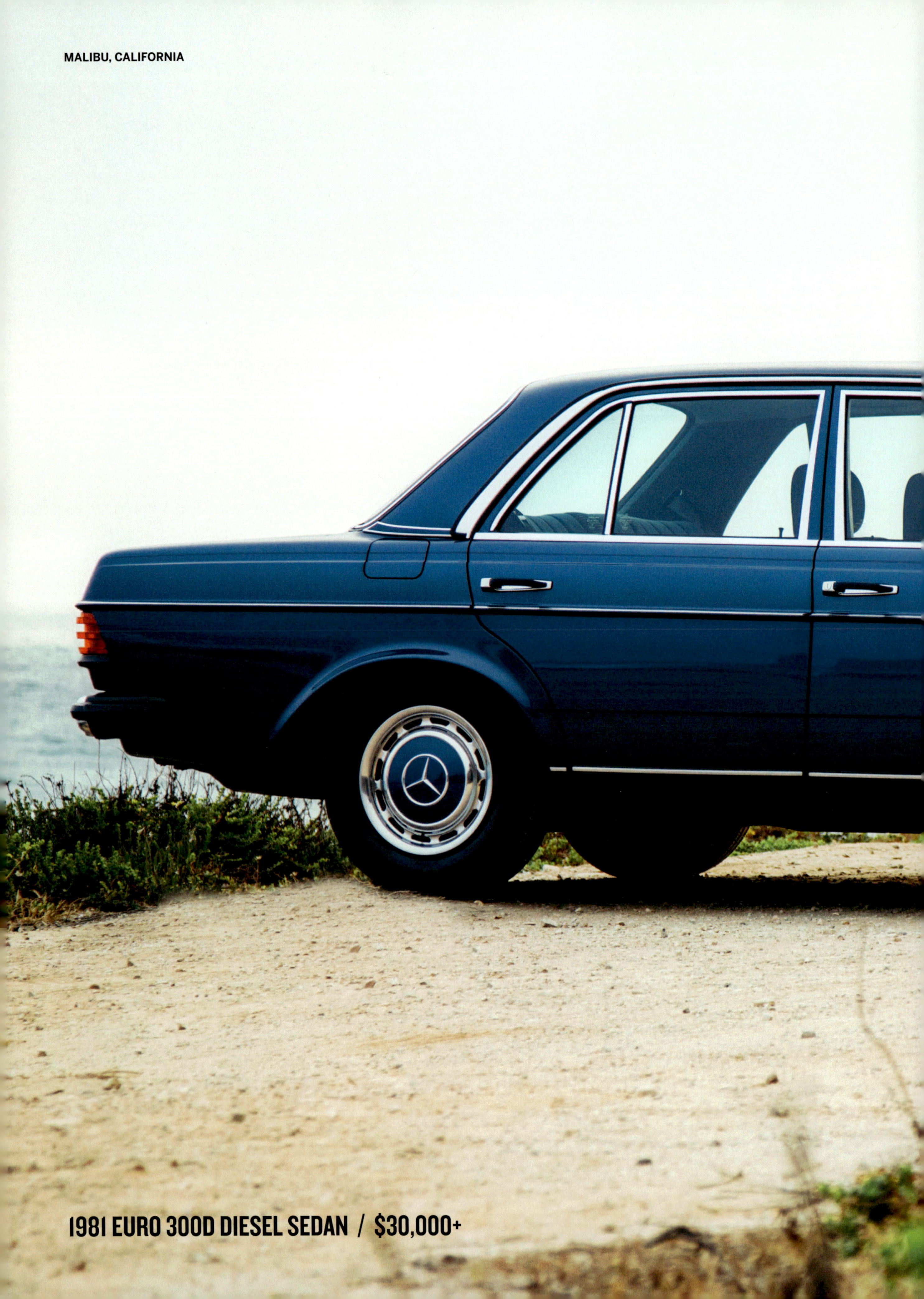

1981 EURO 300D DIESEL SEDAN / $30,000+

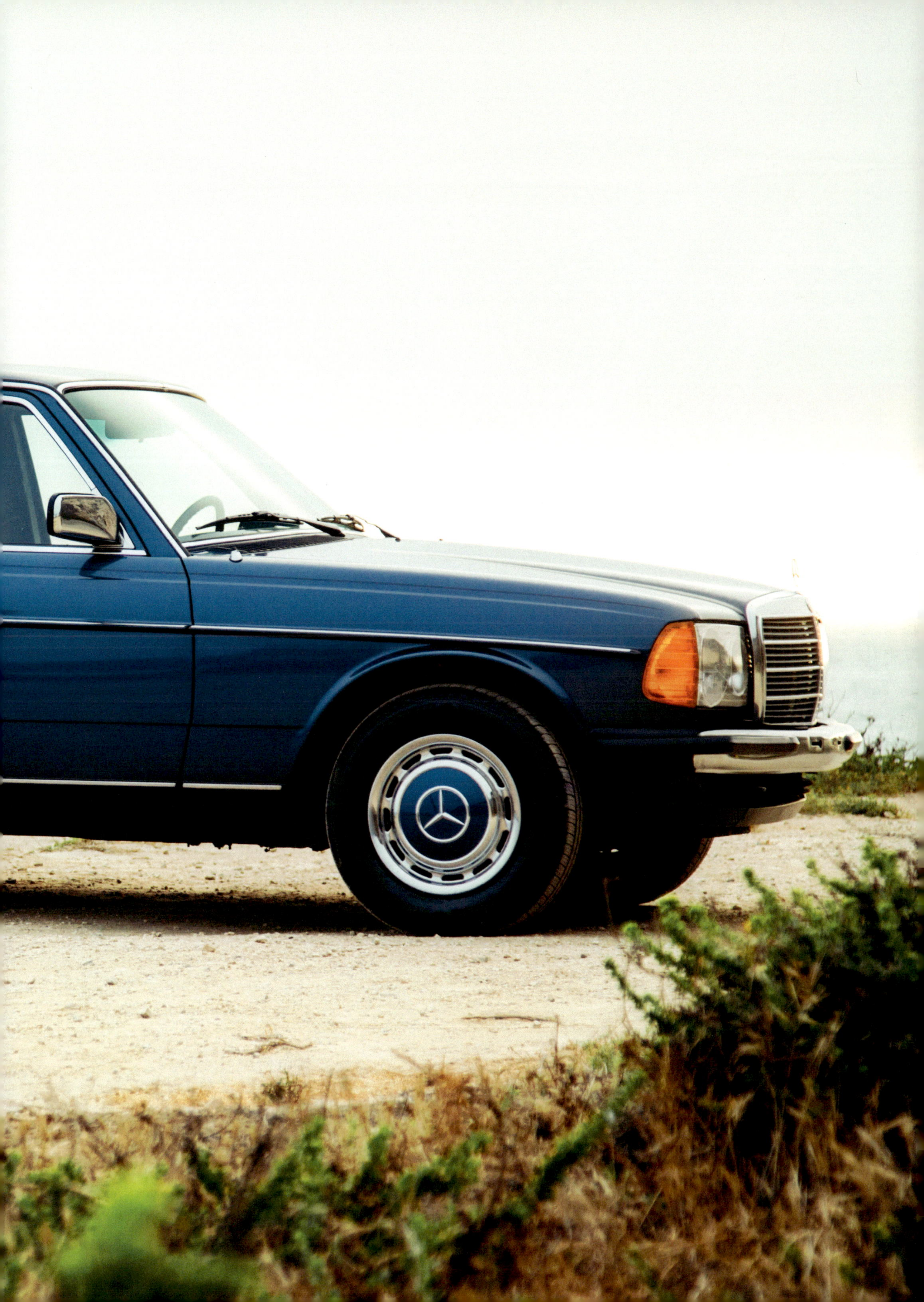

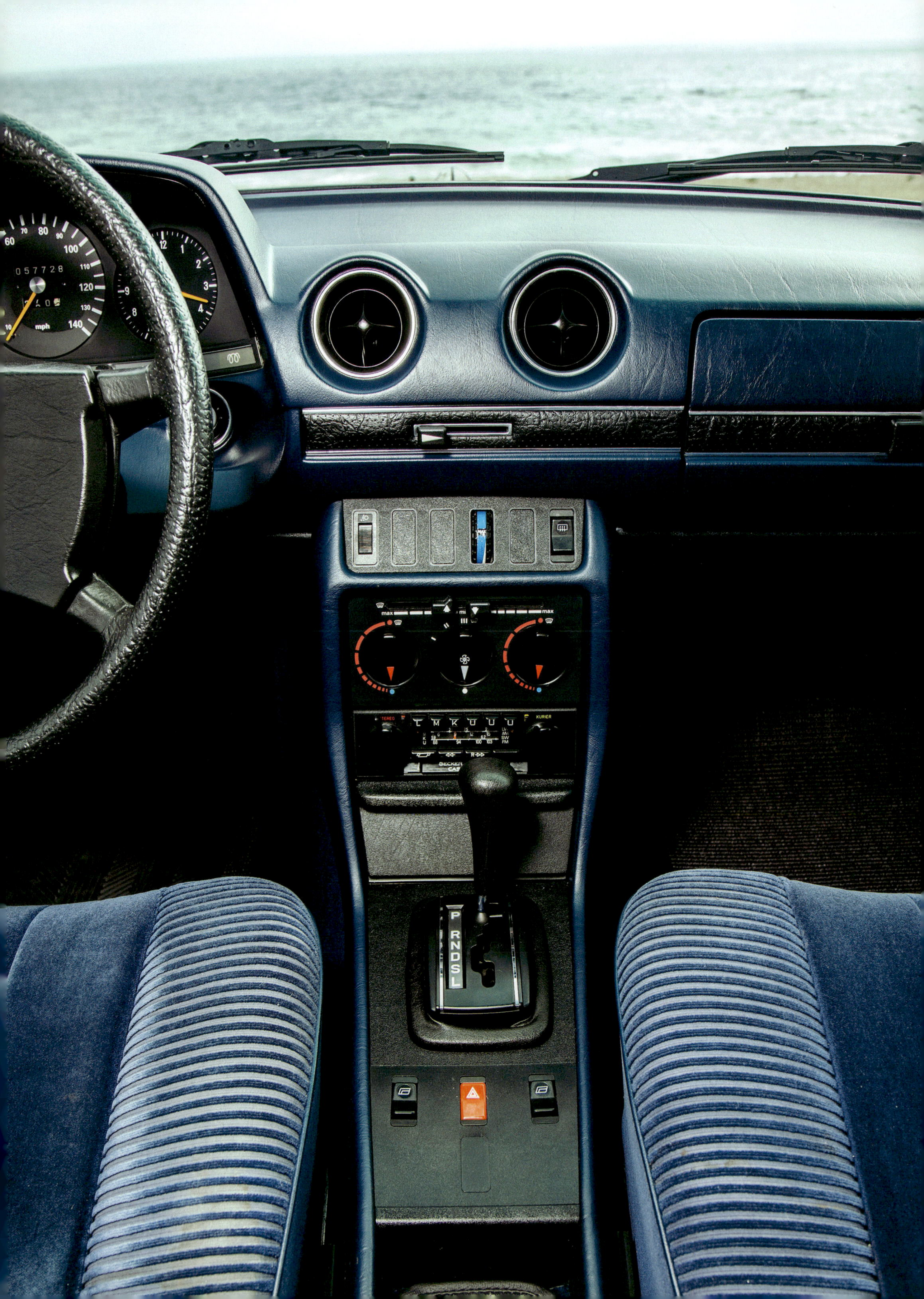
mph
KURIER
P RNDSL

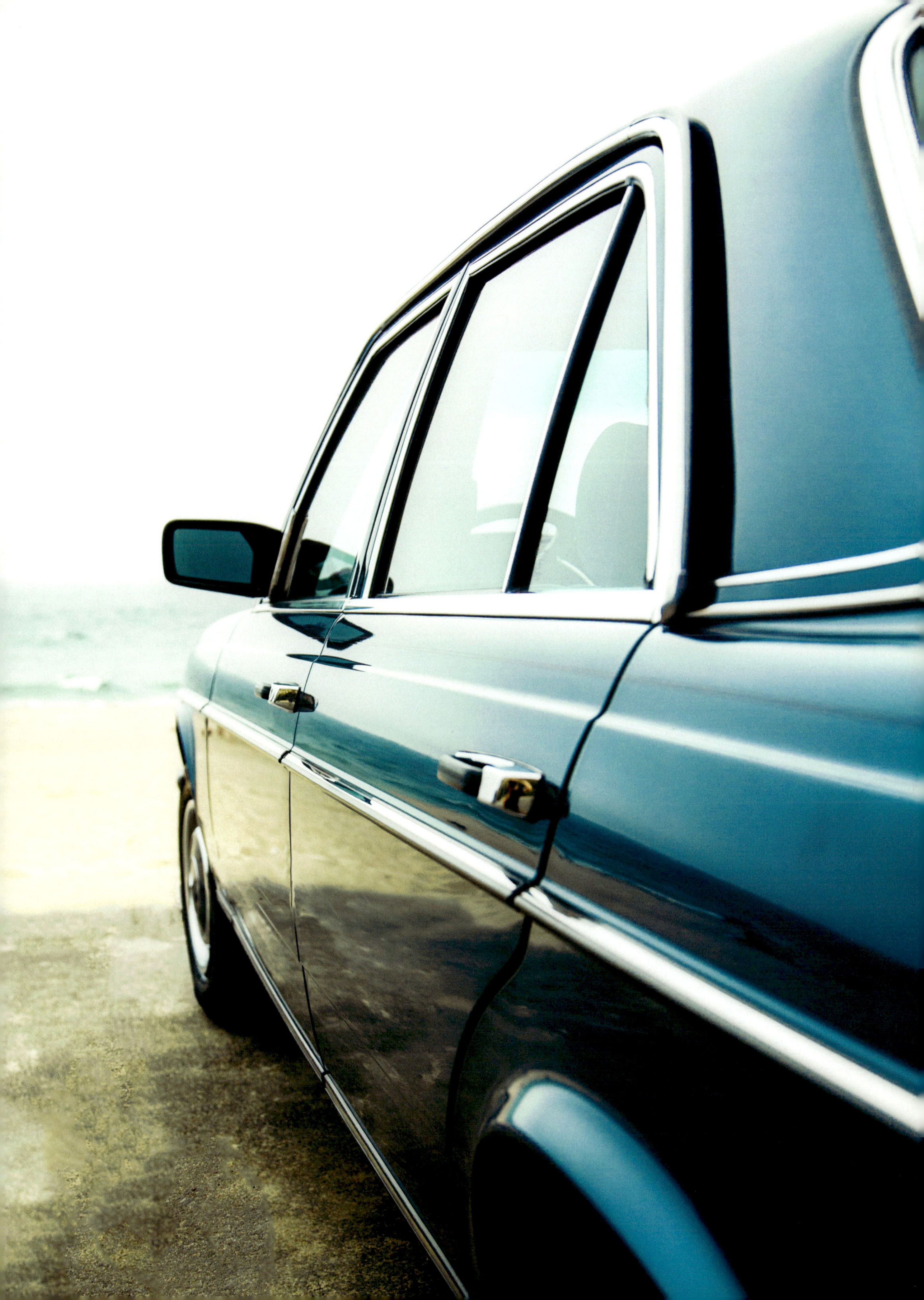

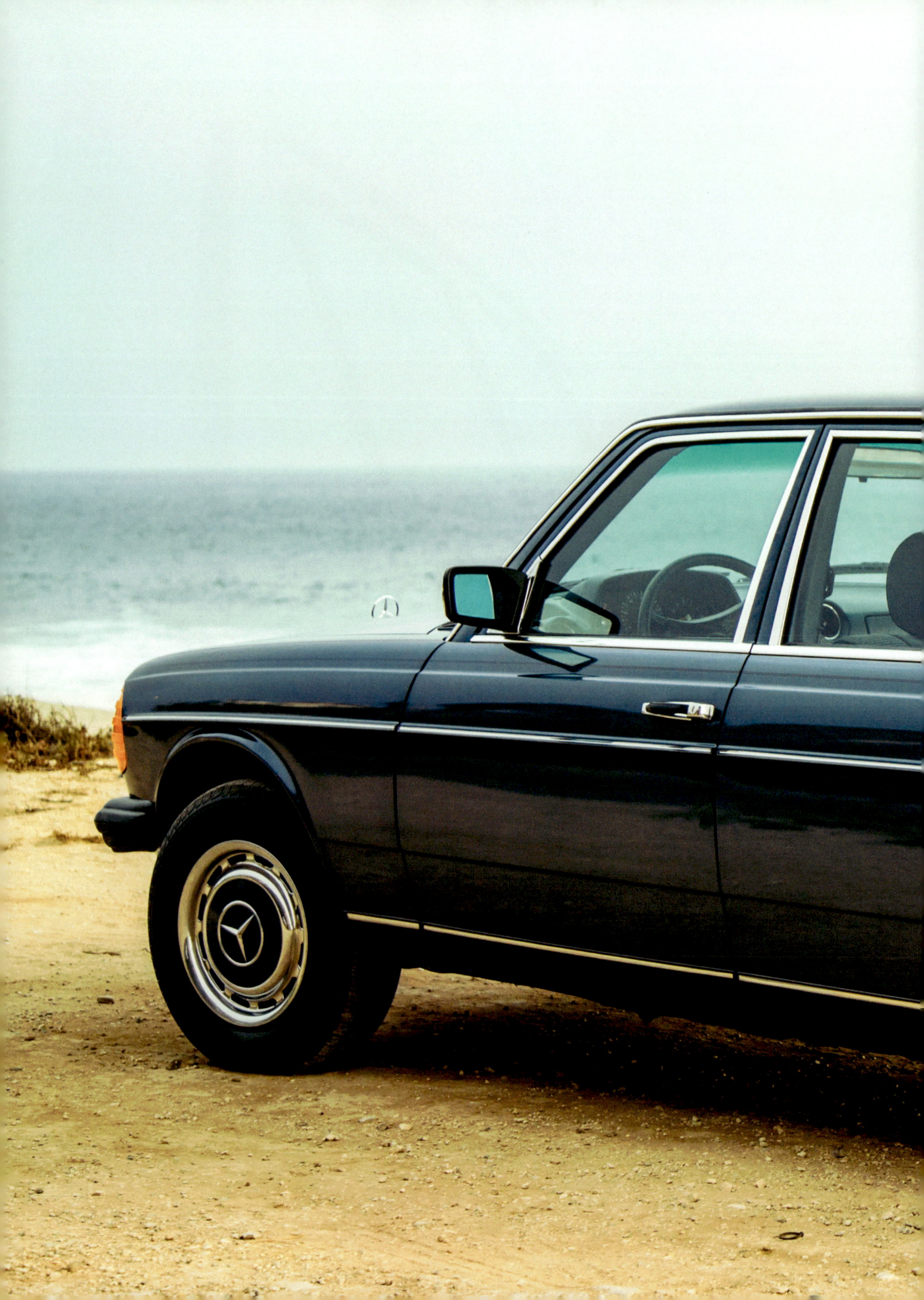

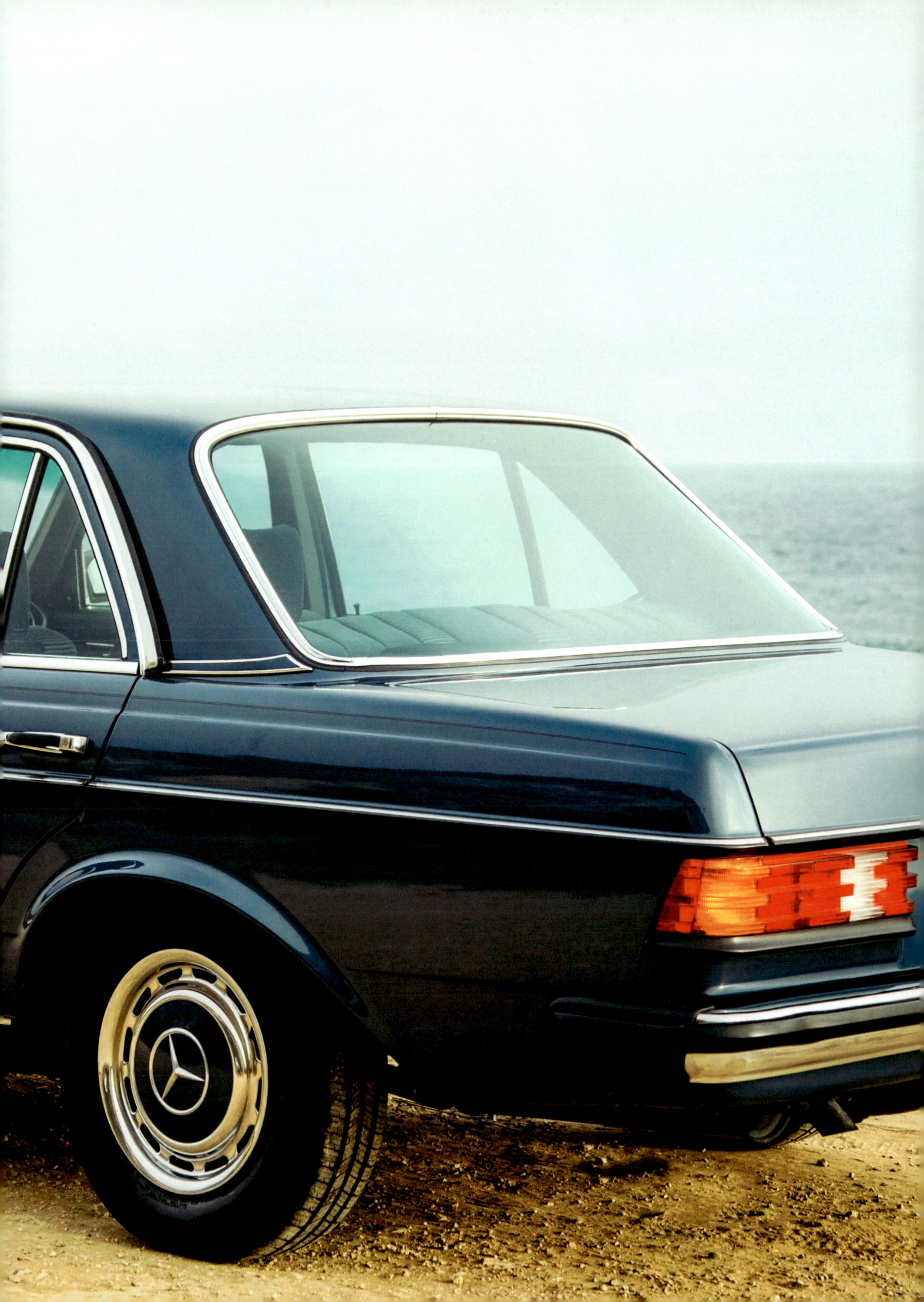

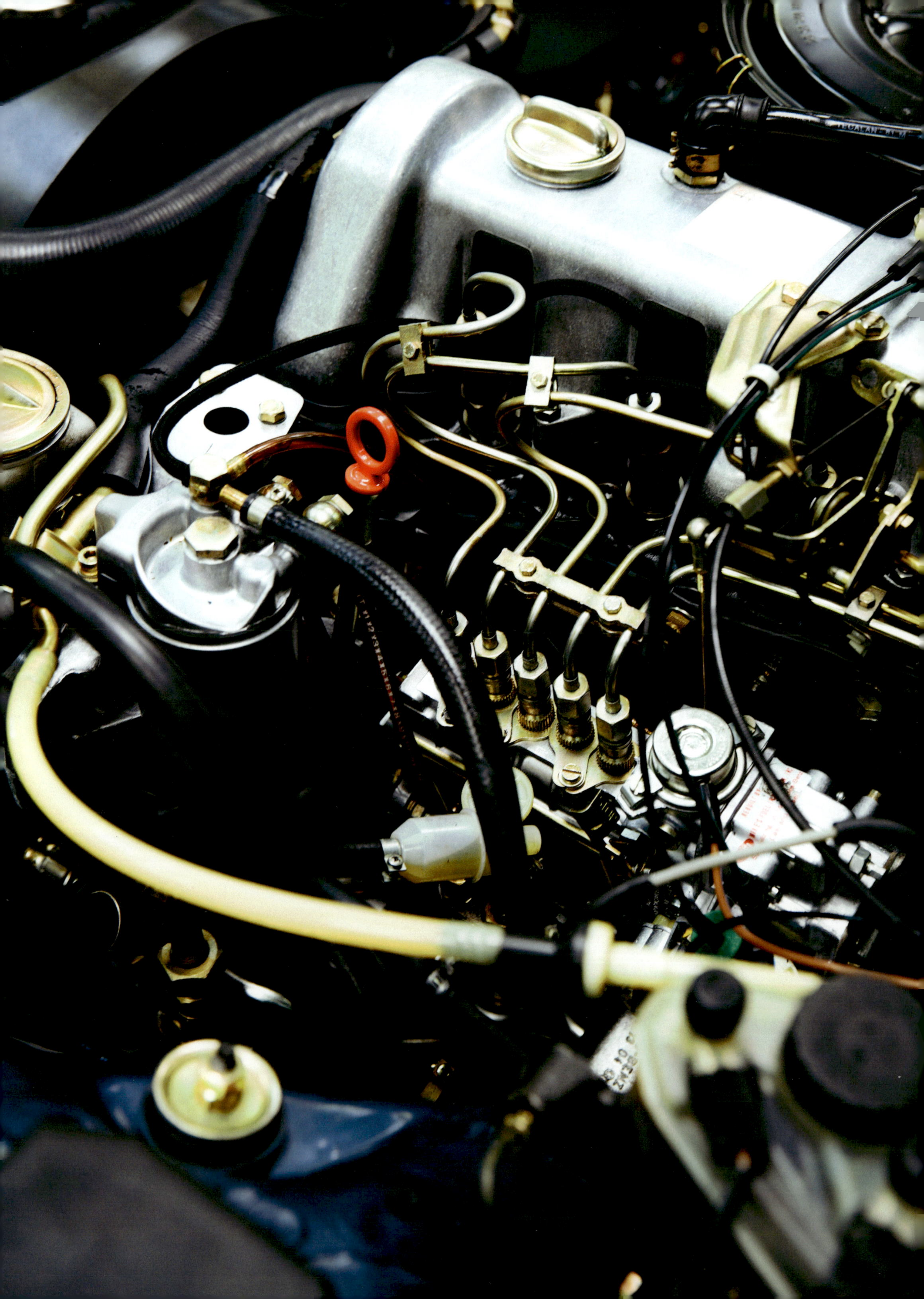

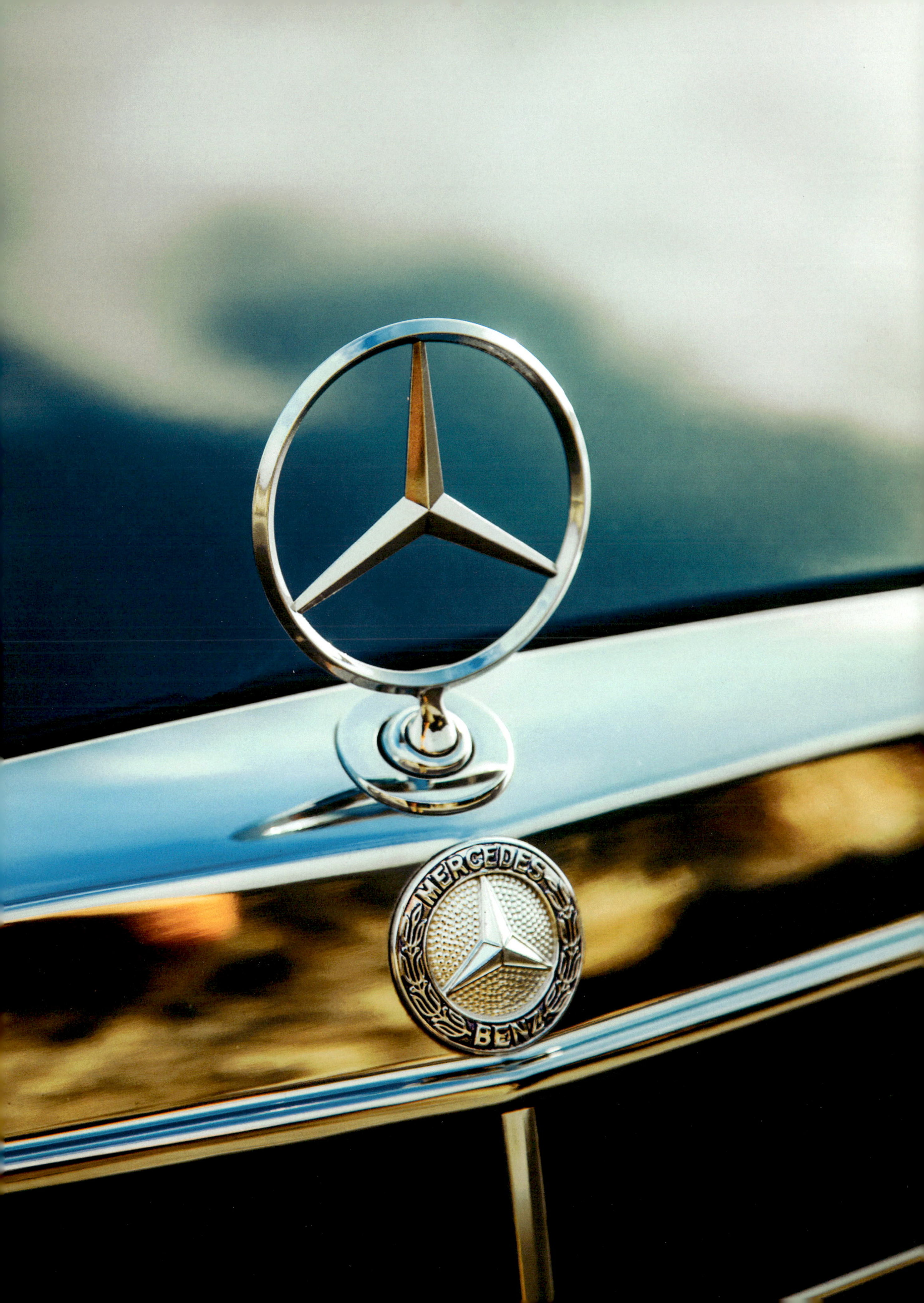
MERCEDES
BENZ

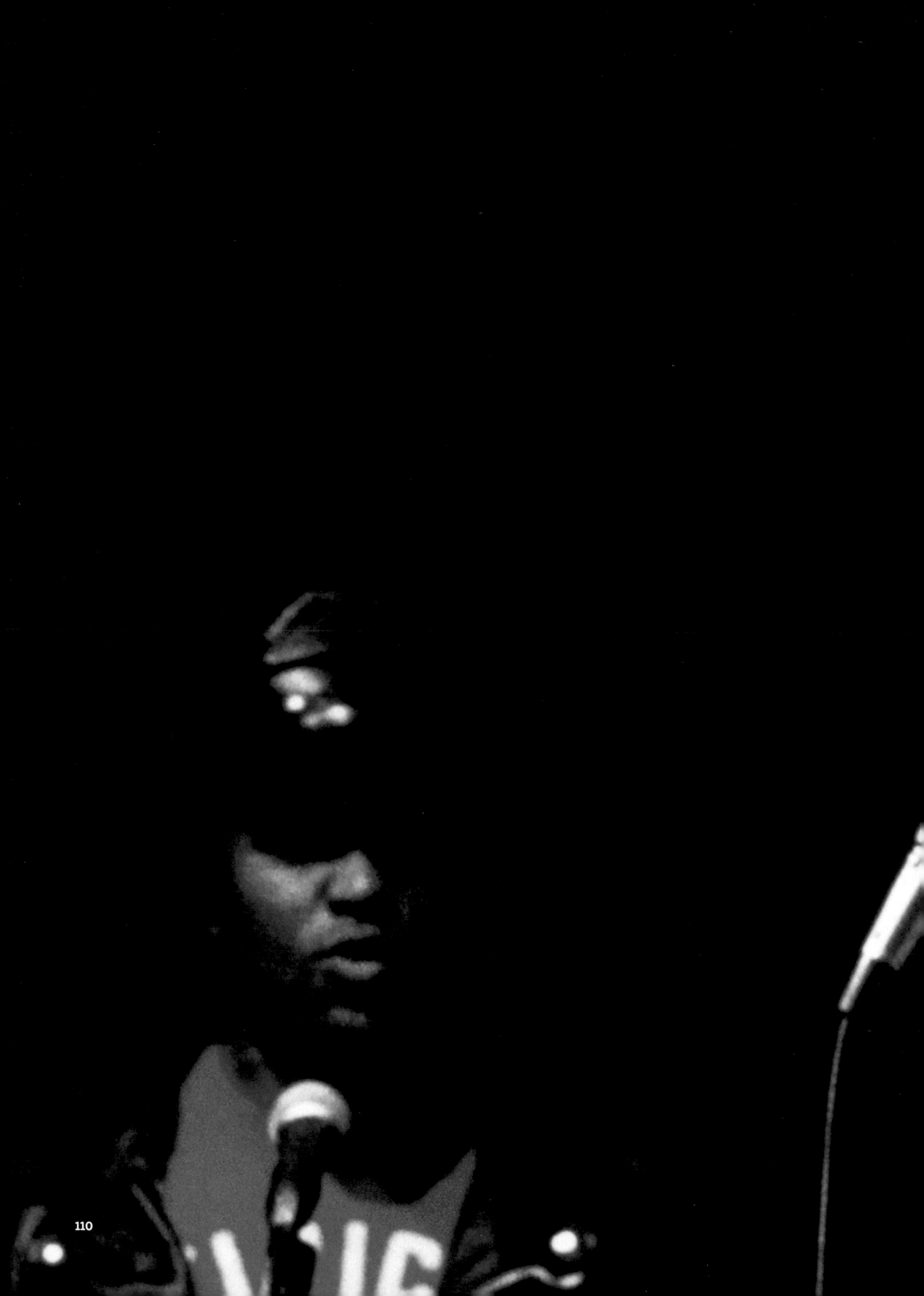

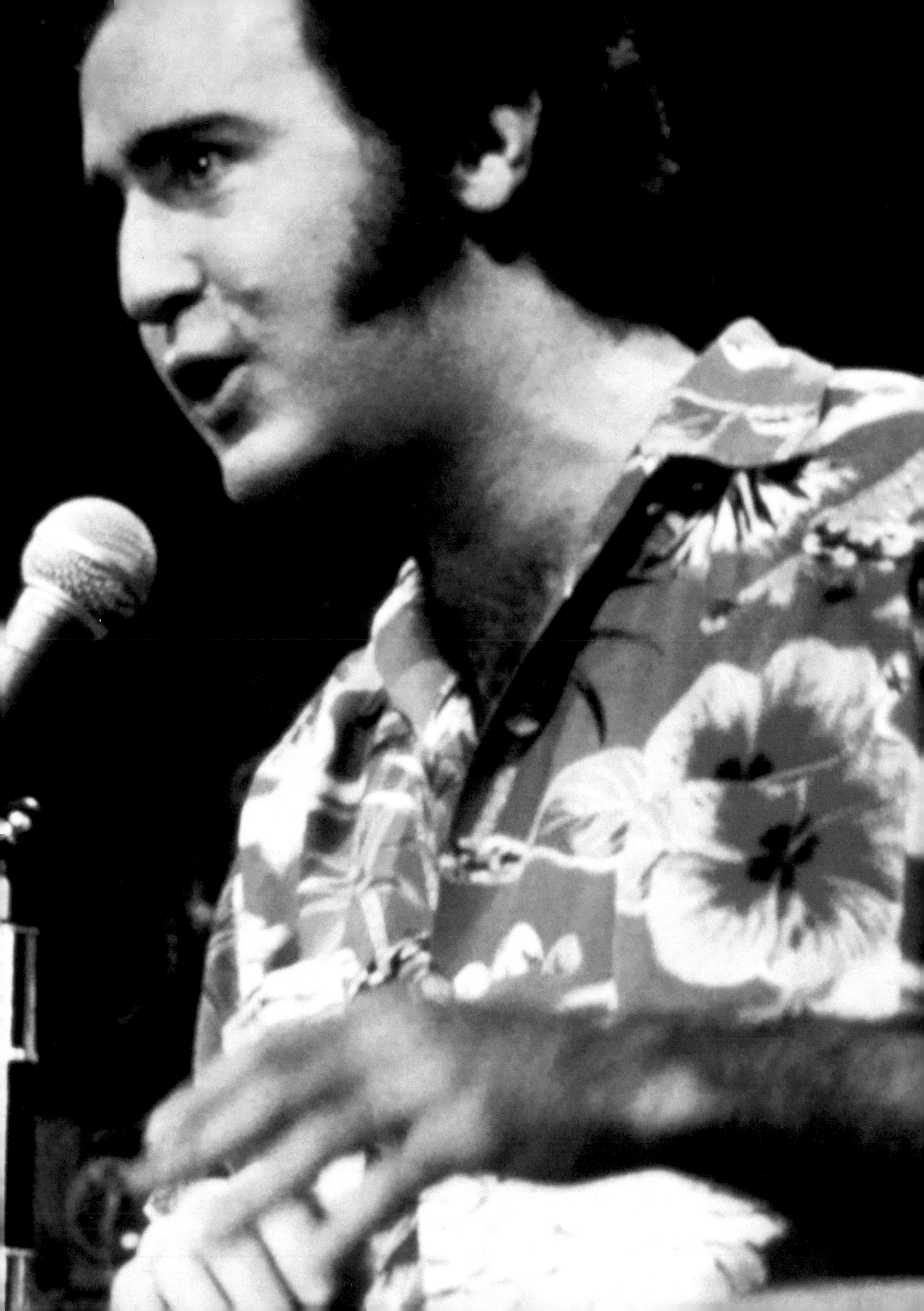

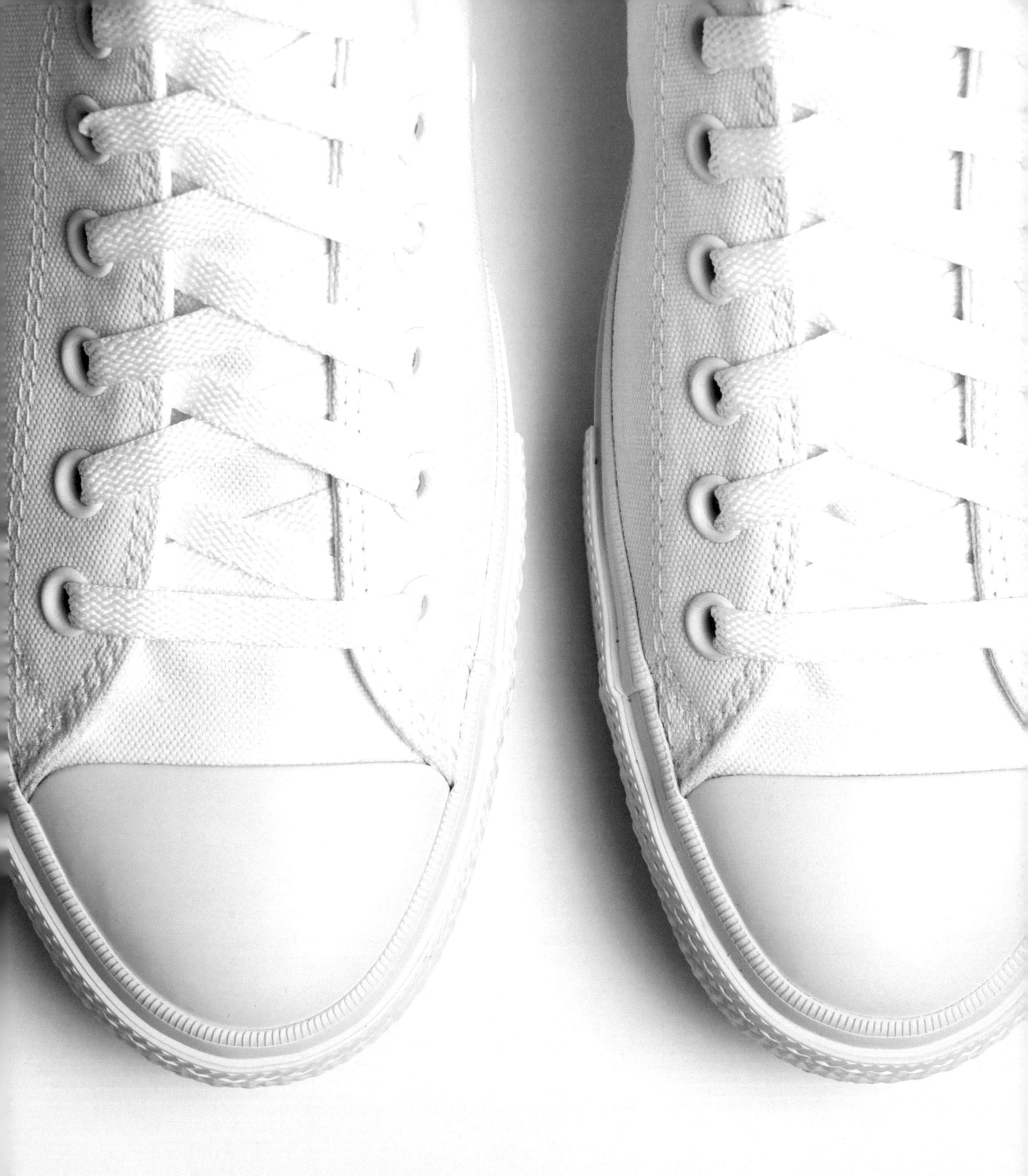

CONVERSE CHUCK TAYLOR ALL STAR II SNEAKER / $75

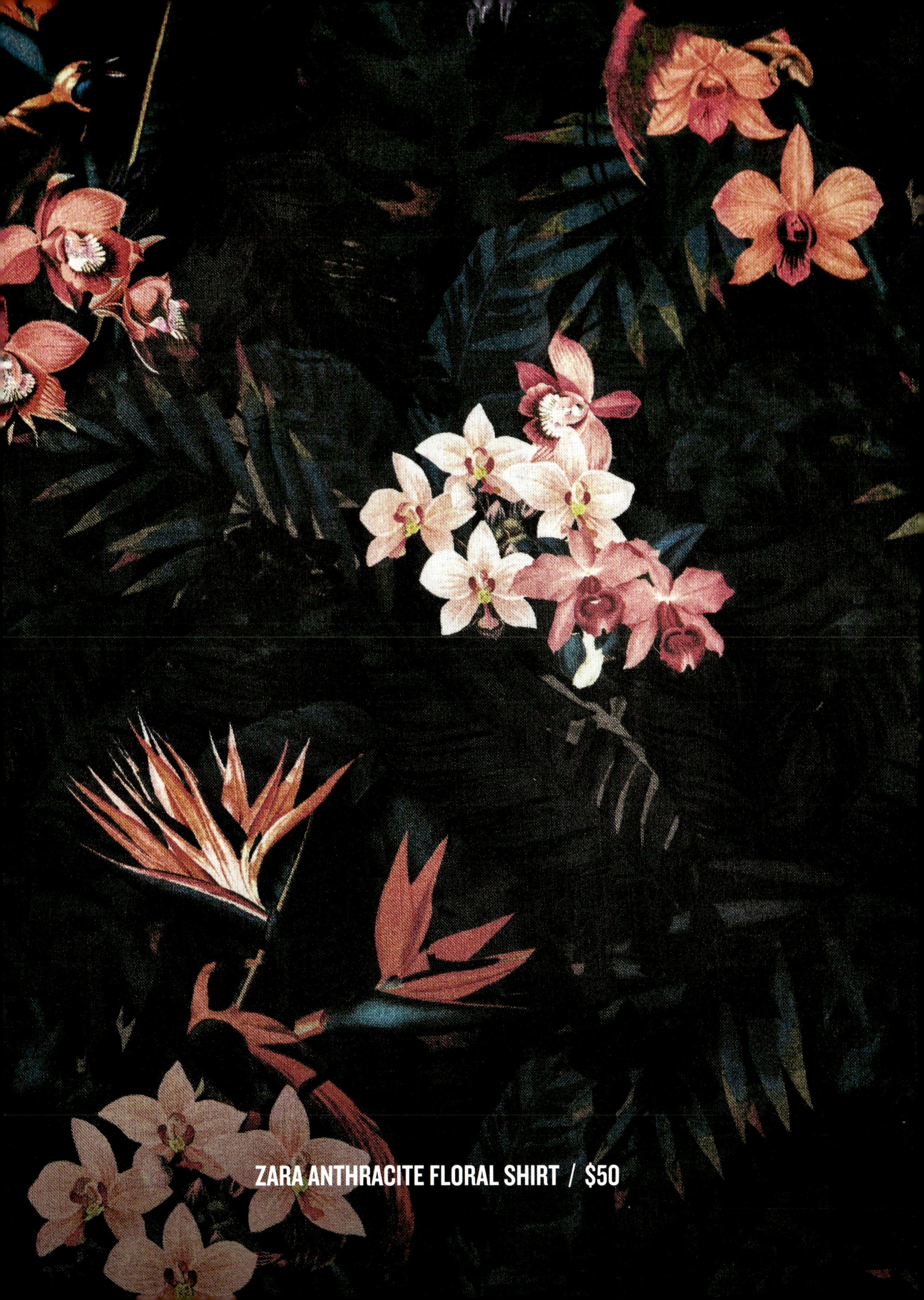

ZARA ANTHRACITE FLORAL SHIRT / $50

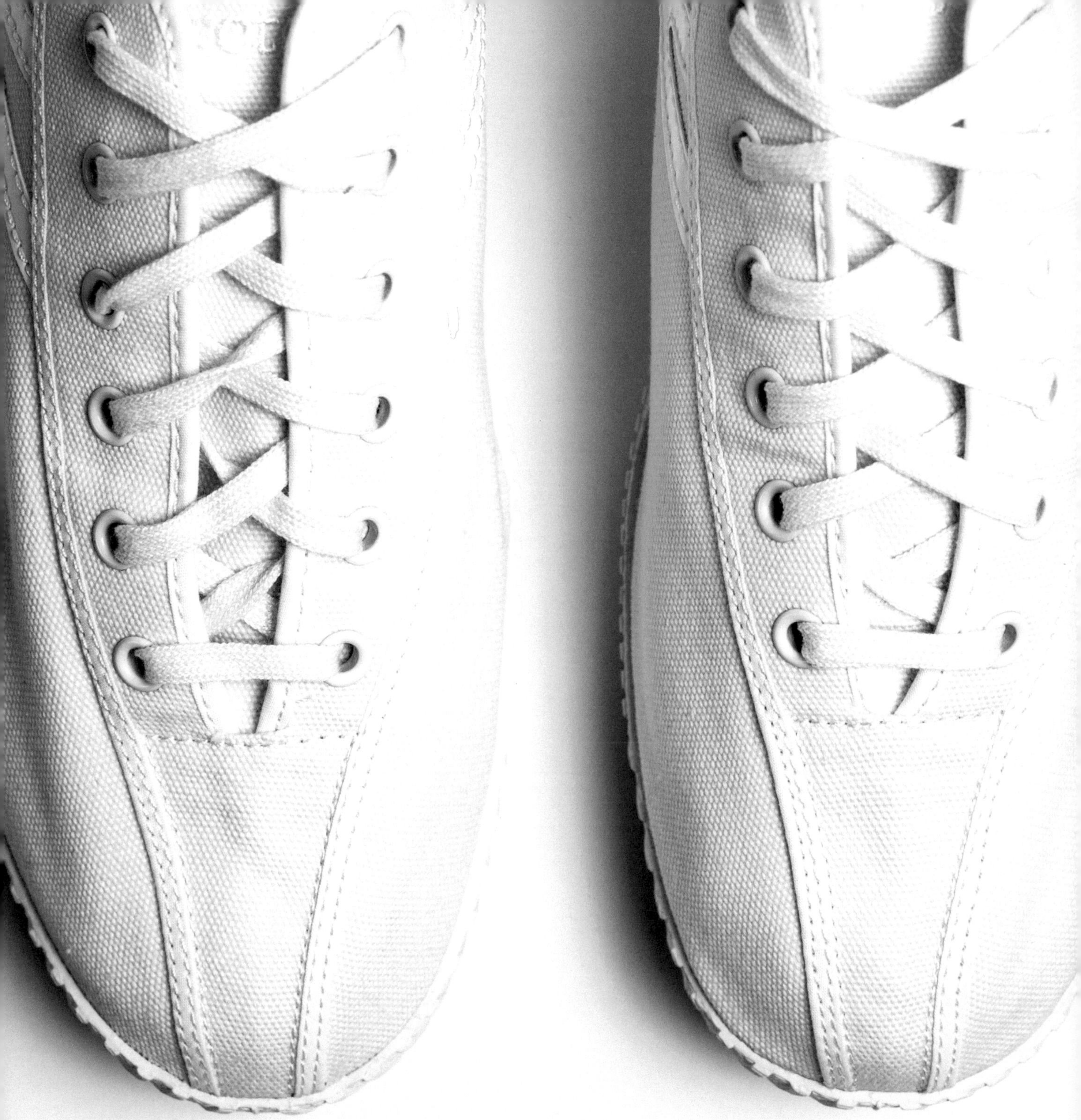

TRETORN NYLITE SNEAKER / $65

FREEMANS SPORTING CLUB FLOWERS SHIRT / $185

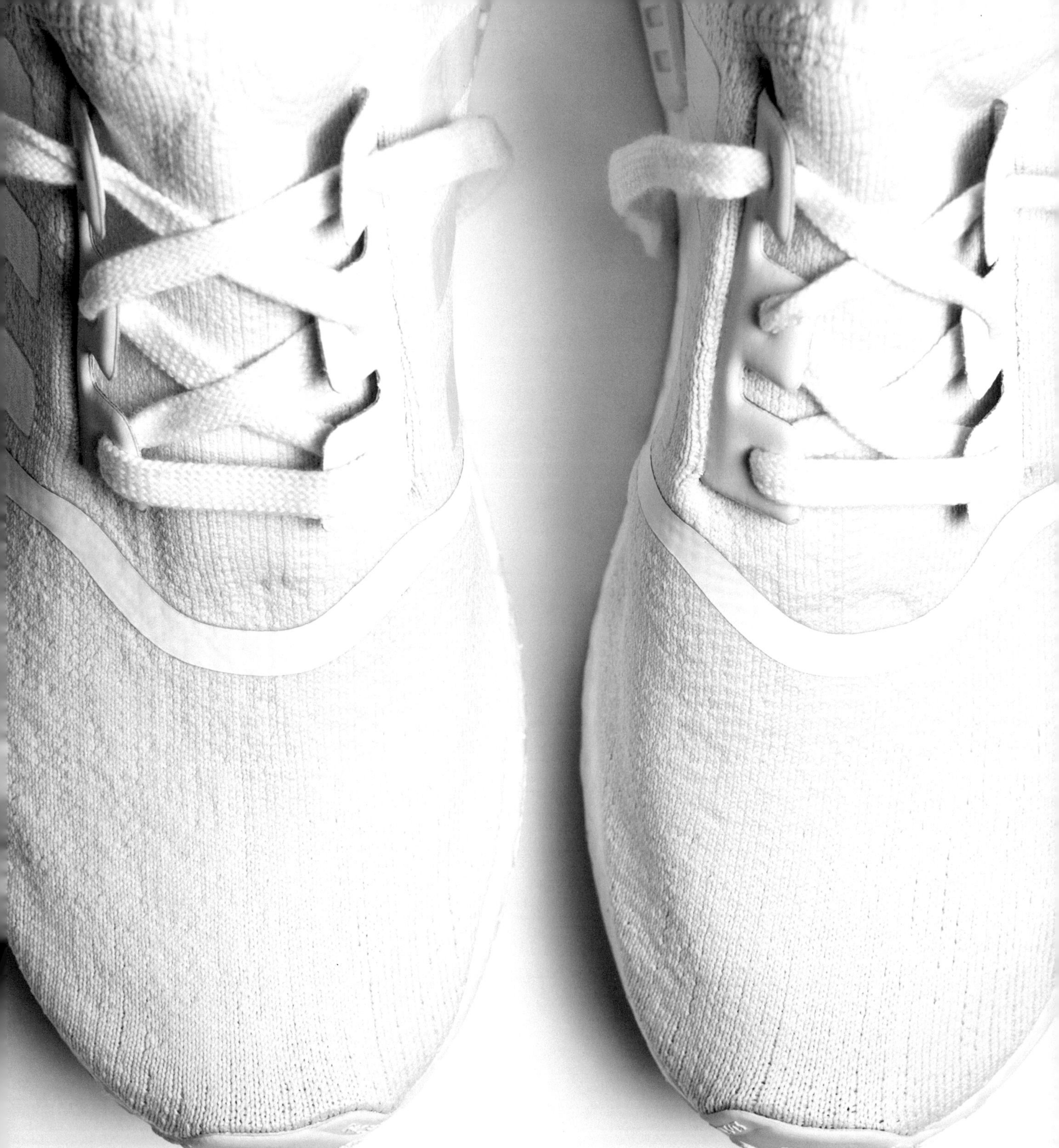

ADIDAS NMD R1 PRIMEKNIT SNEAKER / $250

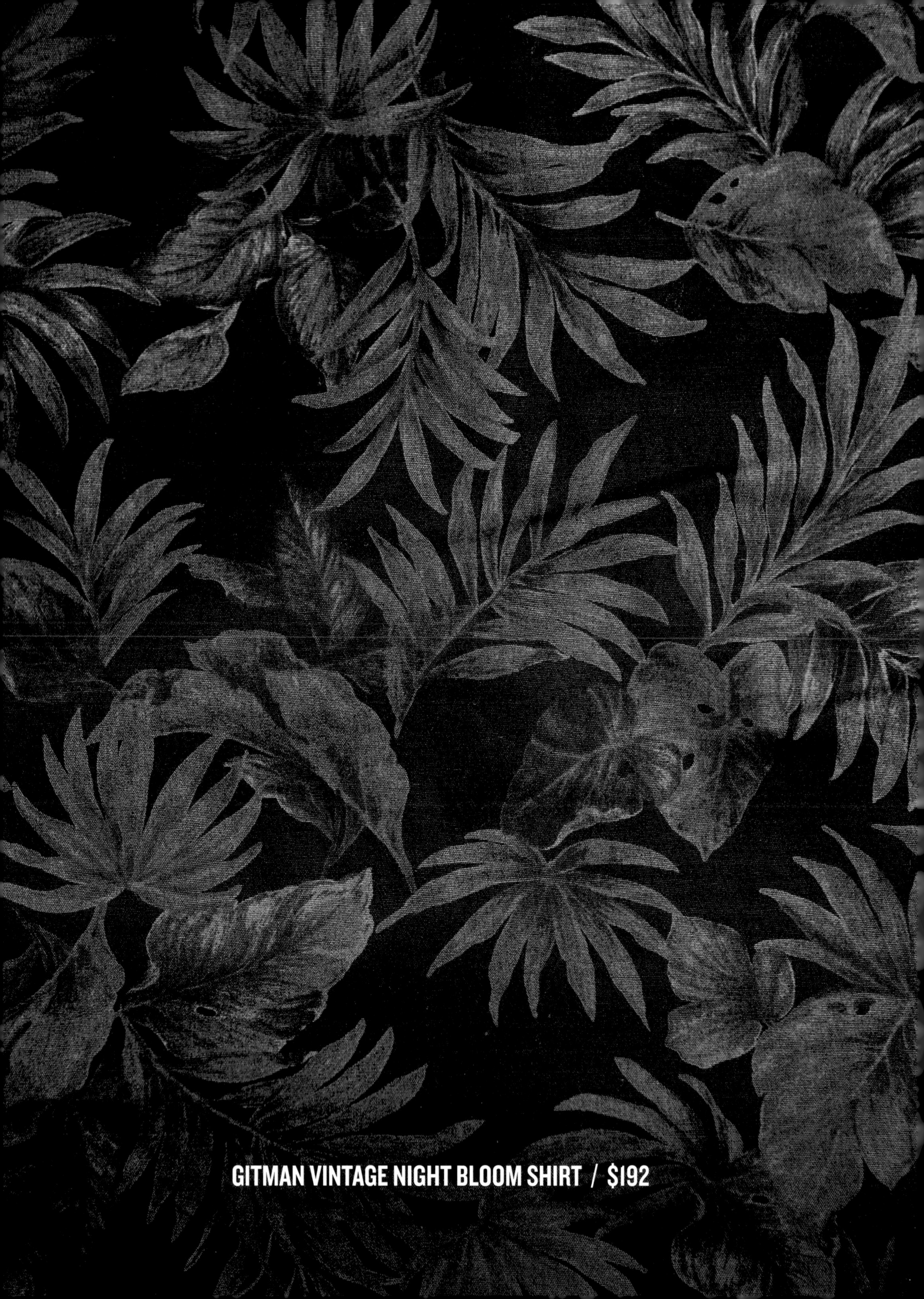

GITMAN VINTAGE NIGHT BLOOM SHIRT / $192

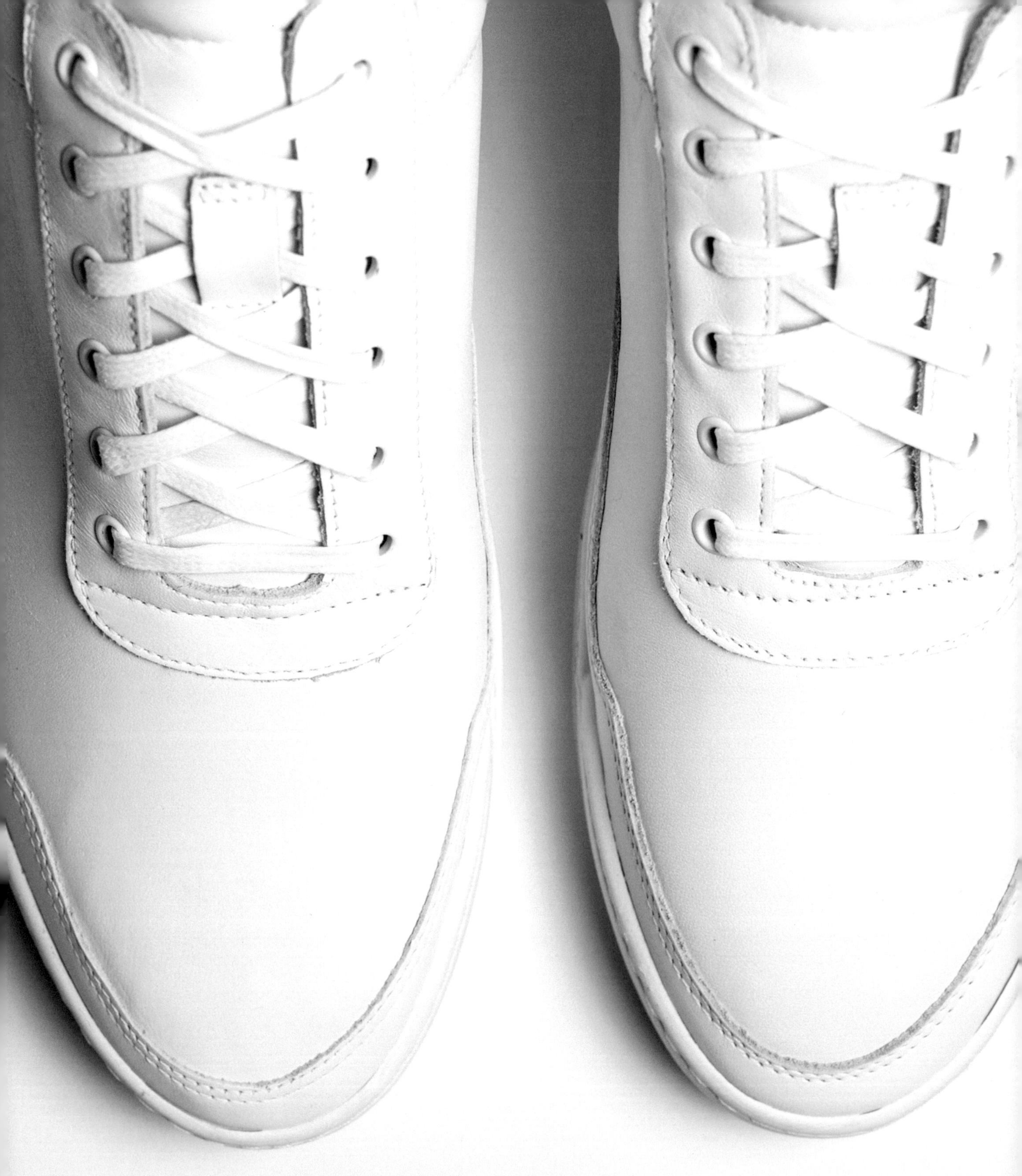

FILLING PIECES LOW TOP GHOST SNEAKER / $235

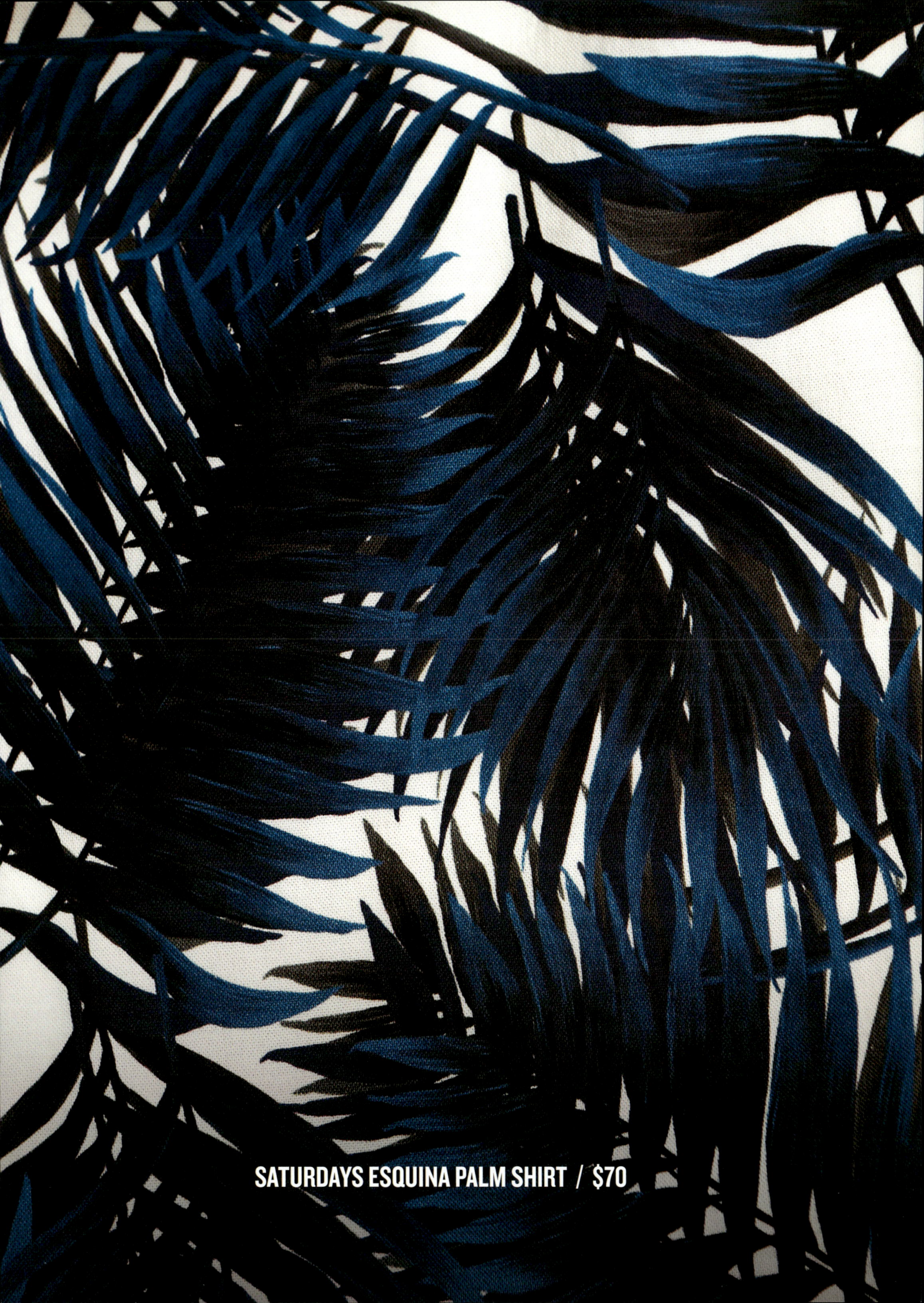
SATURDAYS ESQUINA PALM SHIRT / $70

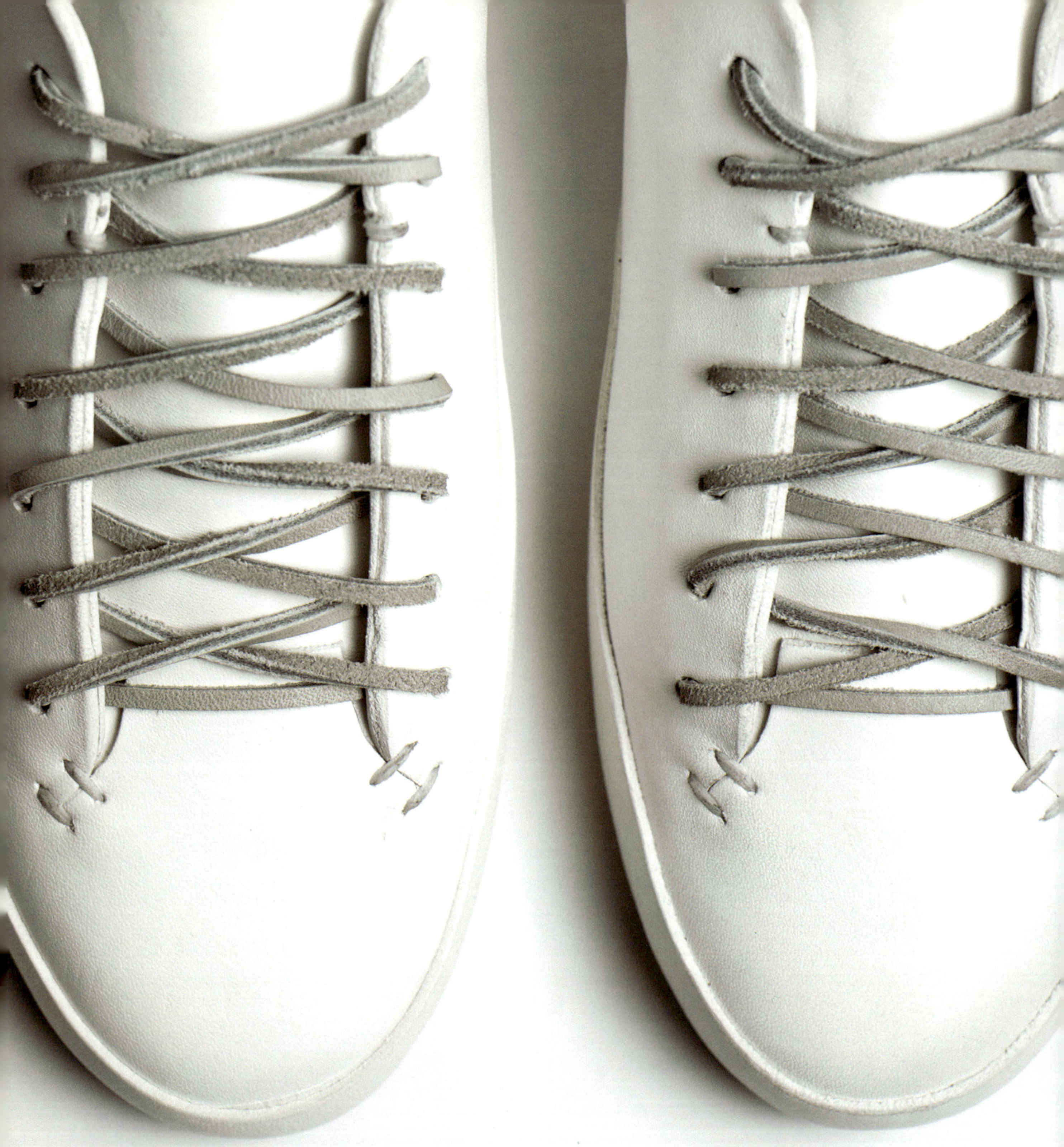

FEIT HAND SEWN SNEAKER / $600

CORRIDOR LILY SHIRT / $195

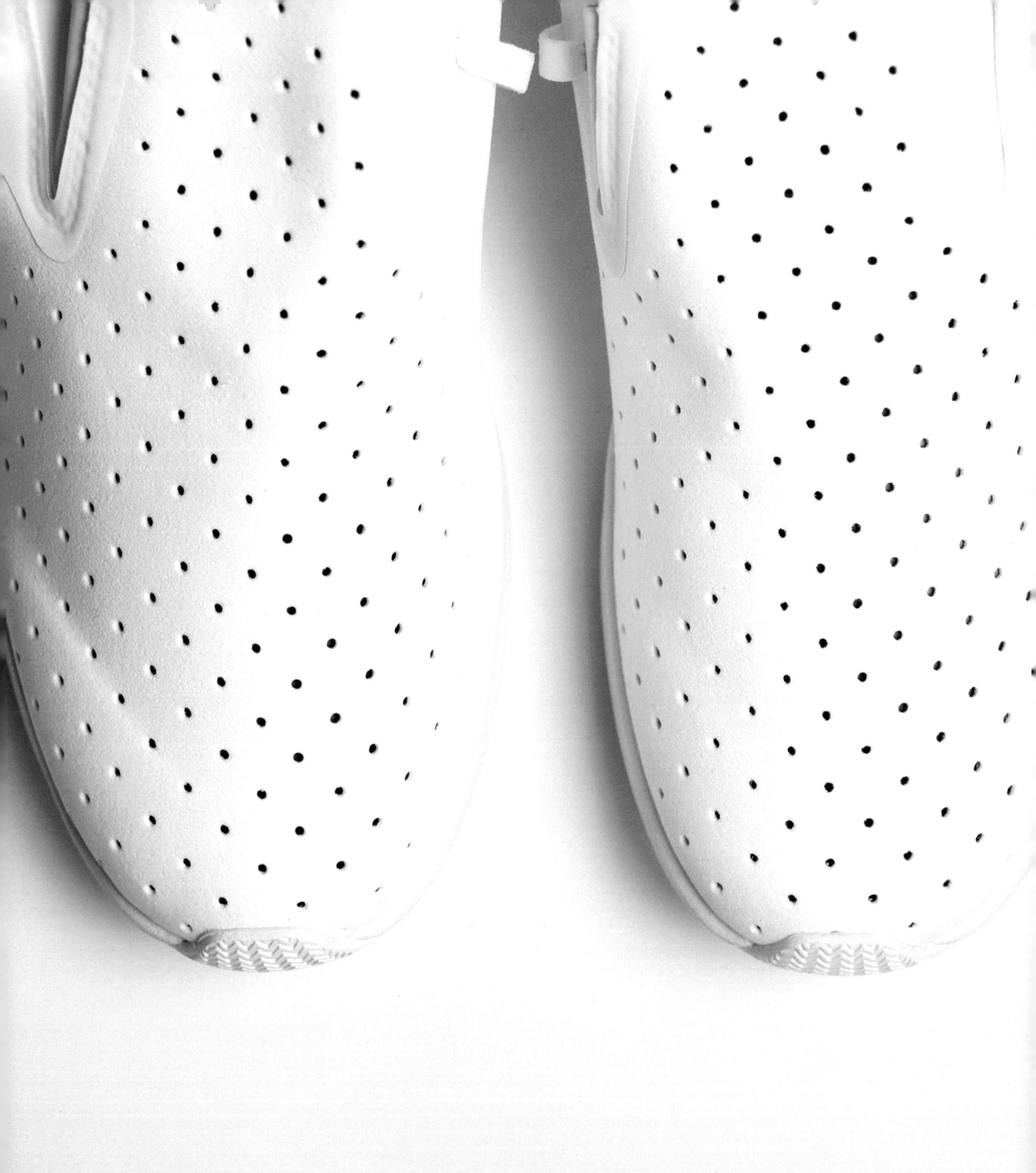

NATIVE APOLLO MOC SNEAKER / $75

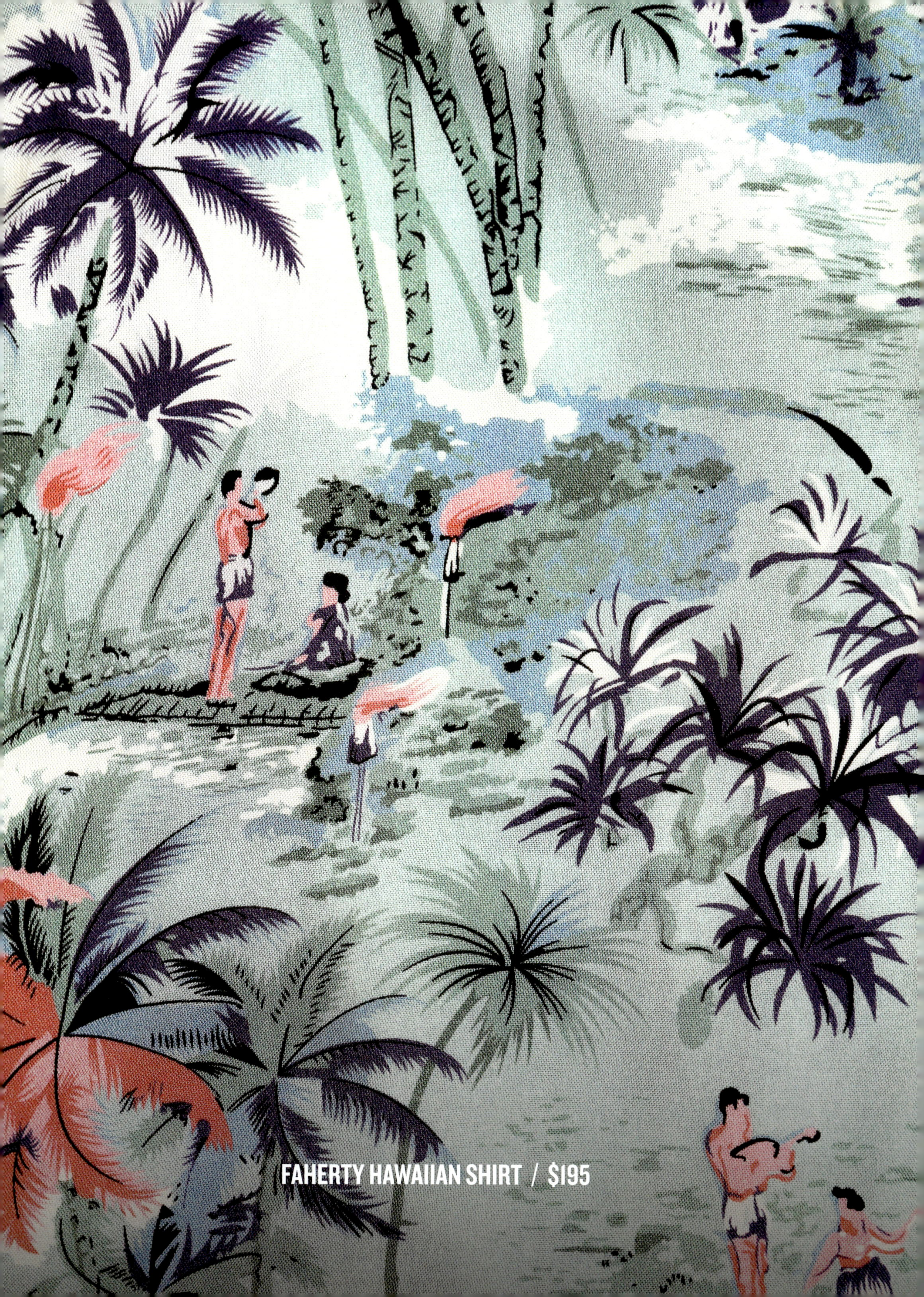

FAHERTY HAWAIIAN SHIRT / $195

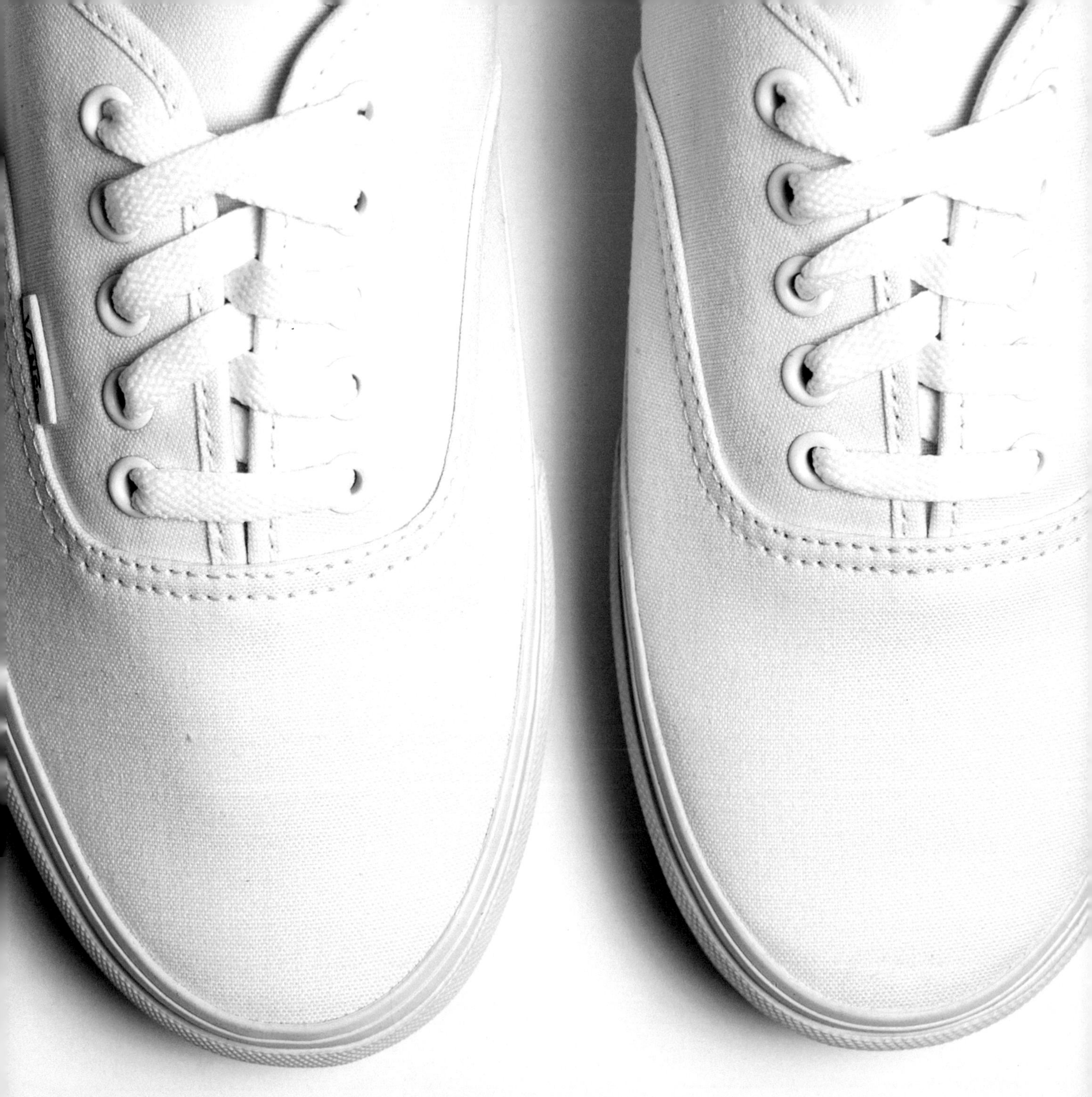

VANS AUTHENTIC SNEAKER / $50

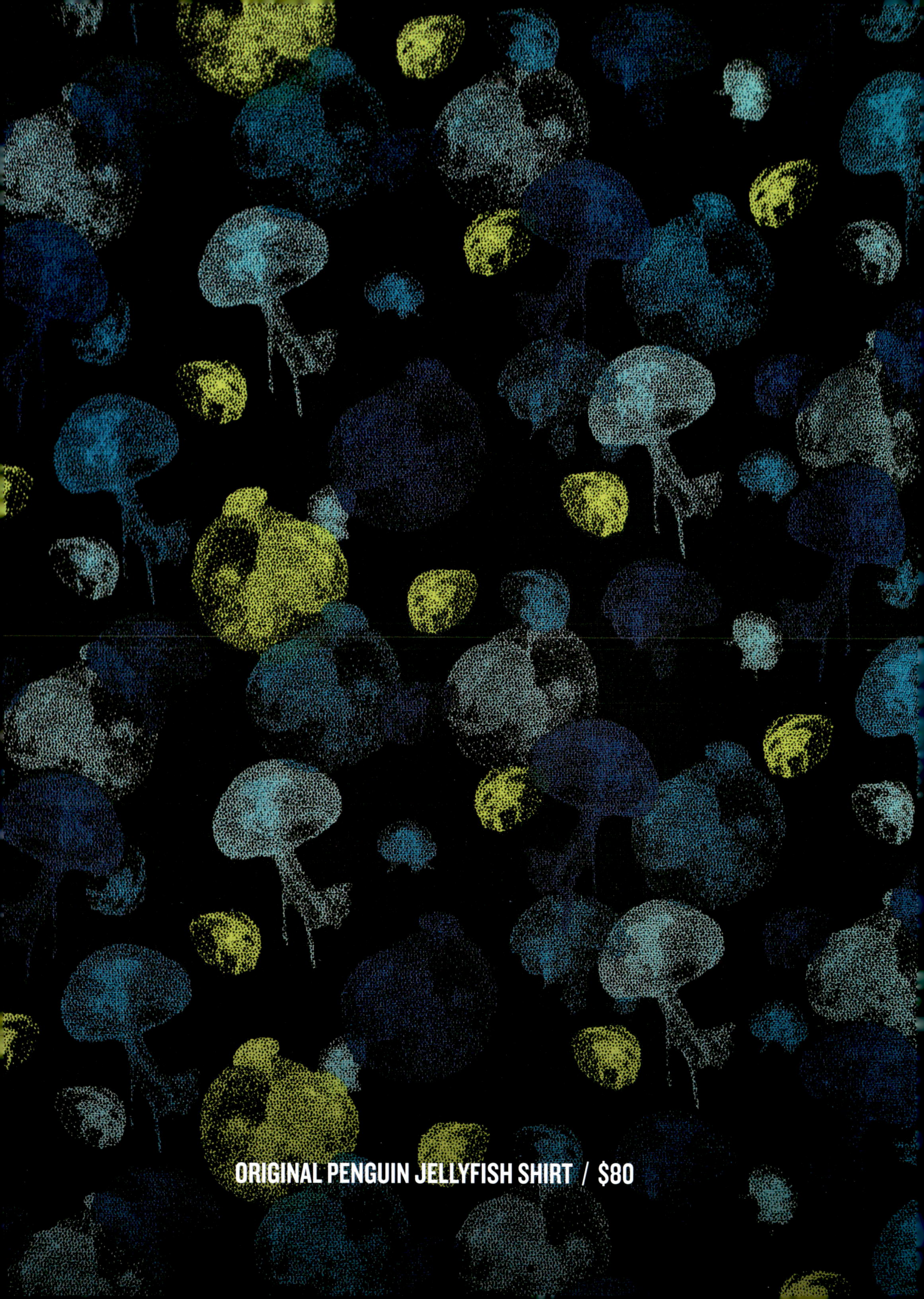

ORIGINAL PENGUIN JELLYFISH SHIRT / $80

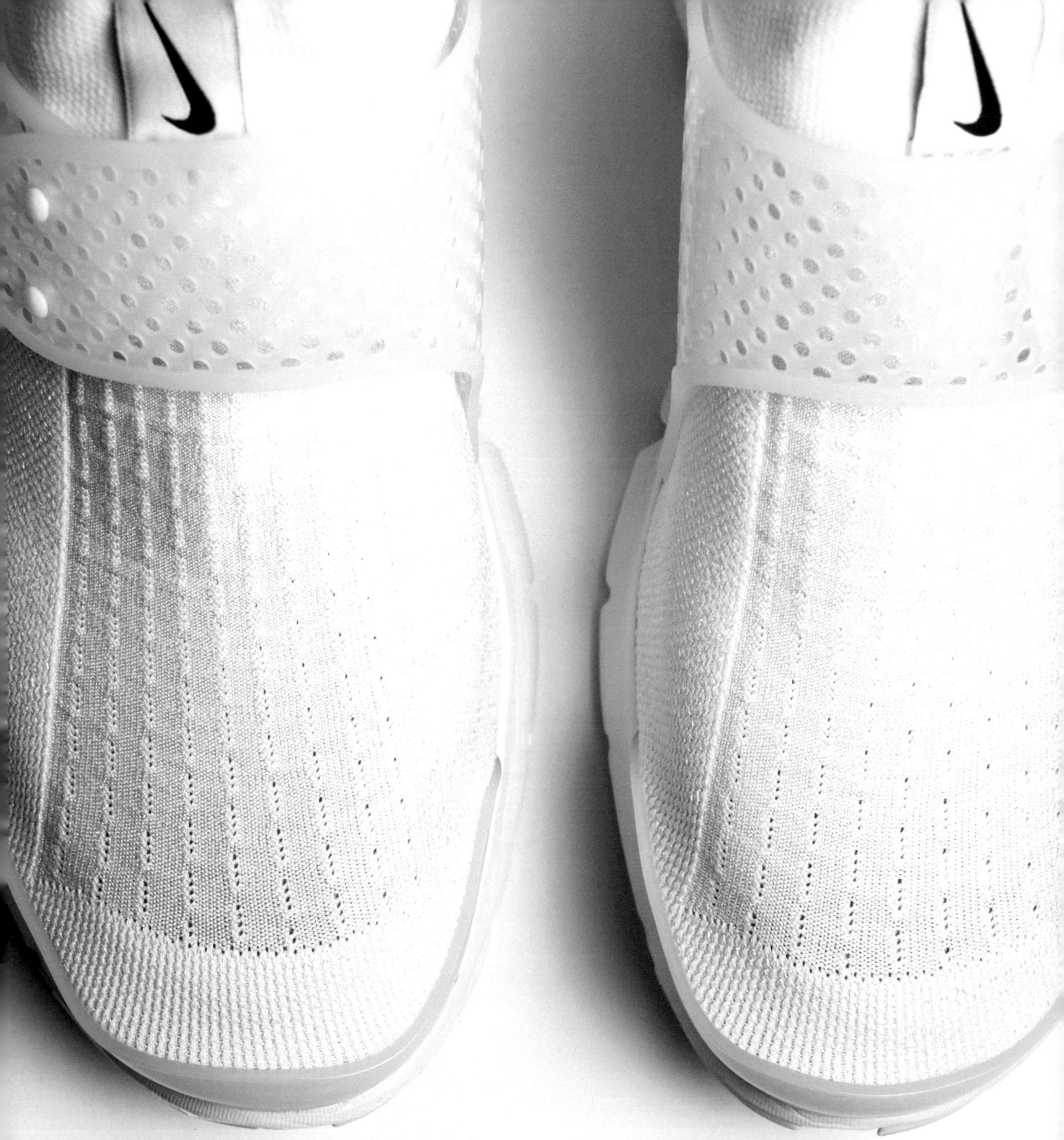

NIKE SOCK DART SP SNEAKER / $210

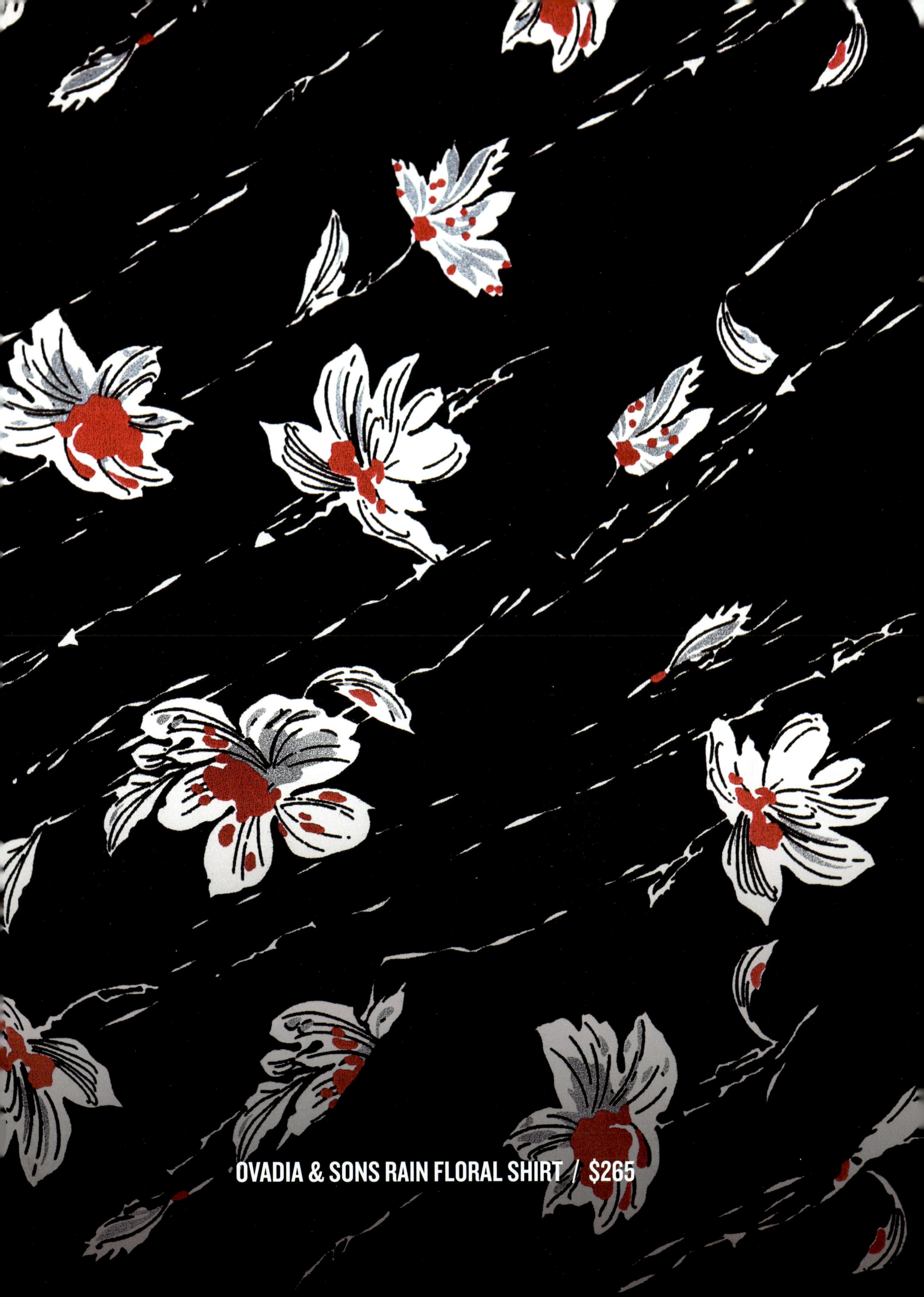

OVADIA & SONS RAIN FLORAL SHIRT / $265

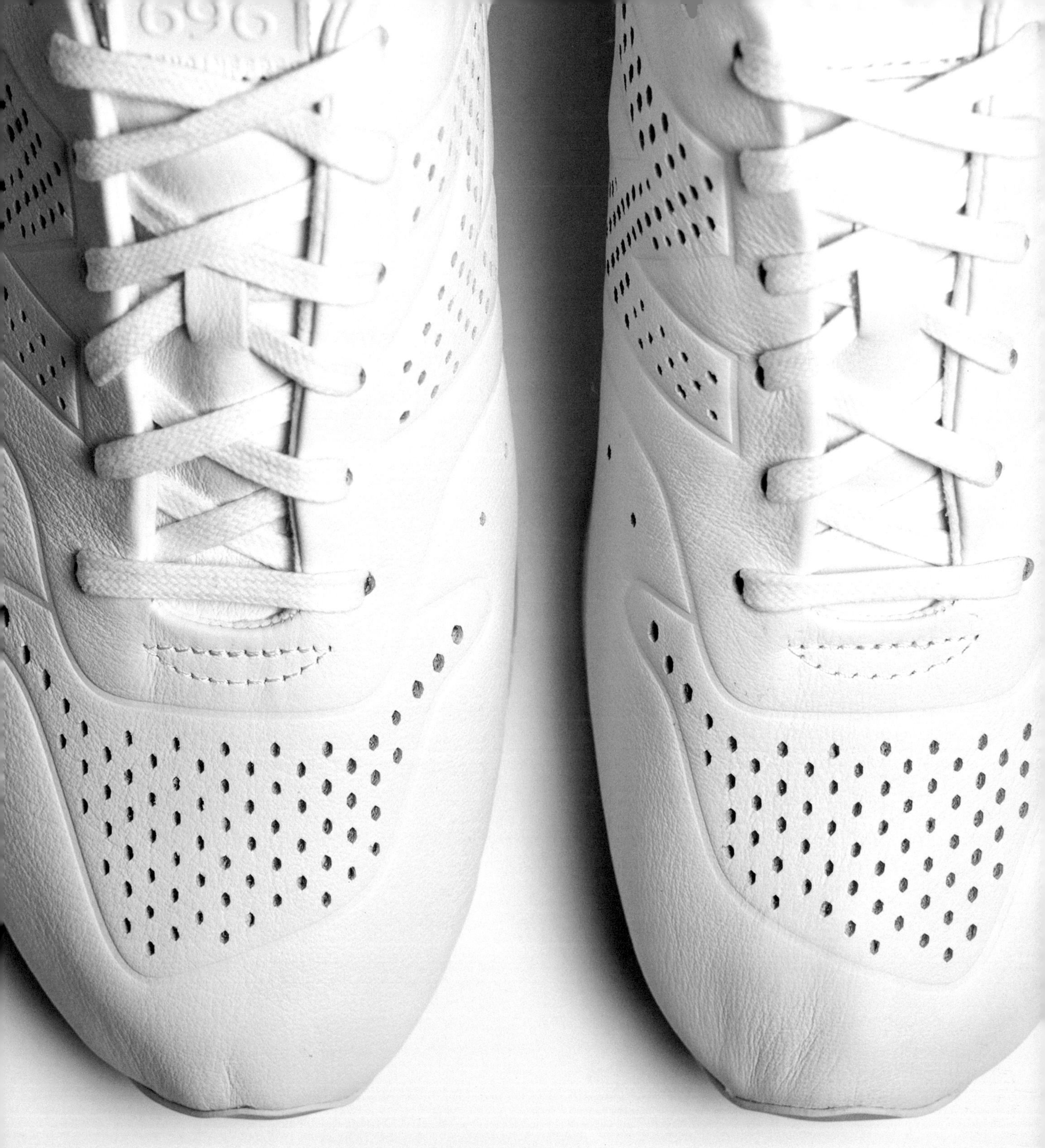

NEW BALANCE MRL696 SNEAKER / $130

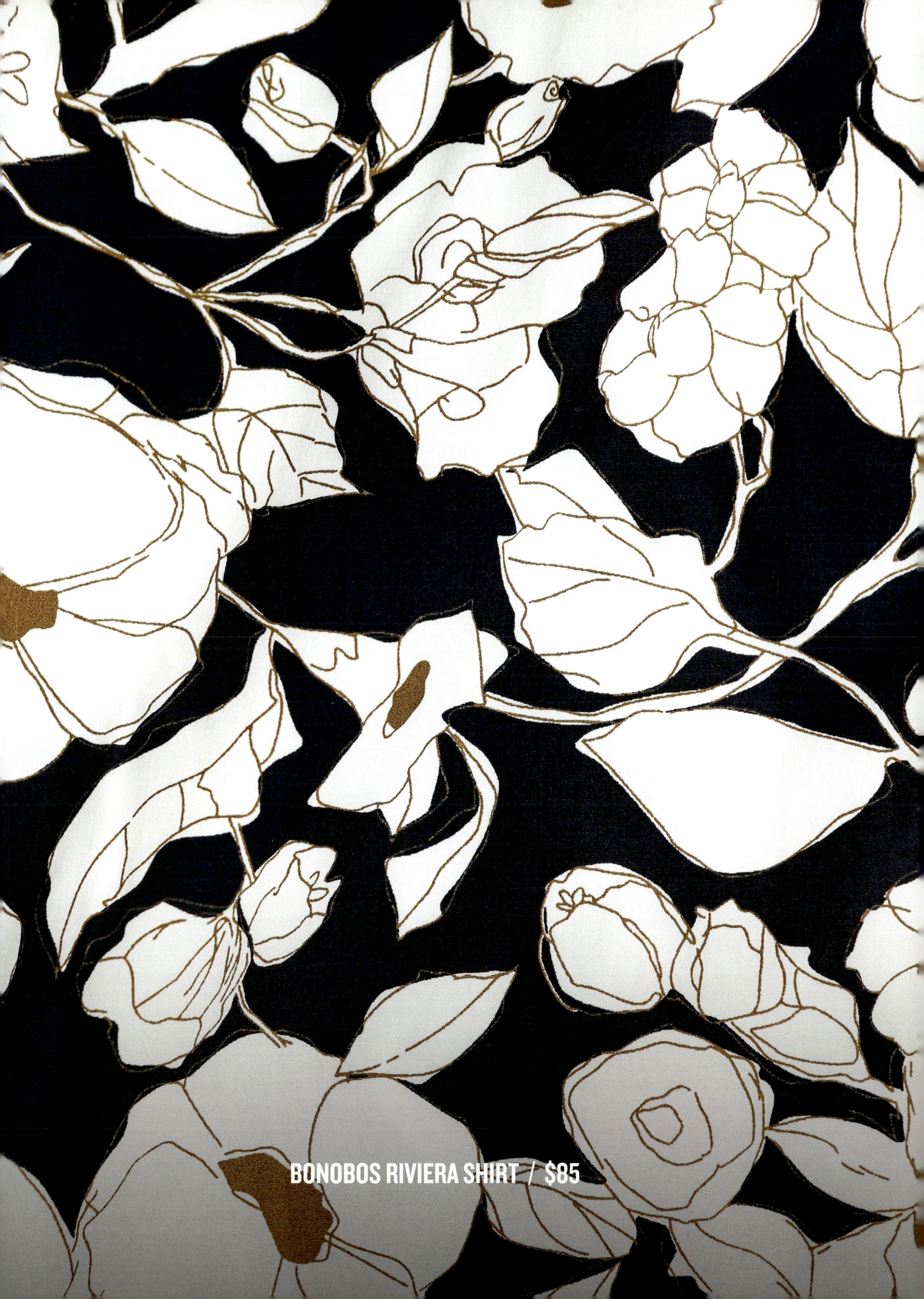
BONOBOS RIVIERA SHIRT / $85

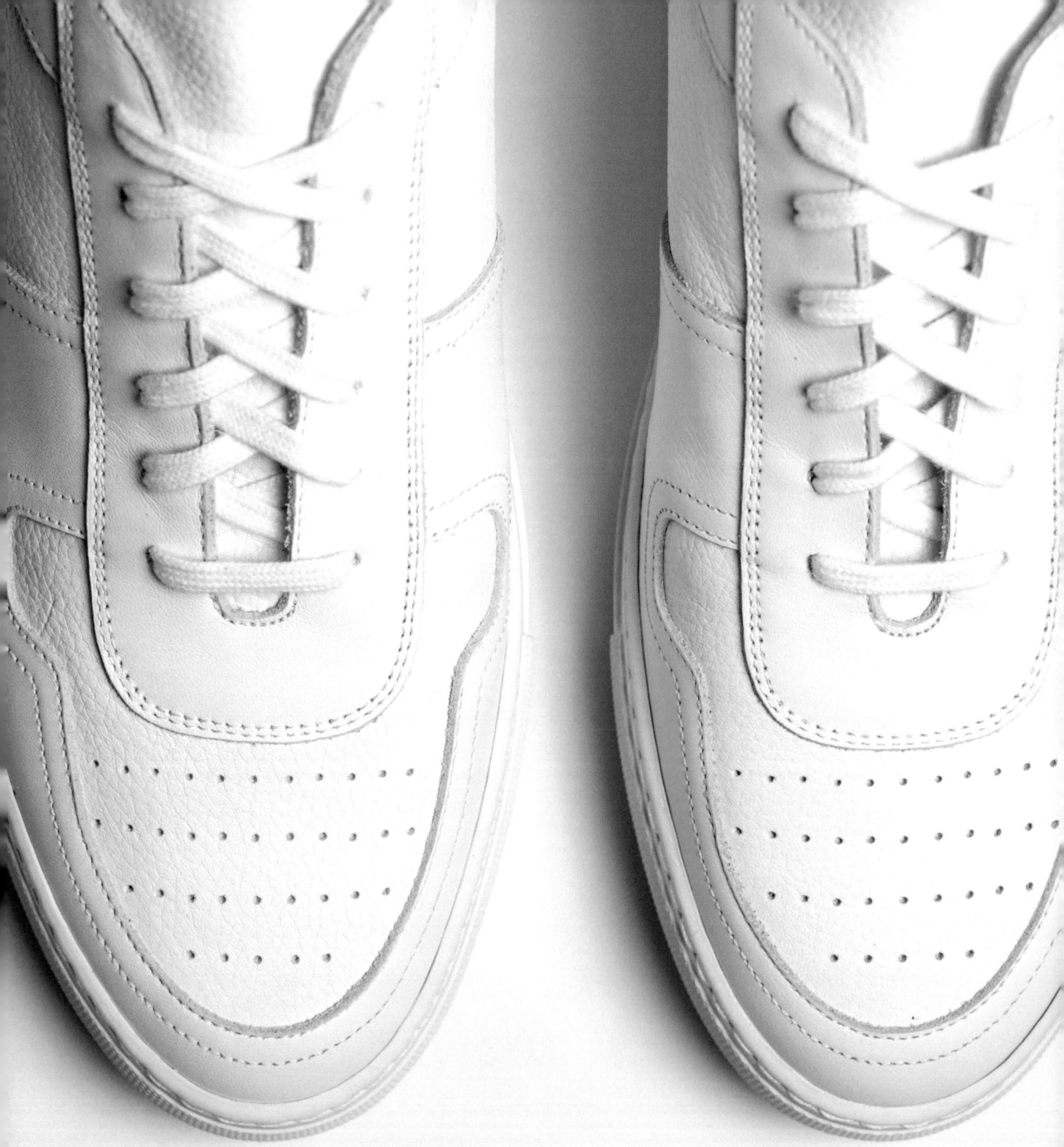

COMMON PROJECTS BBALL SNEAKER / $460

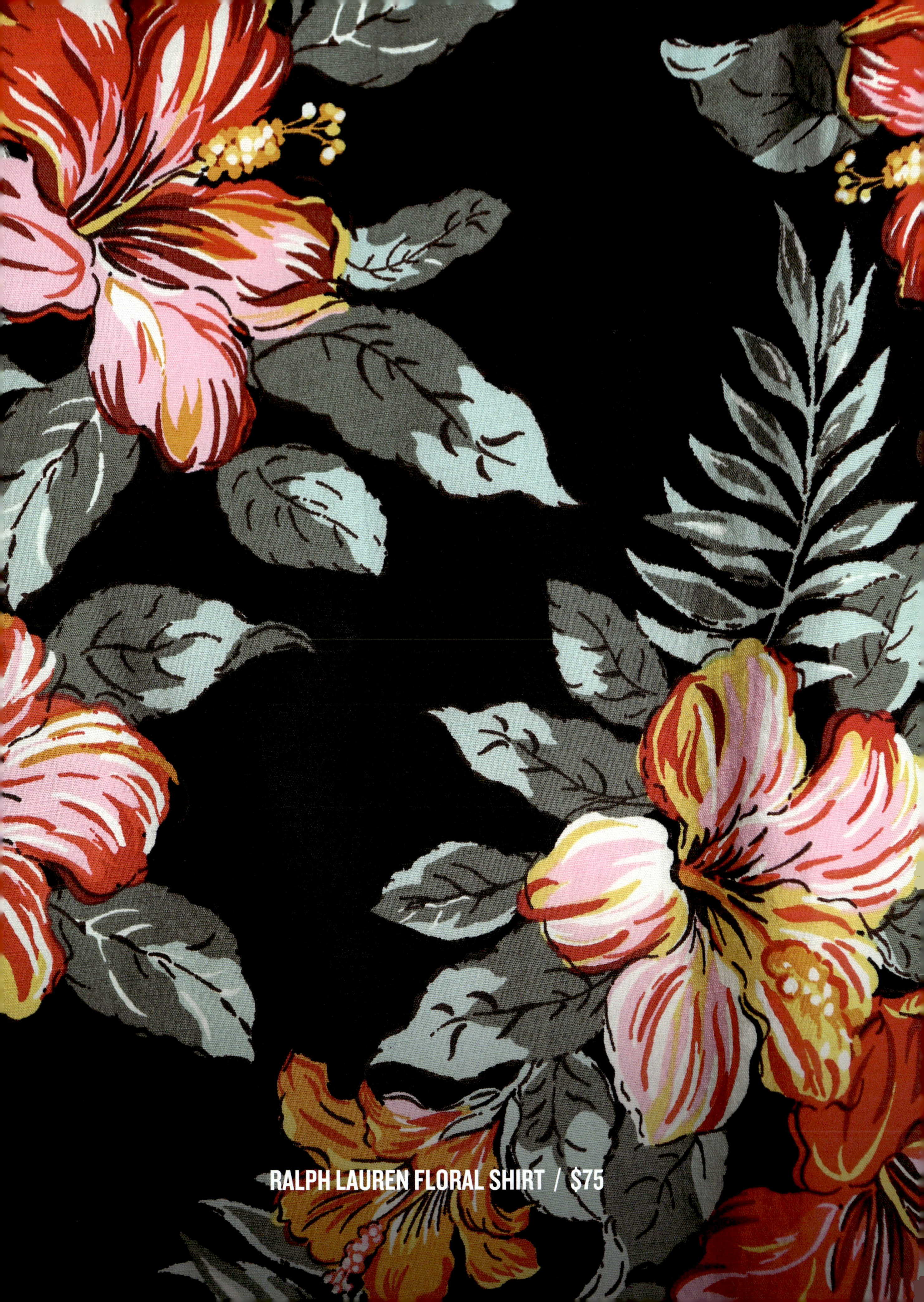

RALPH LAUREN FLORAL SHIRT / $75

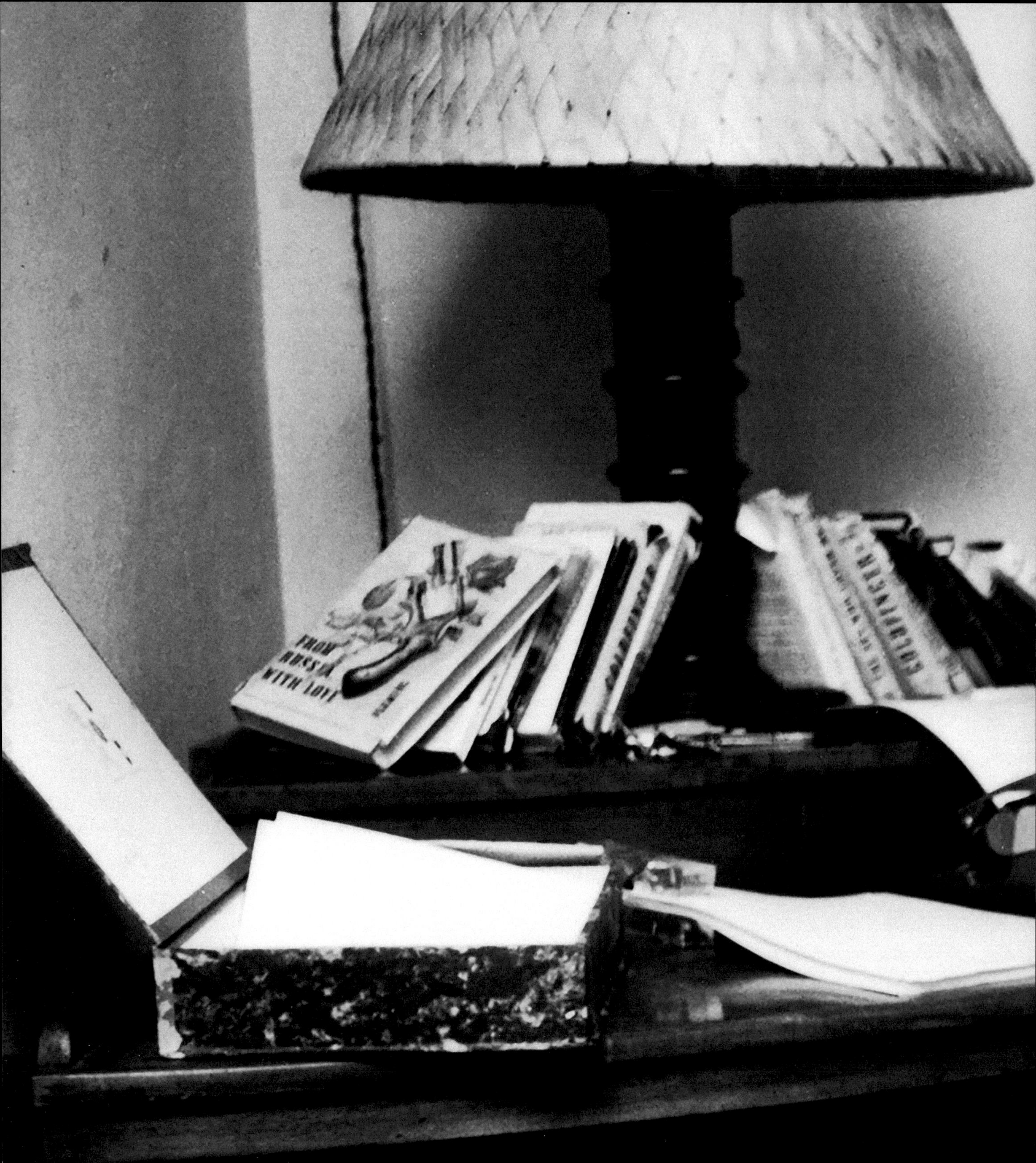
FROM
RUSSIA
WITH LOVE

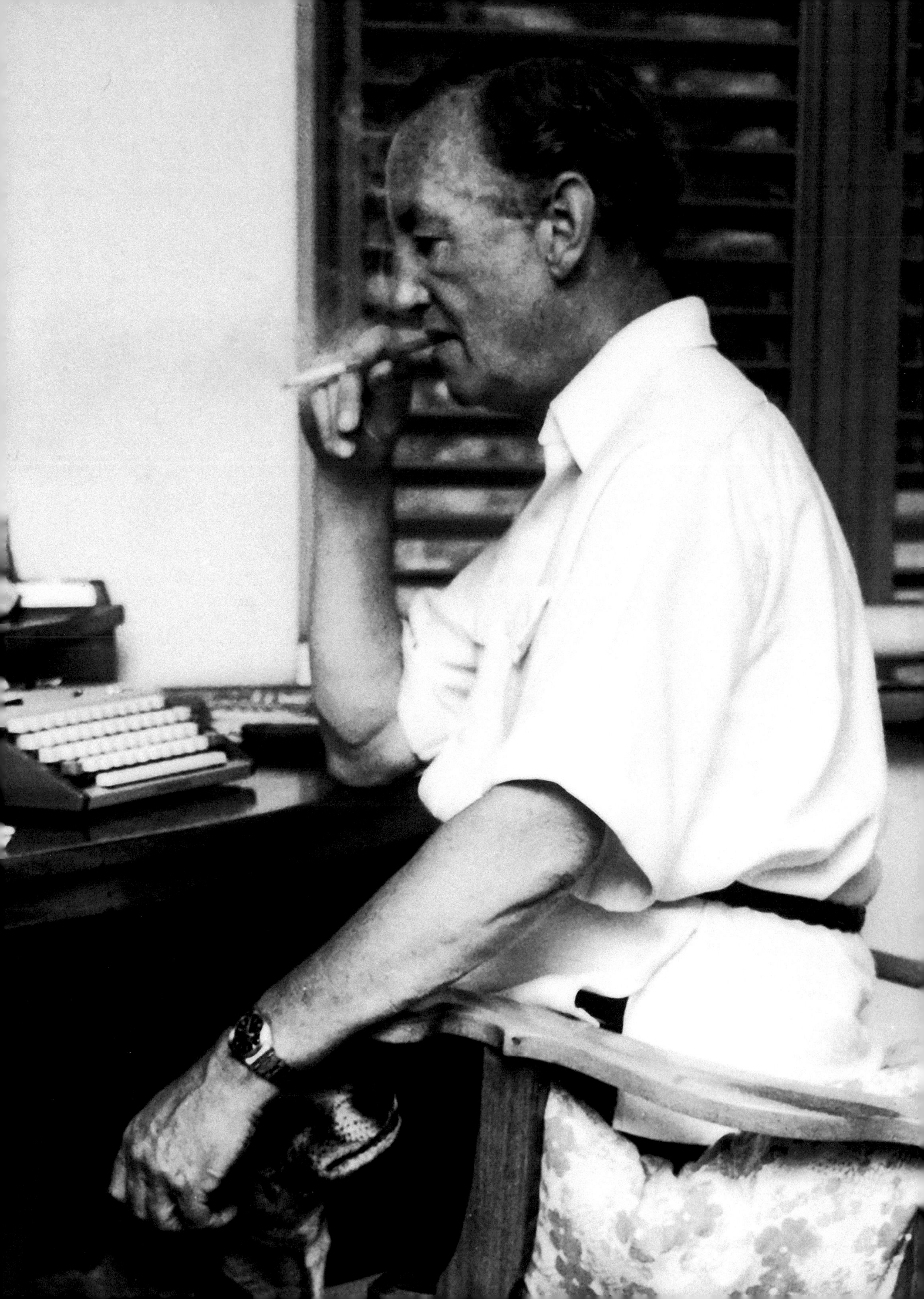

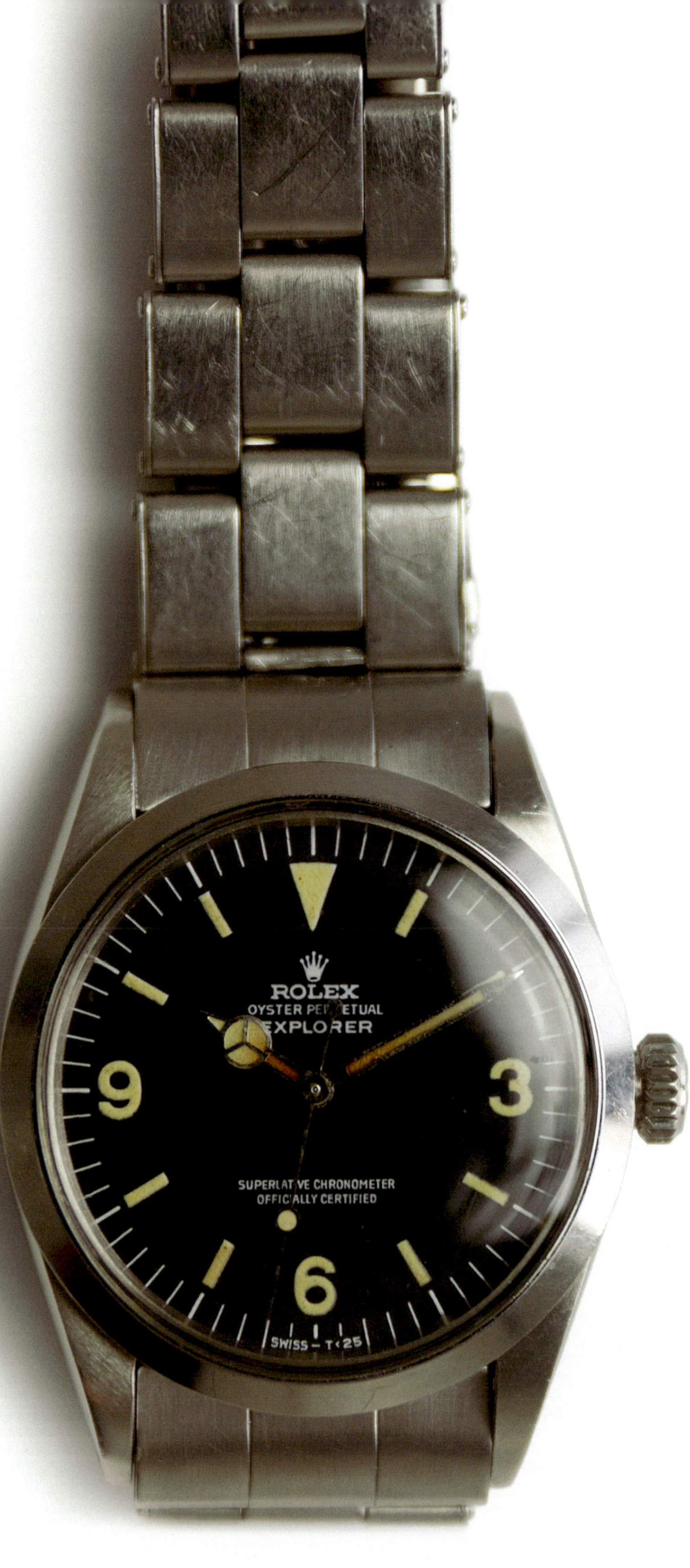

ROLEX EXPLORER 1016 / $7,500+

There were other Explorer models released closer to the prototype's famous trip up Everest, but none had the longevity of the 1016. Produced from 1963 to 1989, this model's combination of a slim 36mm case and basic black dial with its trademark 3/6/9 markings helped it to become one of Rolex's longest-running references.

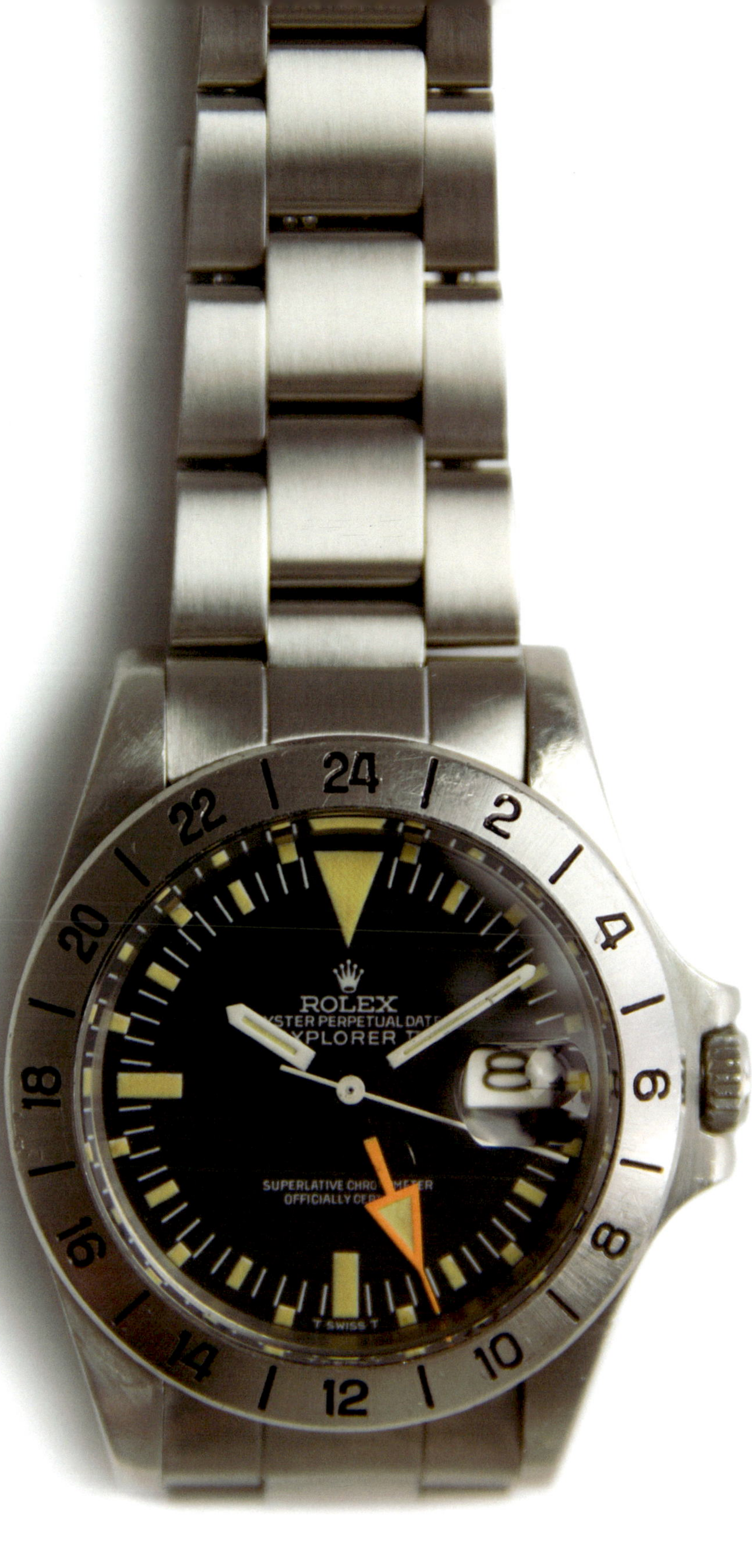

ROLEX EXPLORER II 1655 / $12,000+

Originally designed for cave explorers, the 1655 was the first Explorer II. It's notable for its orange GMT hand, which works in conjunction with the stationary bezel to help distinguish AM from PM. It sports a beefier 40mm build, and remains a bit of a misfit in Rolex's history, making it all the more coveted to those sick of Submariners.

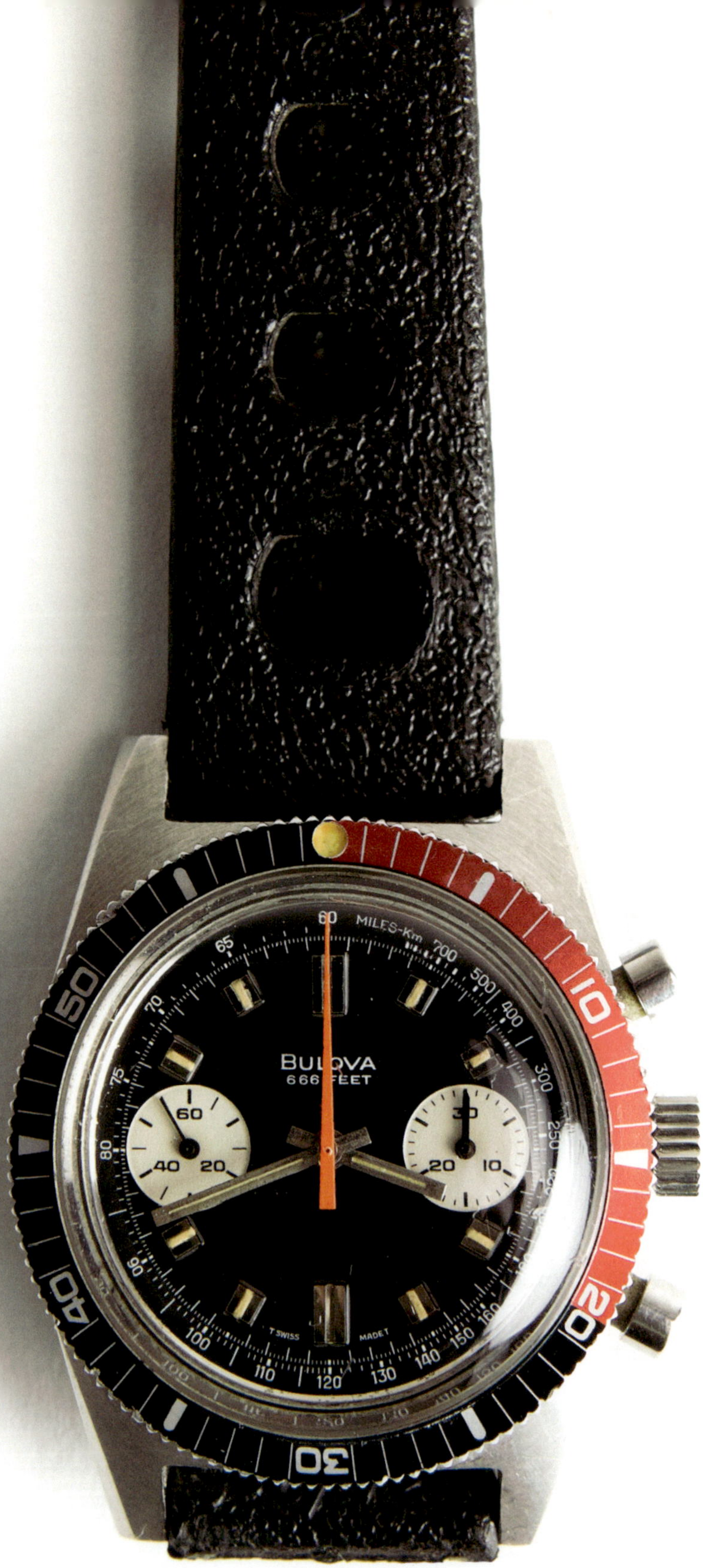

BULOVA DEEP SEA CHRONOGRAPH B / $800+

Known as a "Devil Diver" thanks to its cheeky water resistance rating of 666 feet, this early 1970s timepiece, which measures in at a universal-fitting 38mm size, is powered by the venerable Valjoux 7733 movement, and looks to be close to new old stock (NOS) condition, appearing much as it did when it left the factory.

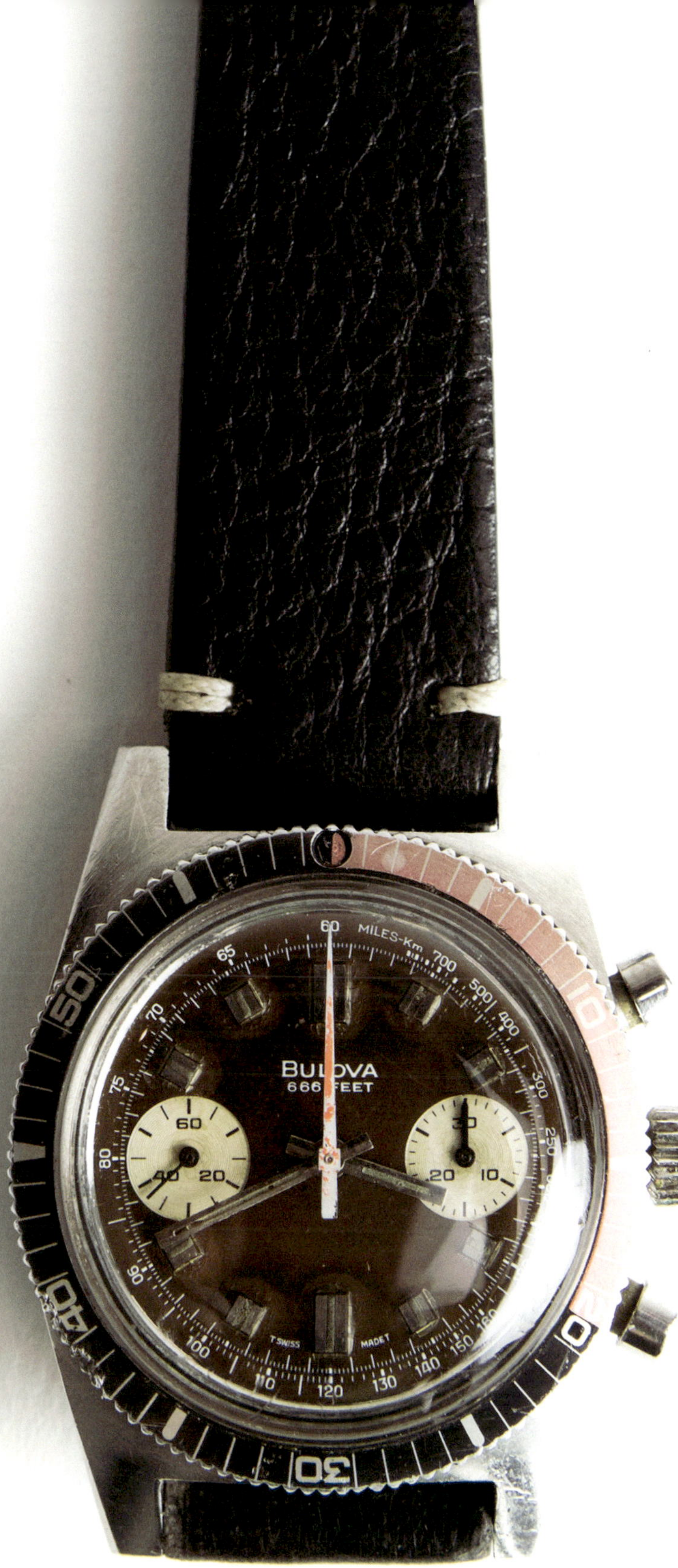

PATINA

TROPICAL BULOVA DEEP SEA CHRONOGRAPH B / $600+

As opposed to the crisp watch on the left, this timepiece wears its age with pride. Its rotating black and red "Coke" bezel is heavily faded, just like the reverse panda dial that has seen its black change to a bubbled "tropical" brown. This is the wrist equivalent of a pair of perfectly broken-in jeans. That you've had for 45 years.

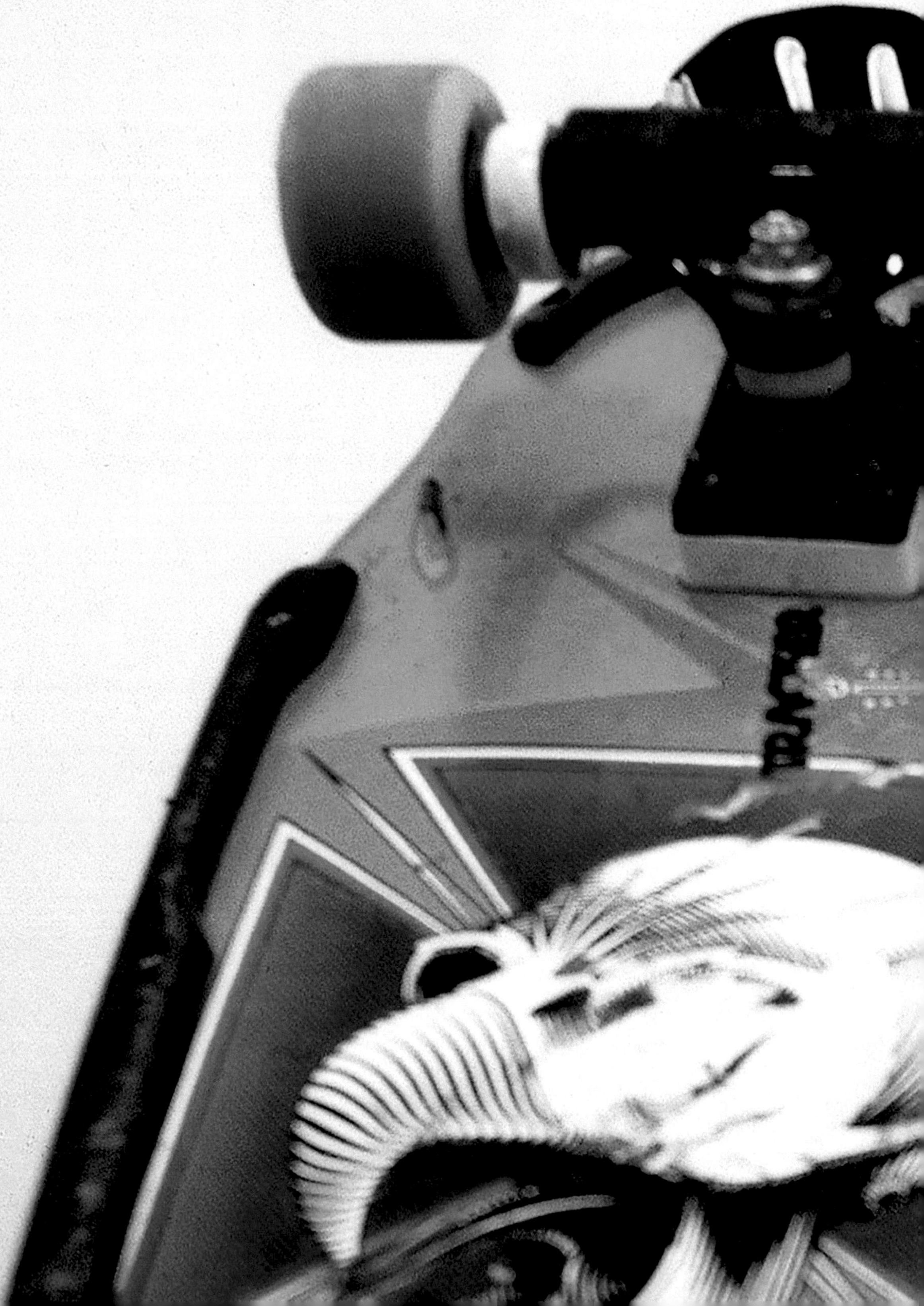

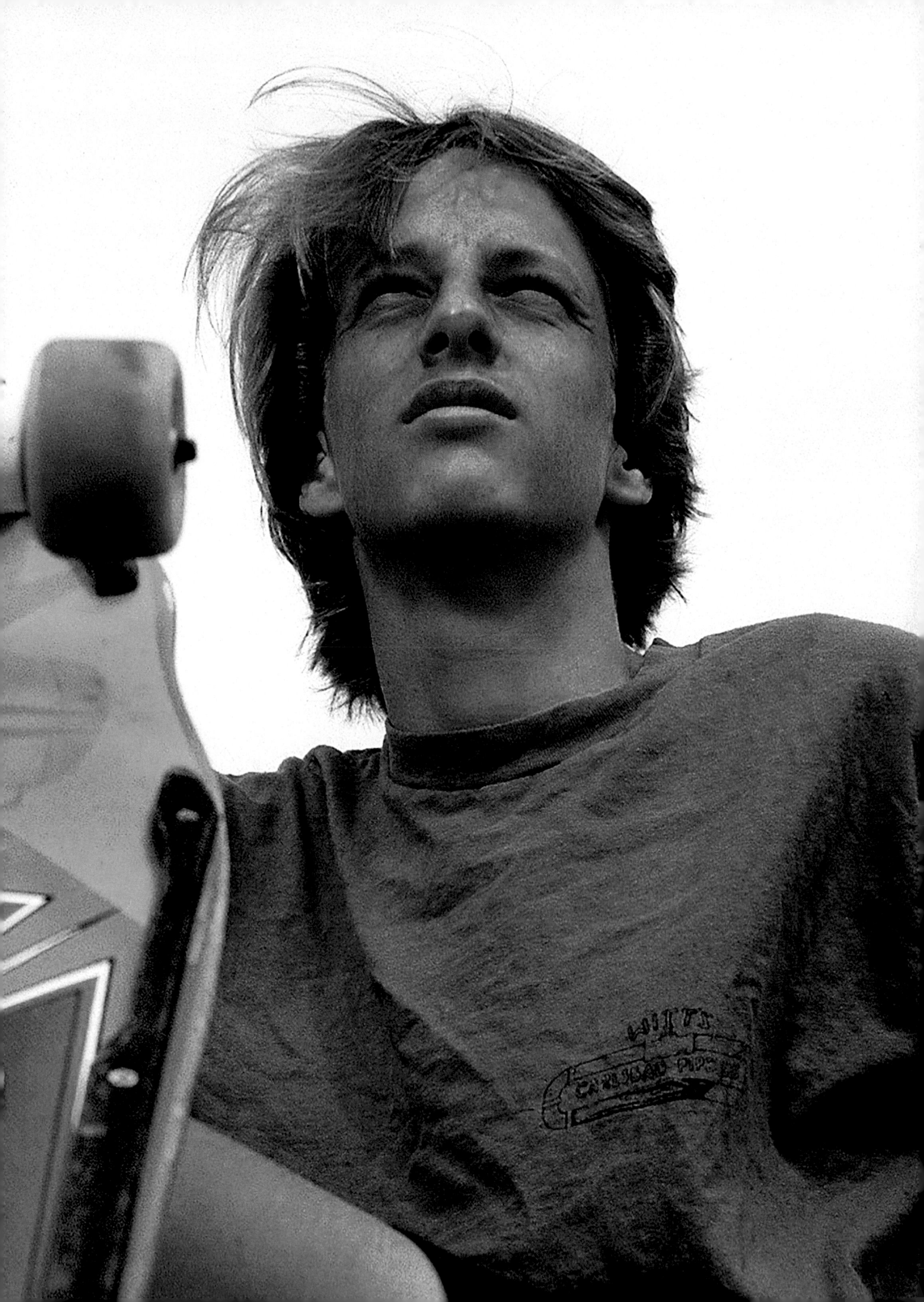

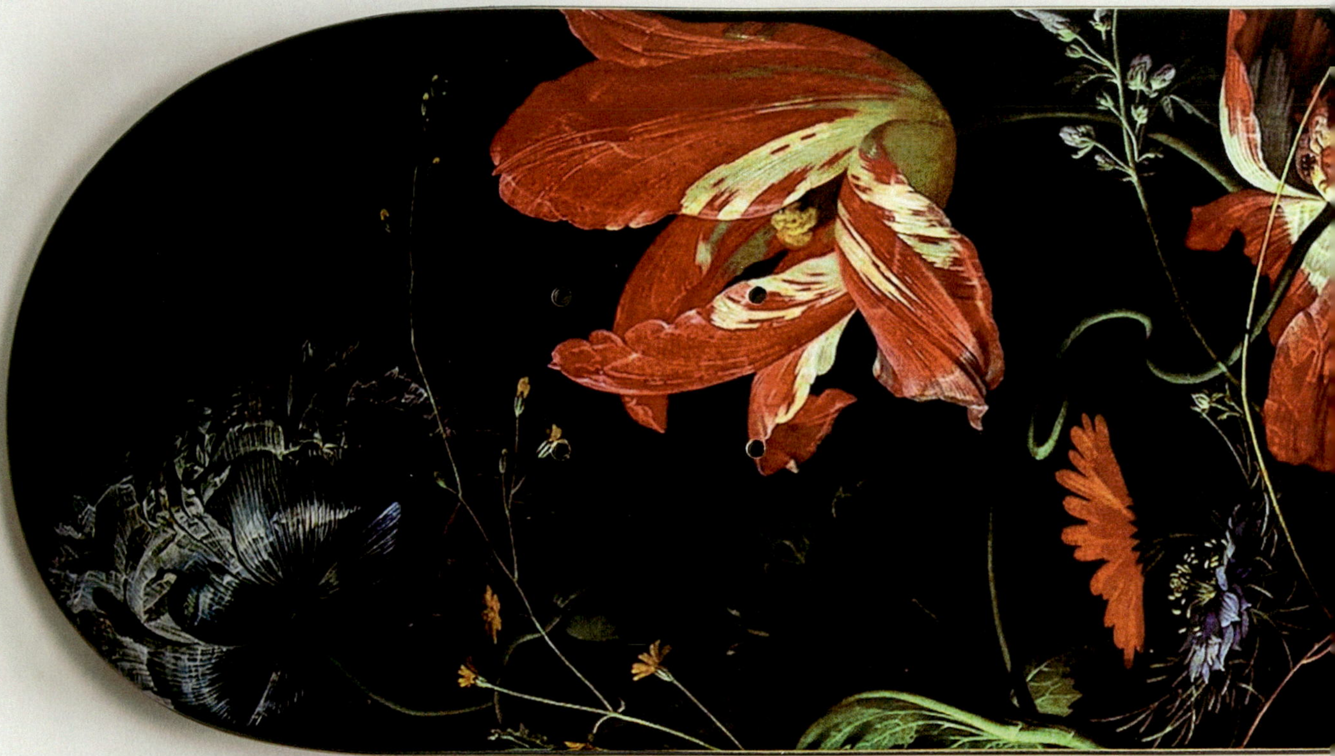

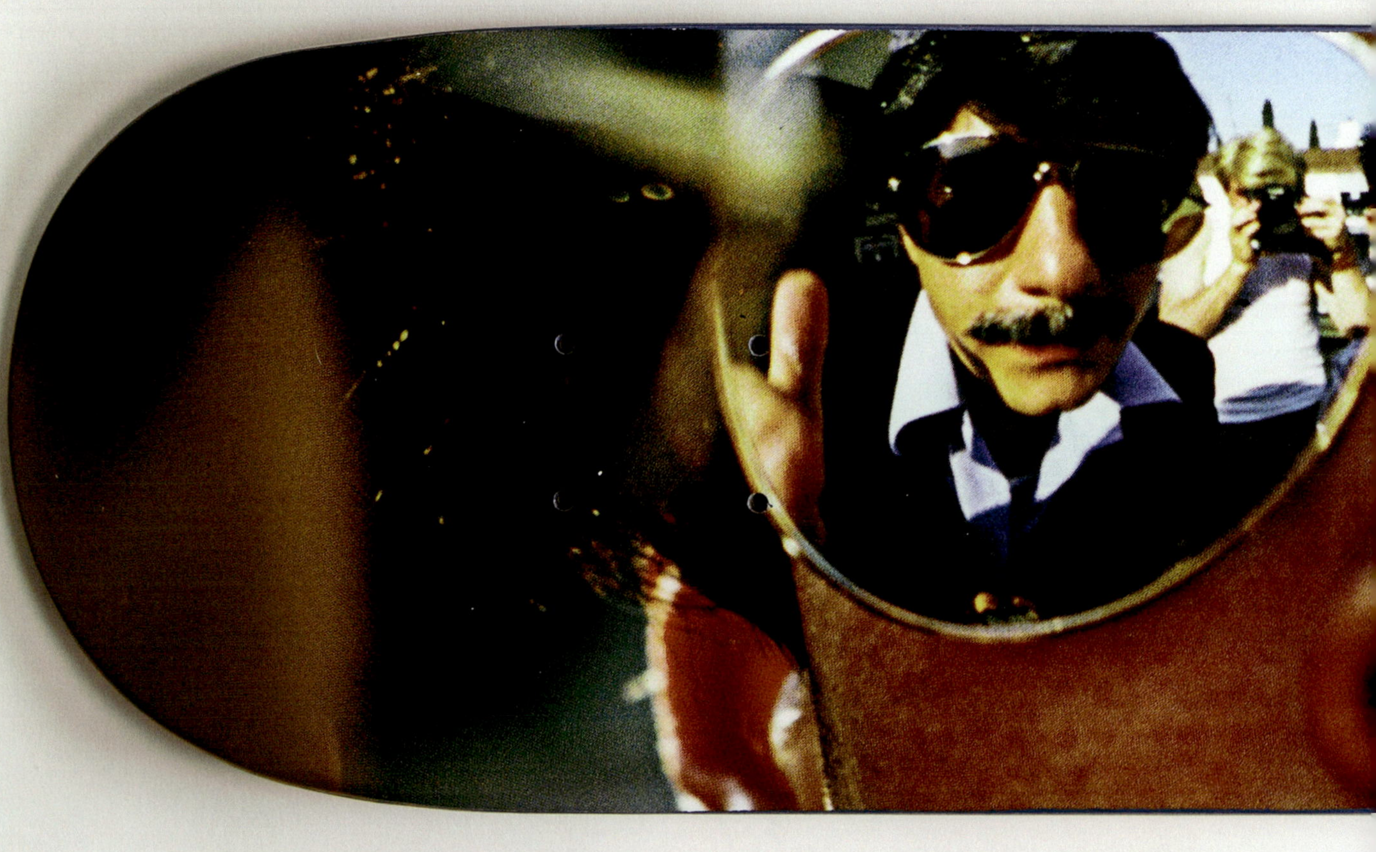

BOOM-ART DIPTYCH FLOWERS / $200 / SET OF 2

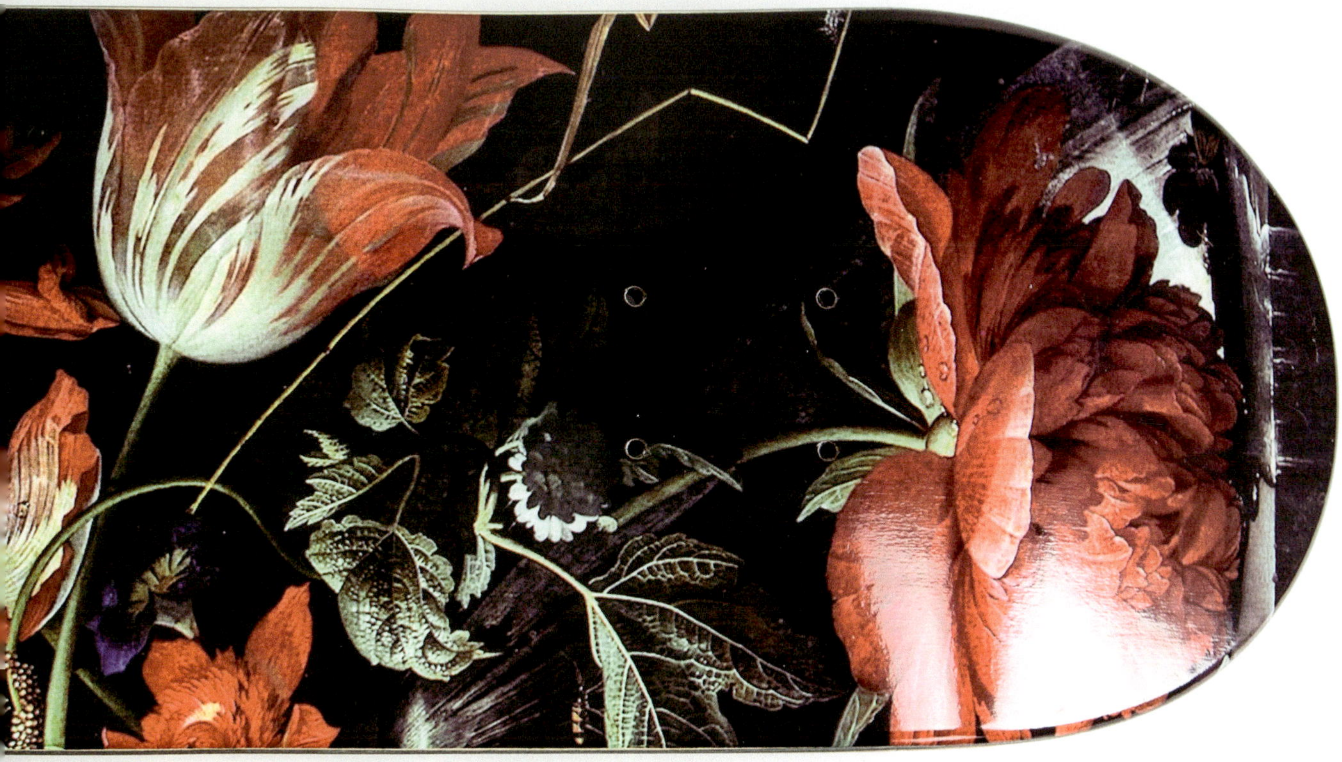

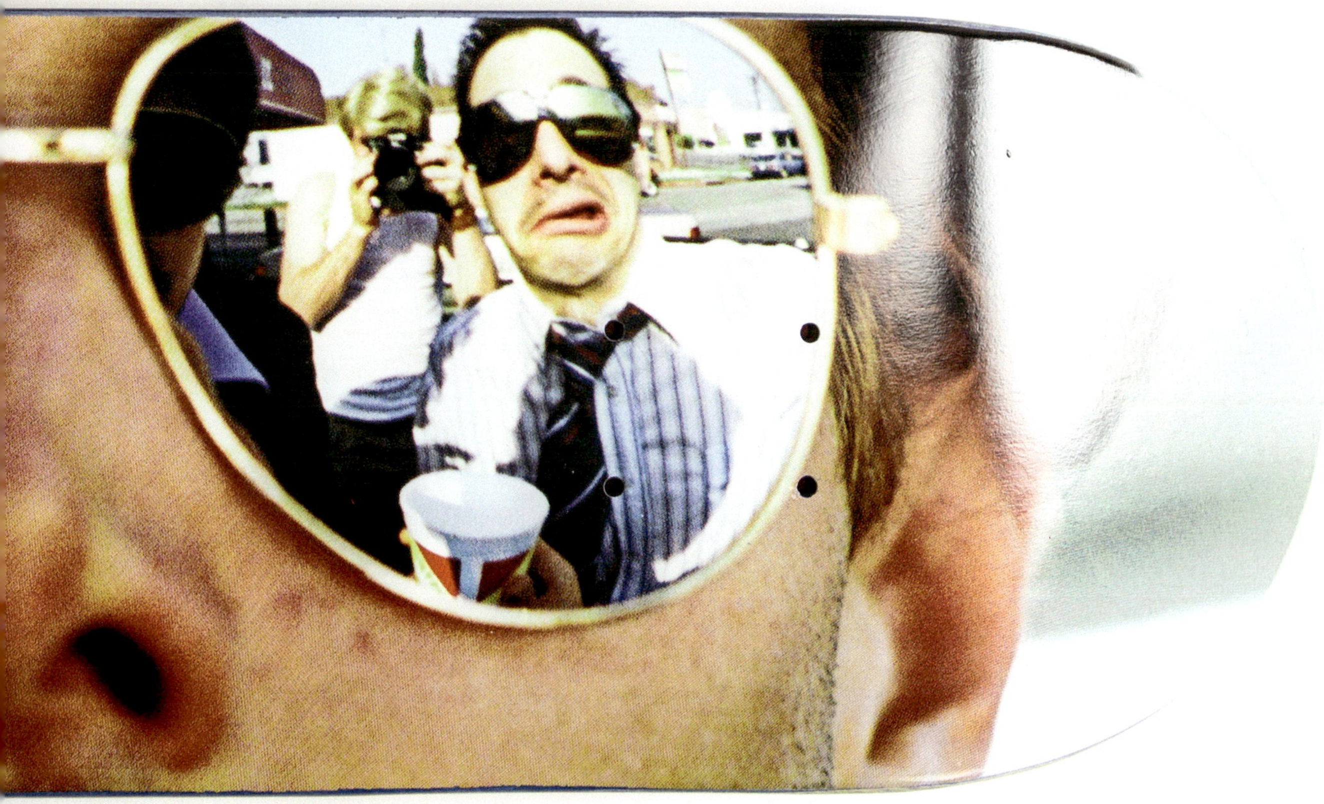

GIRL SKATEBOARDS SPIKE JONZE BEASTIE BOYS / $50

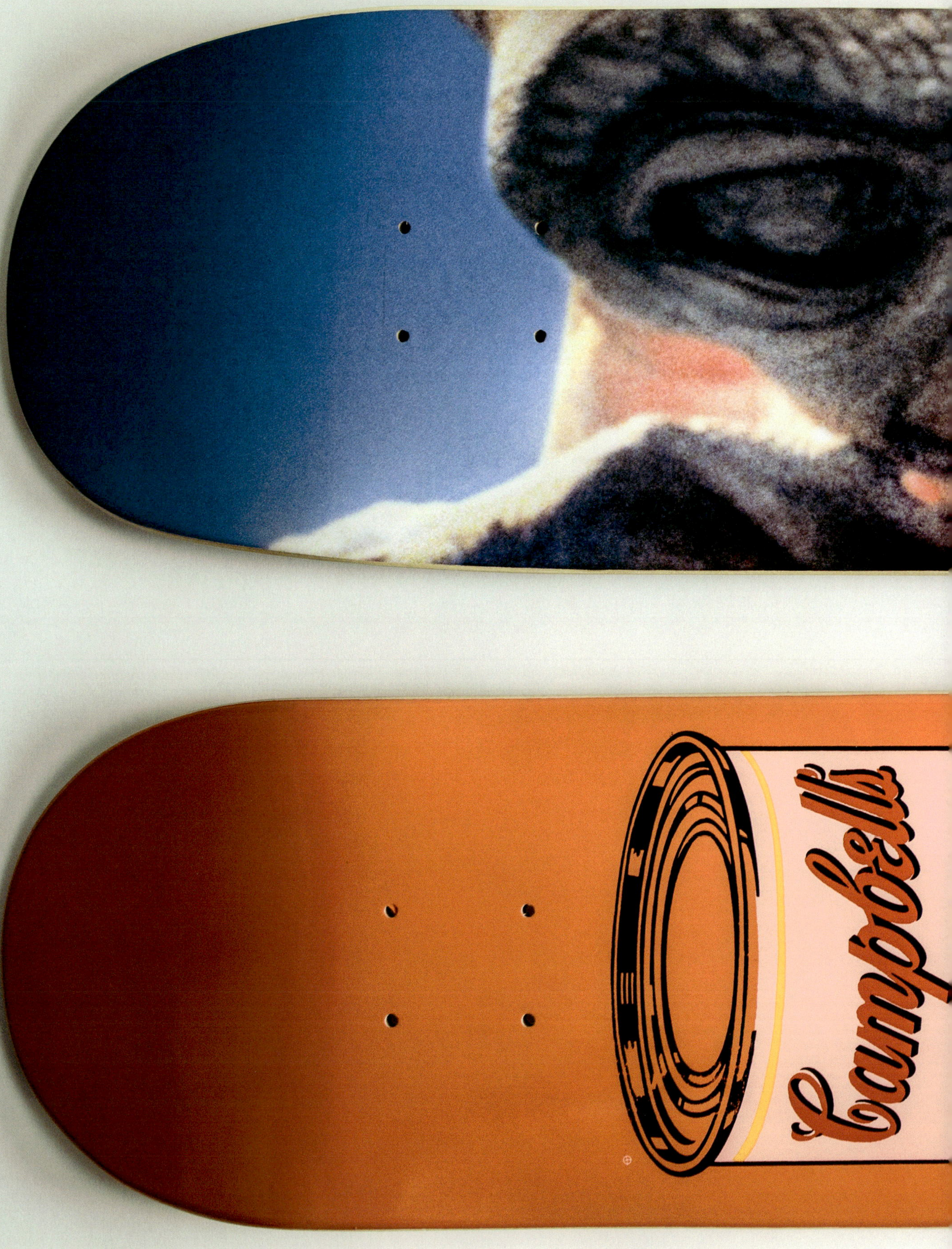
Campbell's

SUPREME E.T. / $150

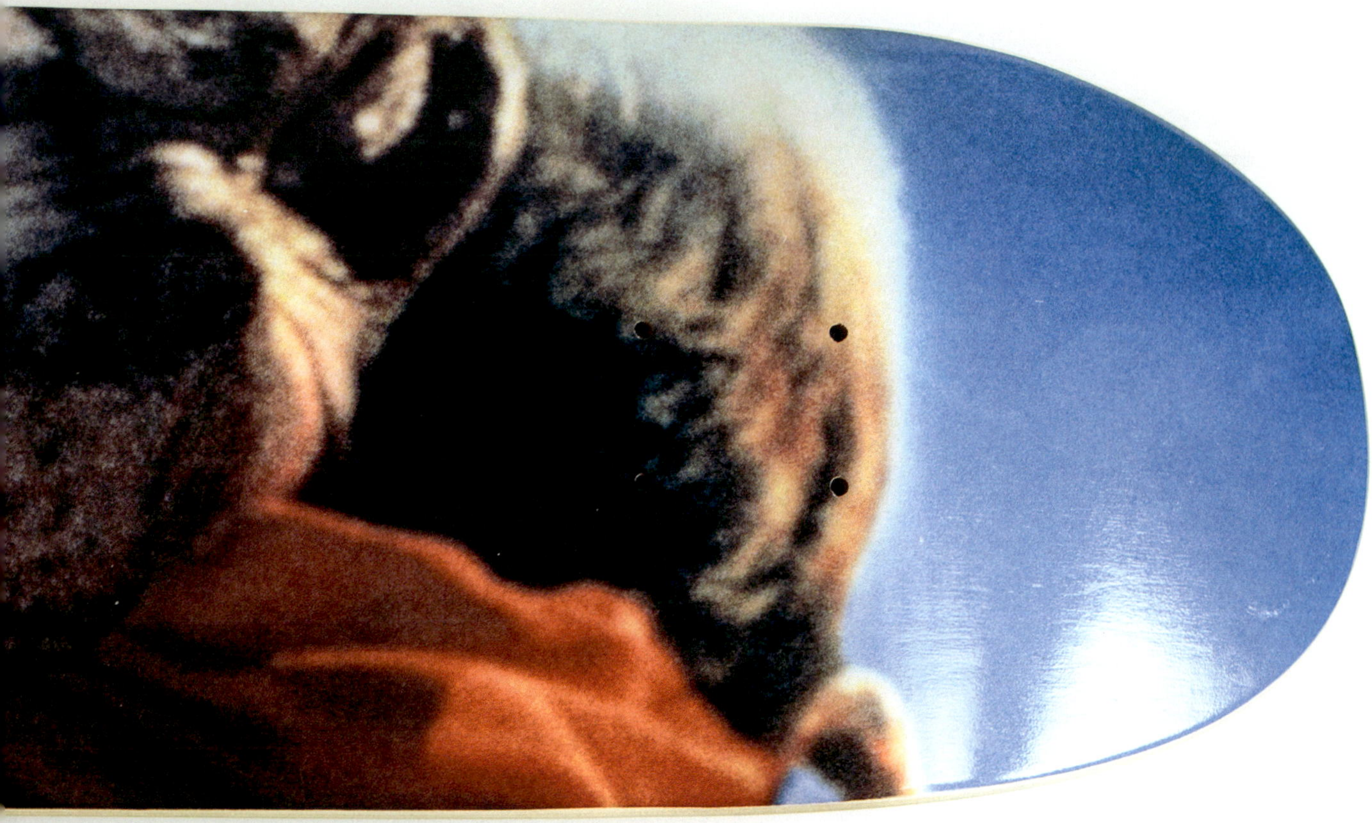

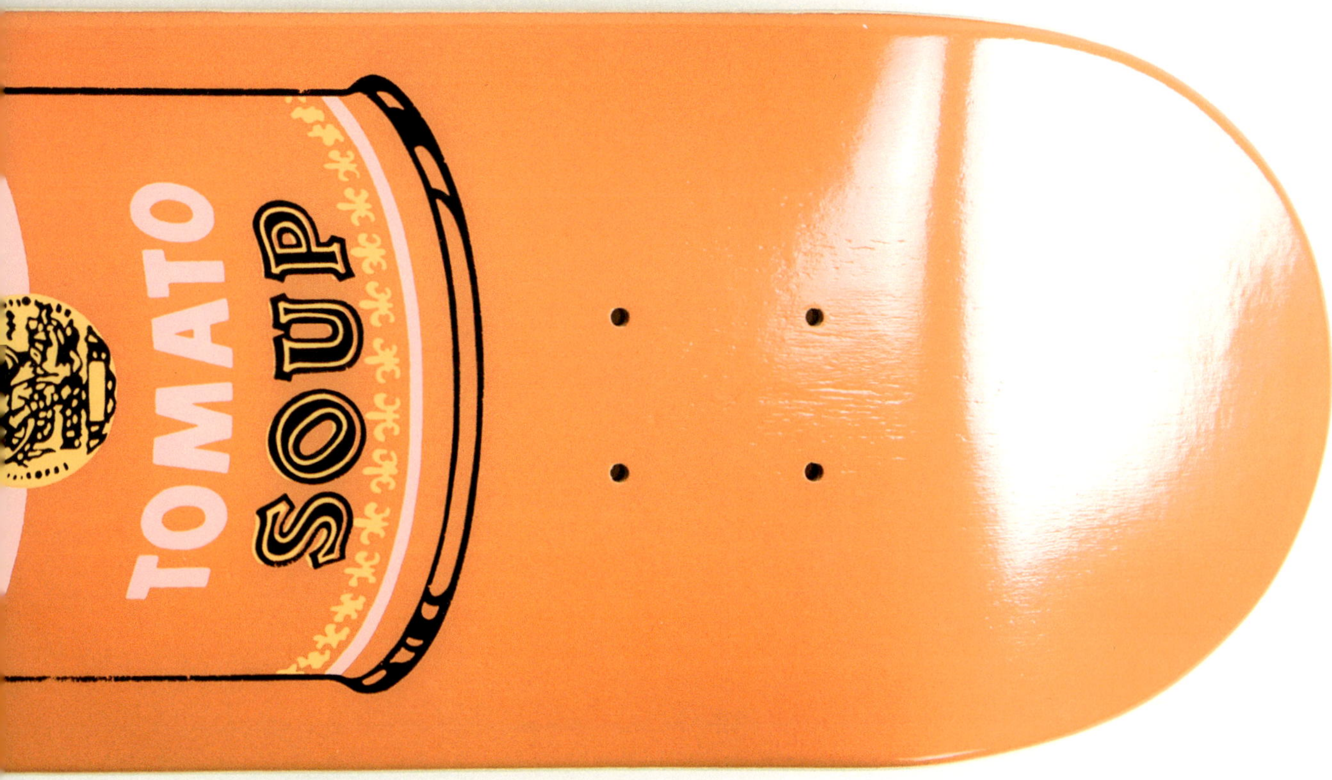

THE SKATEROOM X ANDY WARHOL / $1,570 / SET OF 8

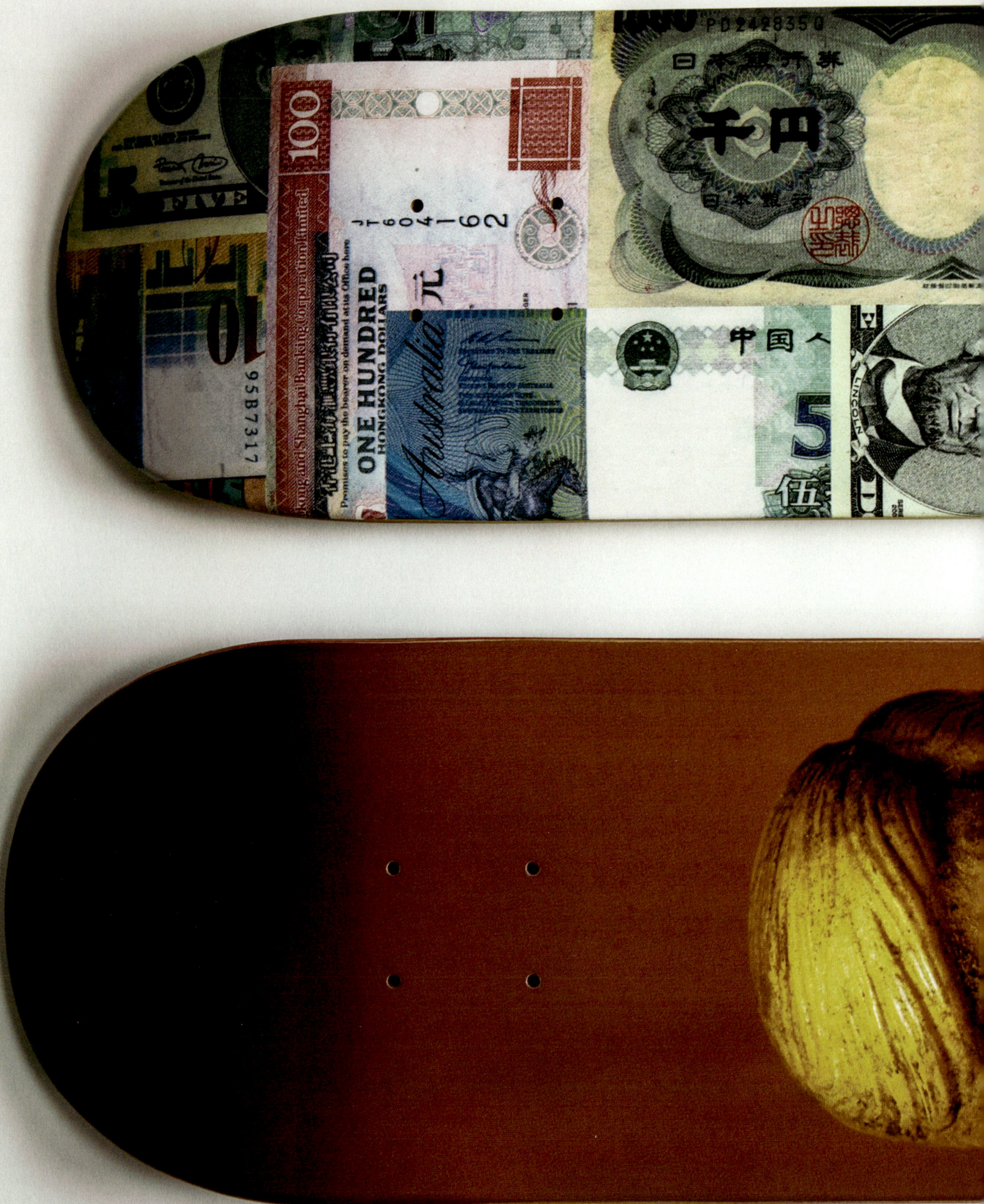

PD2428350
日本銀行券
千円
日本銀行
100
ONE HUNDRED
HONGKONG DOLLARS
元
Australia
中国人
伍
FIVE
LINCOLN
95B7317

FUTURA 2000 FOR LOVE OR MONEY / $500

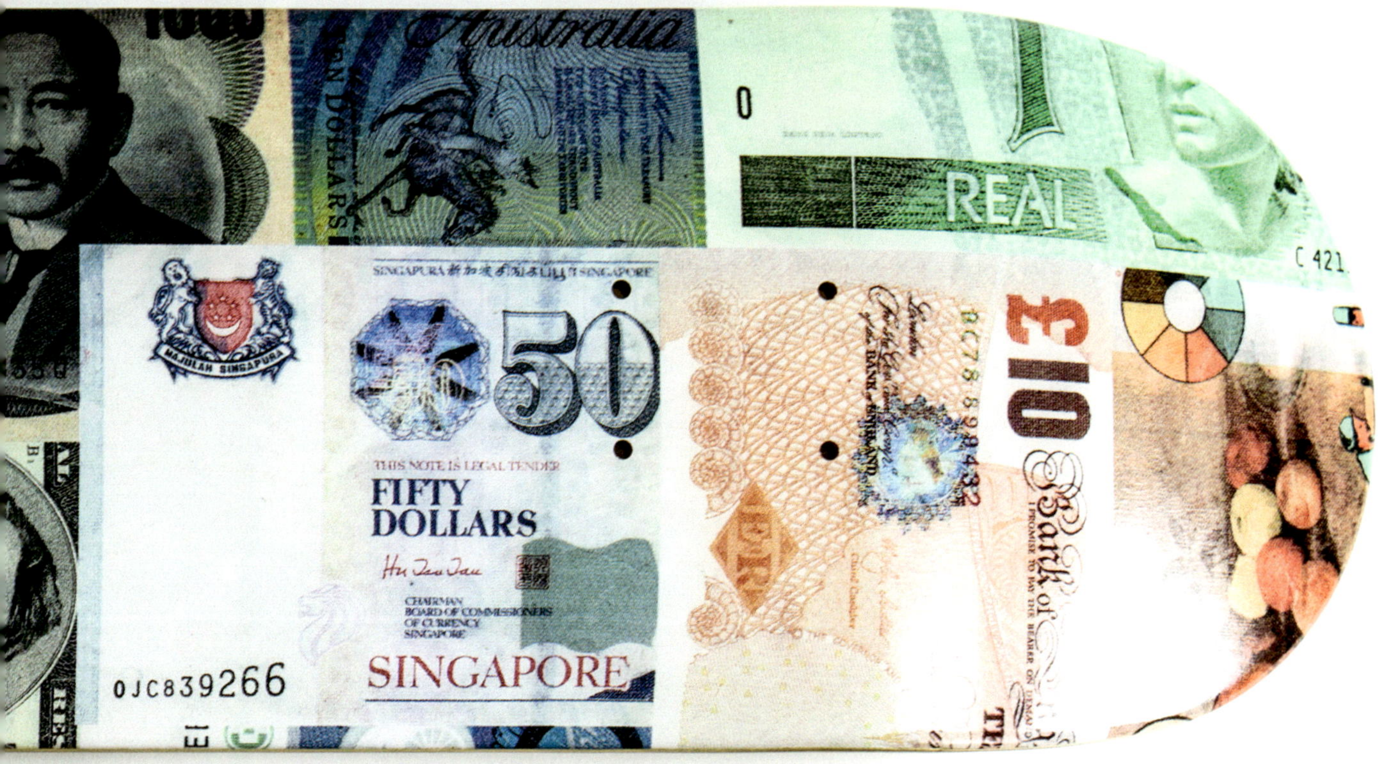

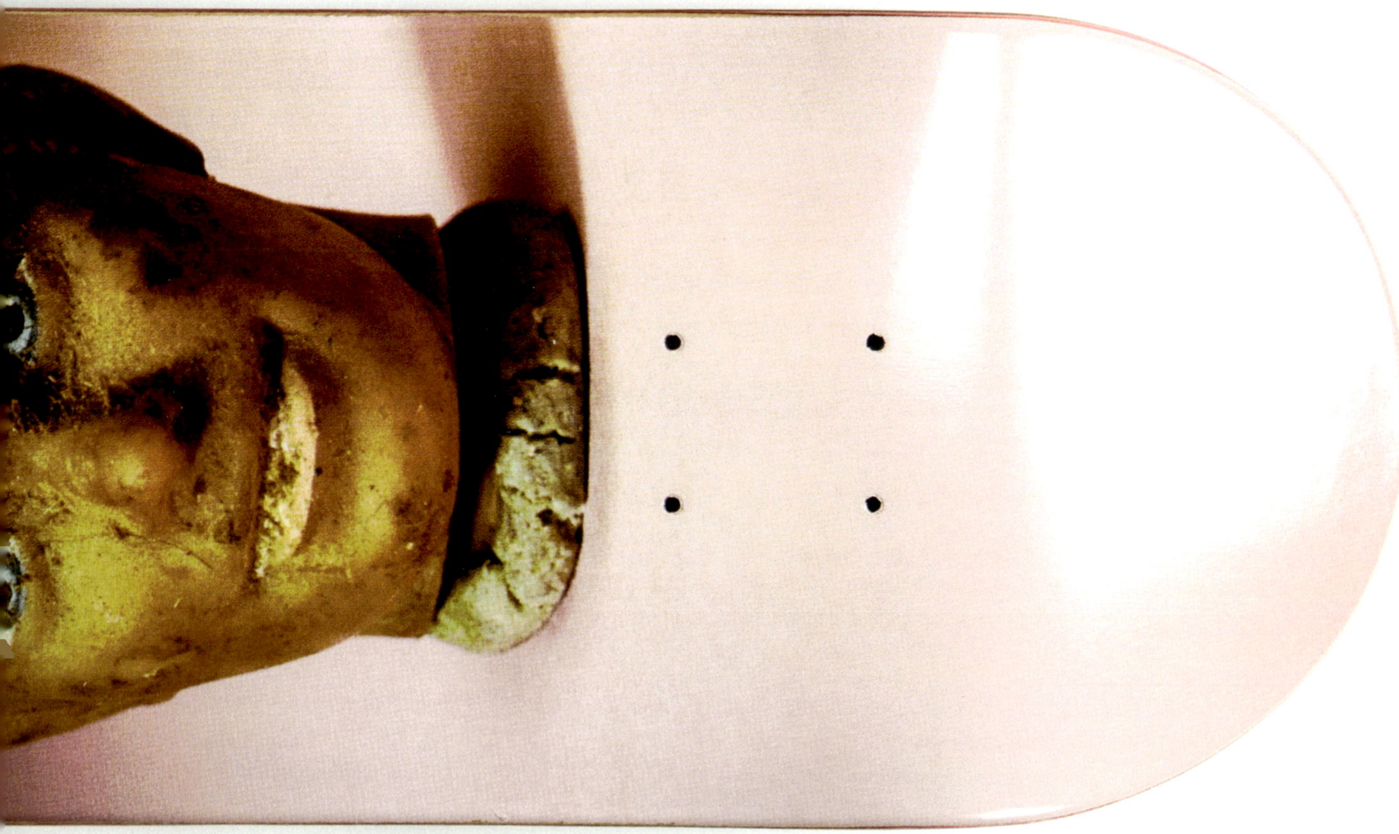

THE SKATEROOM X PAUL MCCARTHY KEN / $355

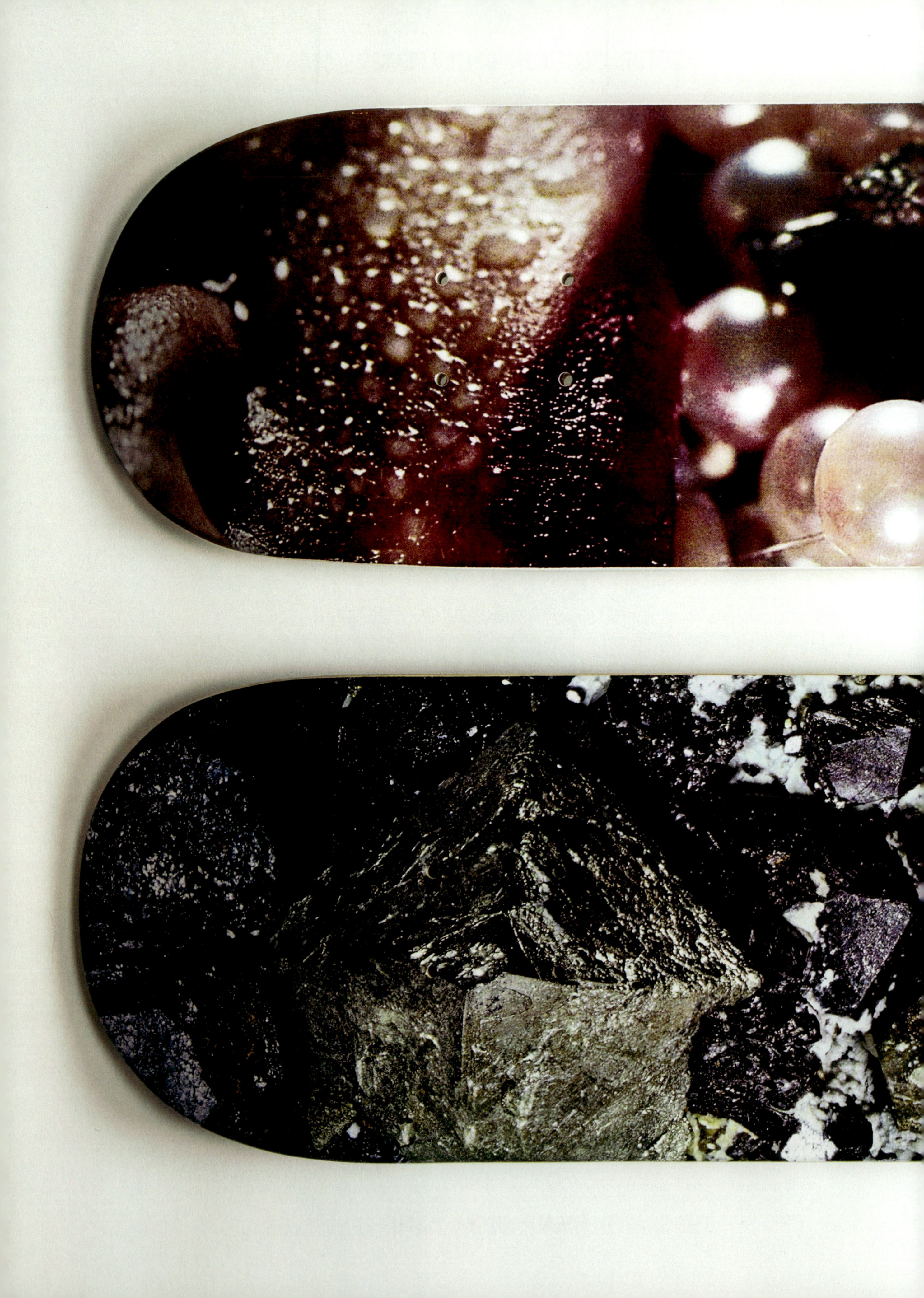

SUPREME X MARILYN MINTER PEARLS / $1,000

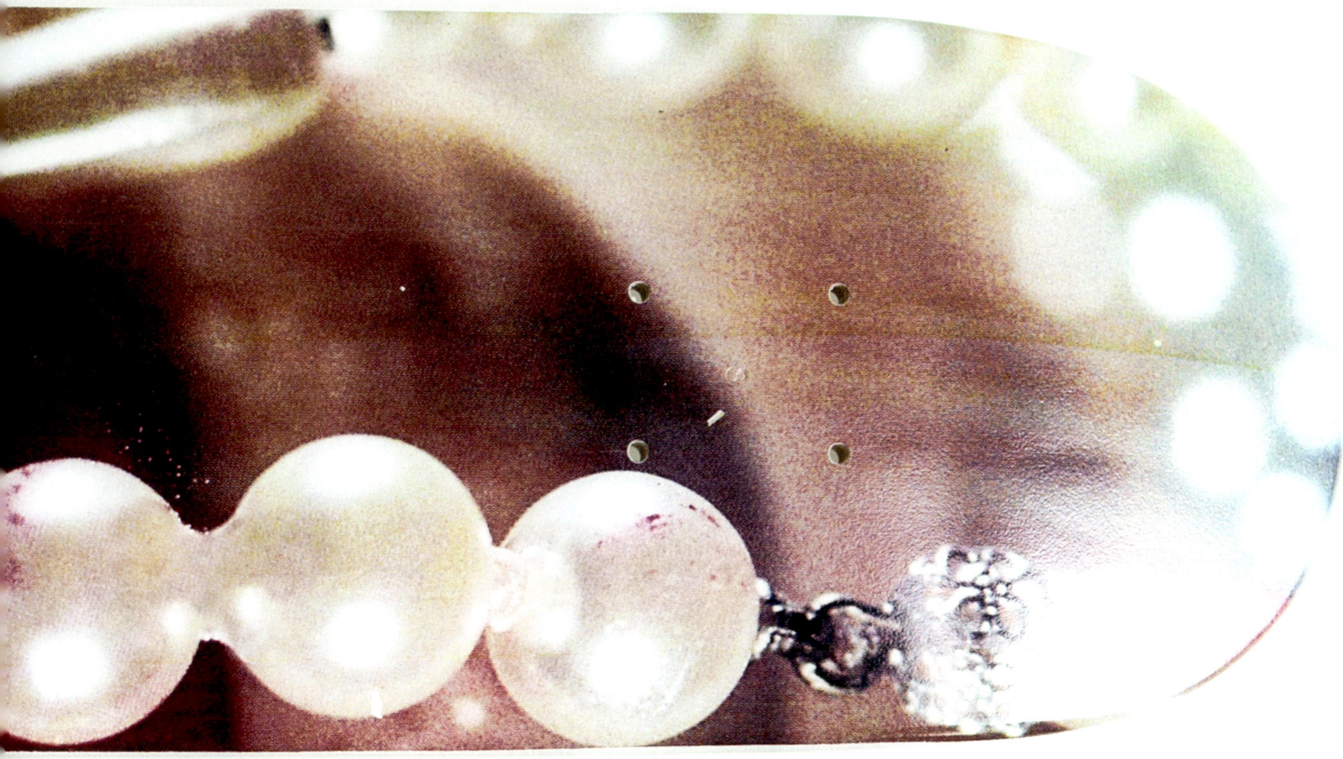

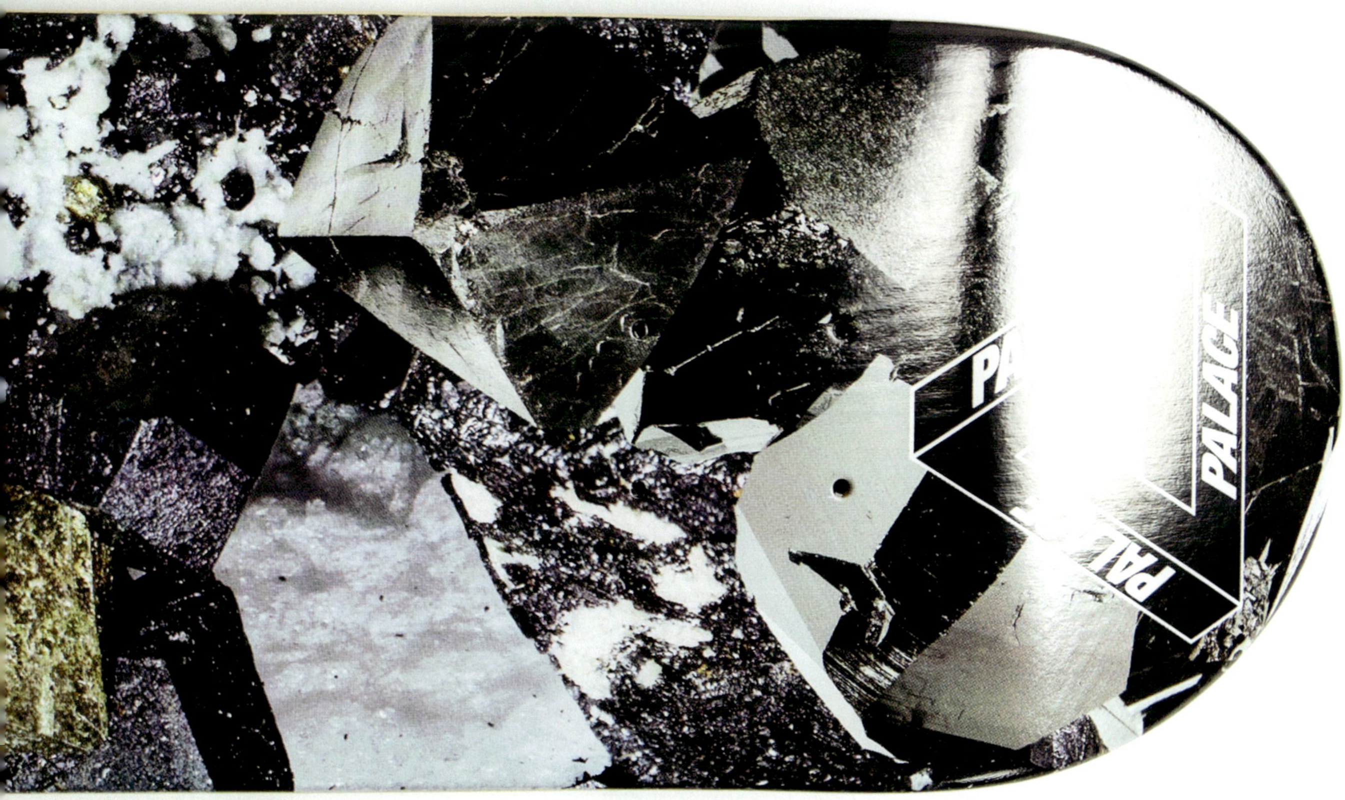

PALACE ROCKY / $50

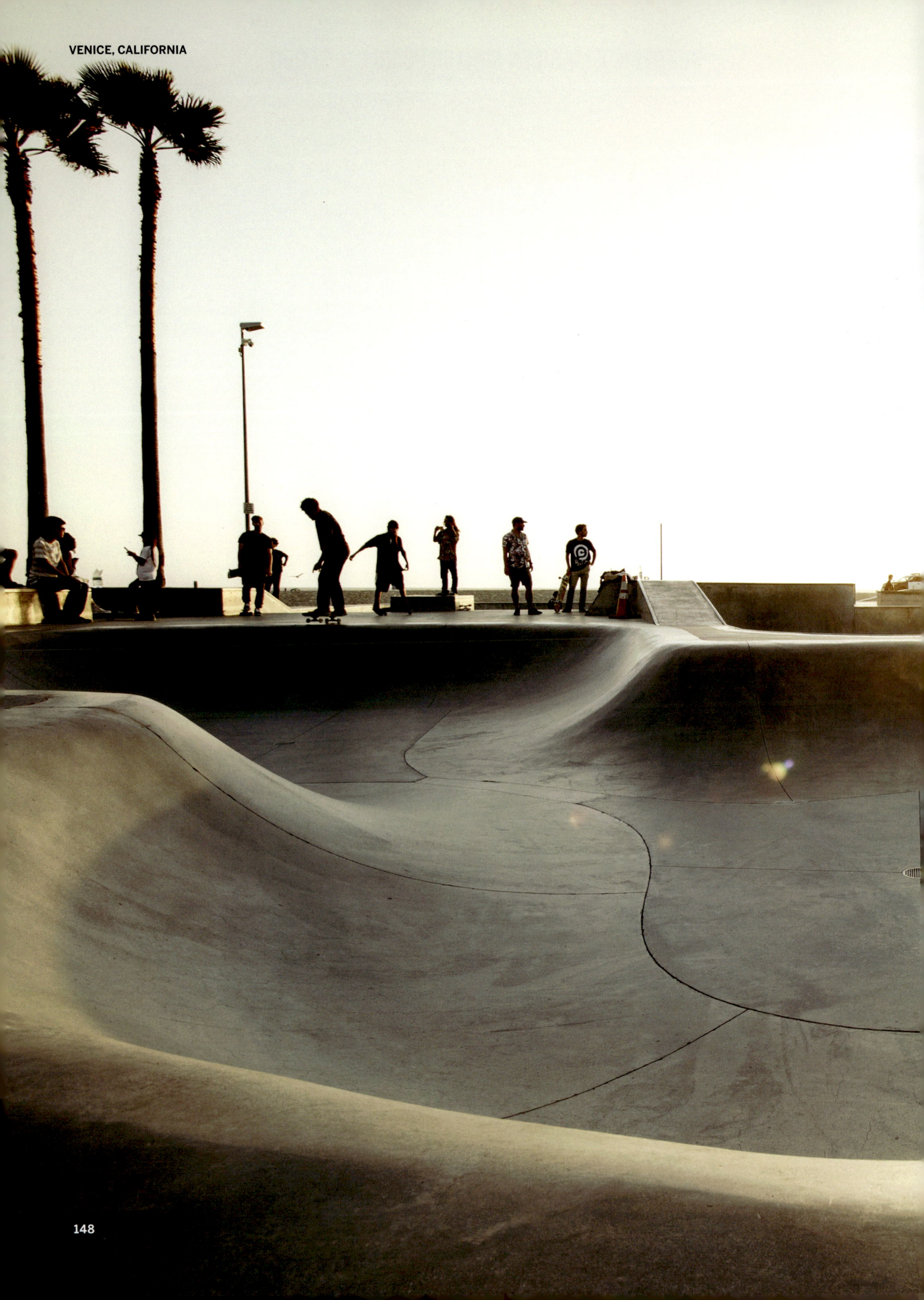

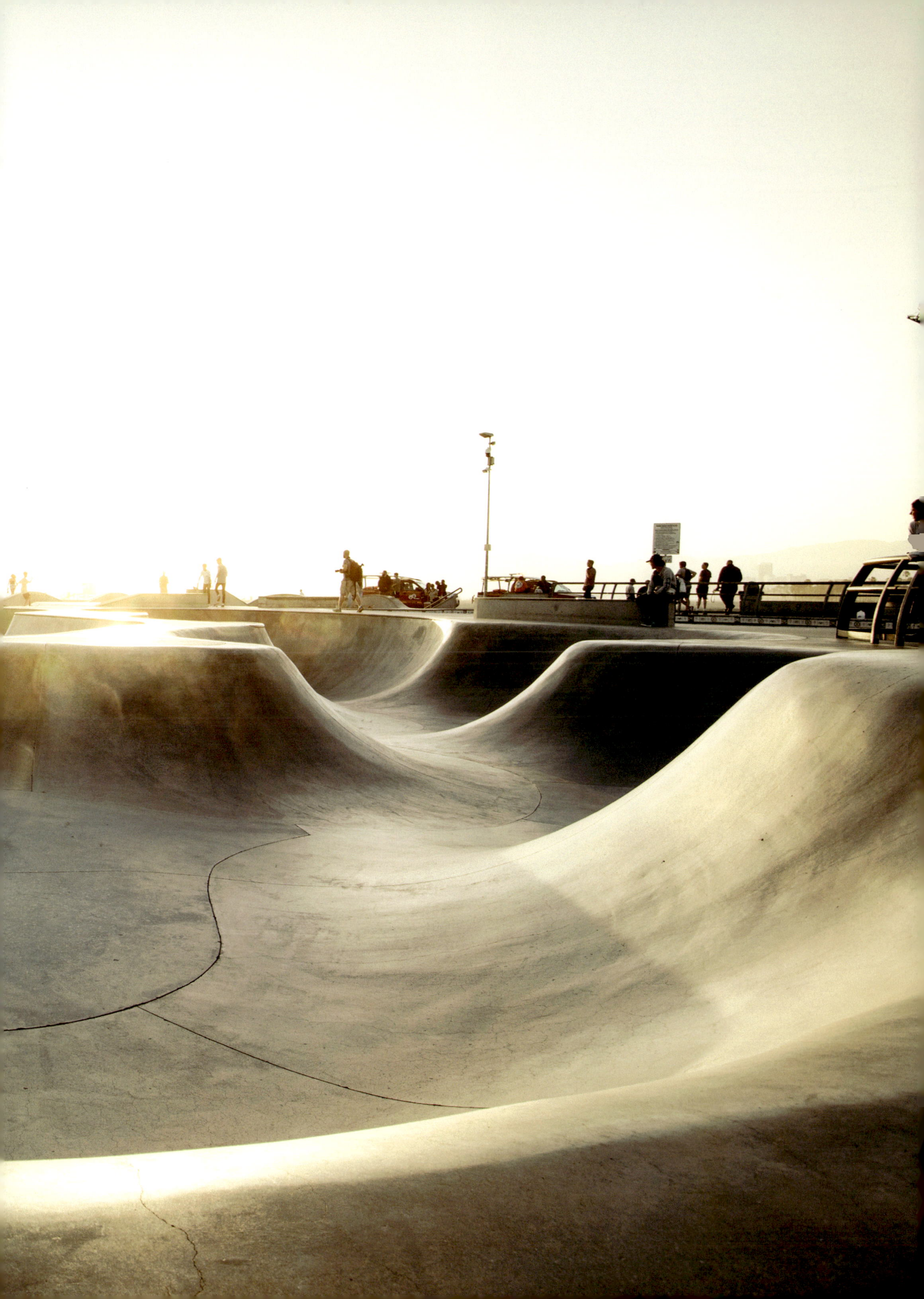

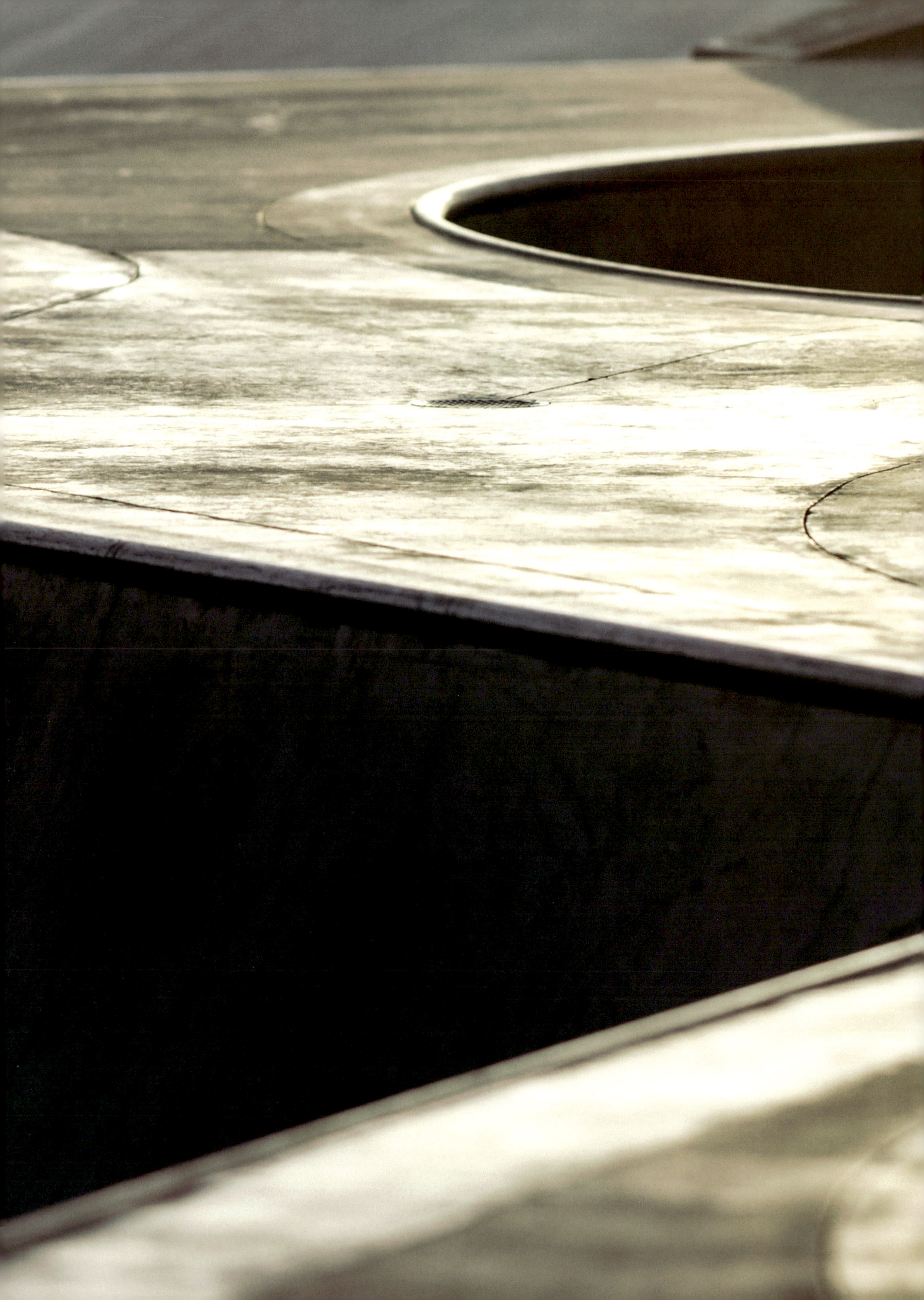

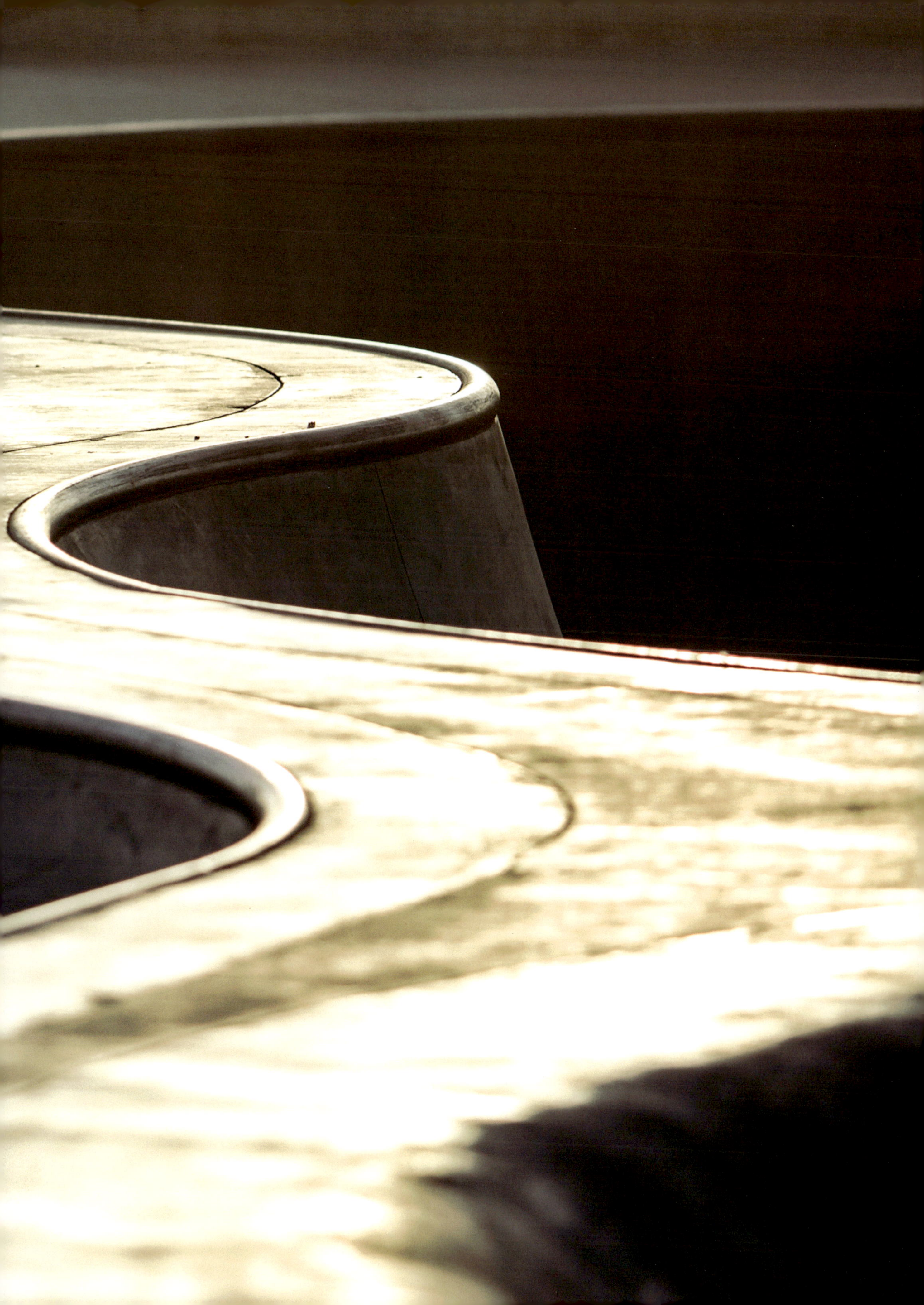

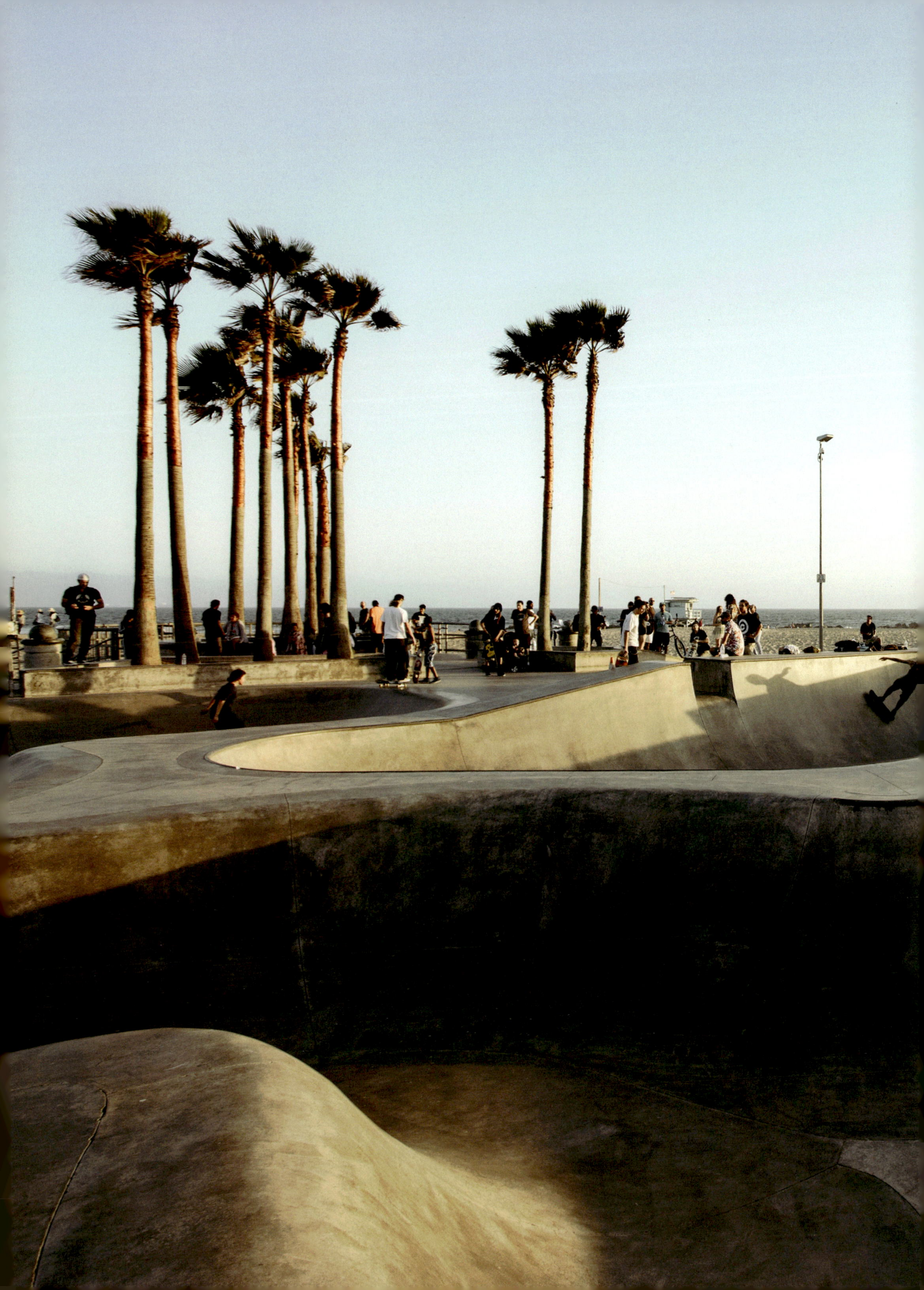

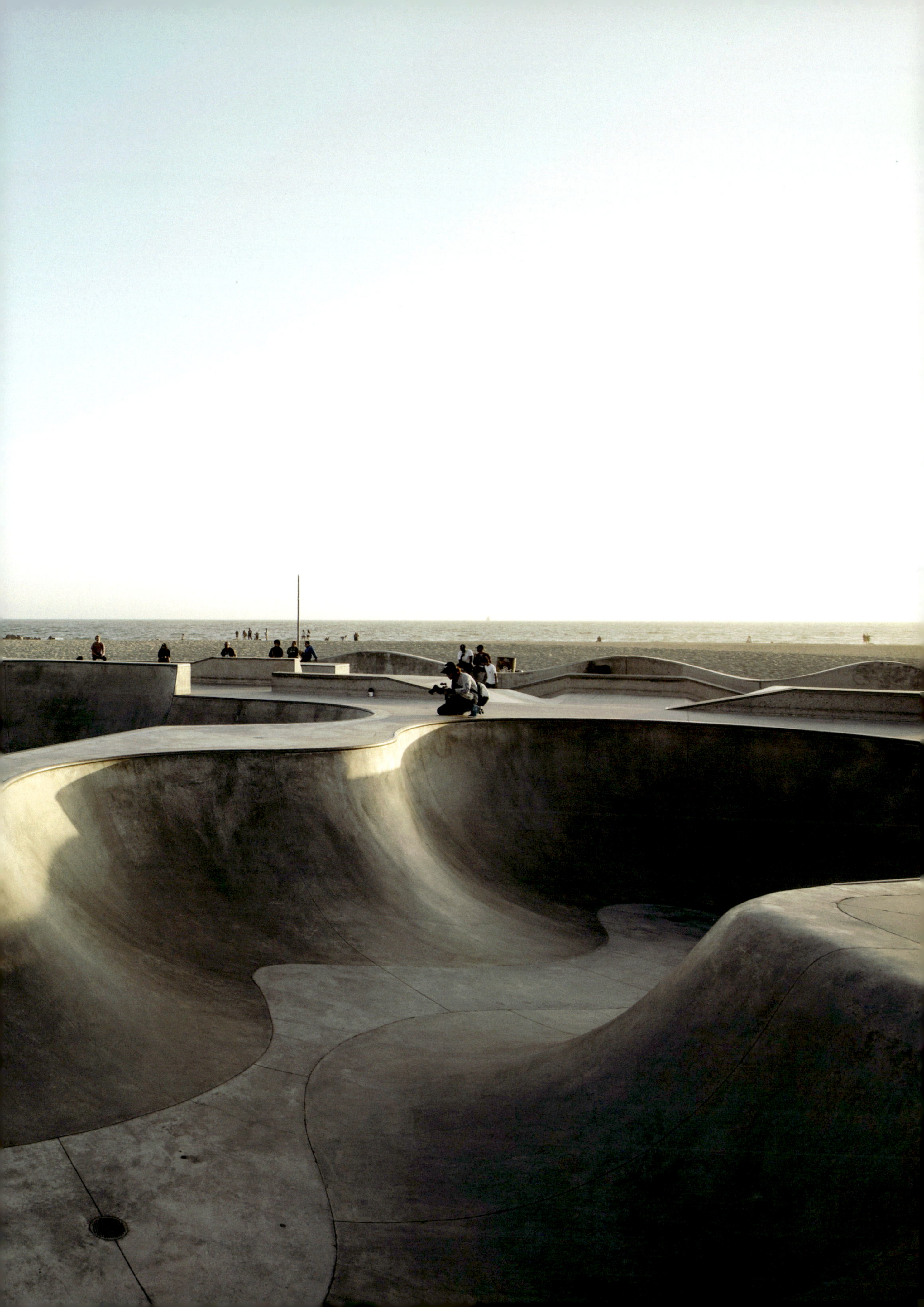

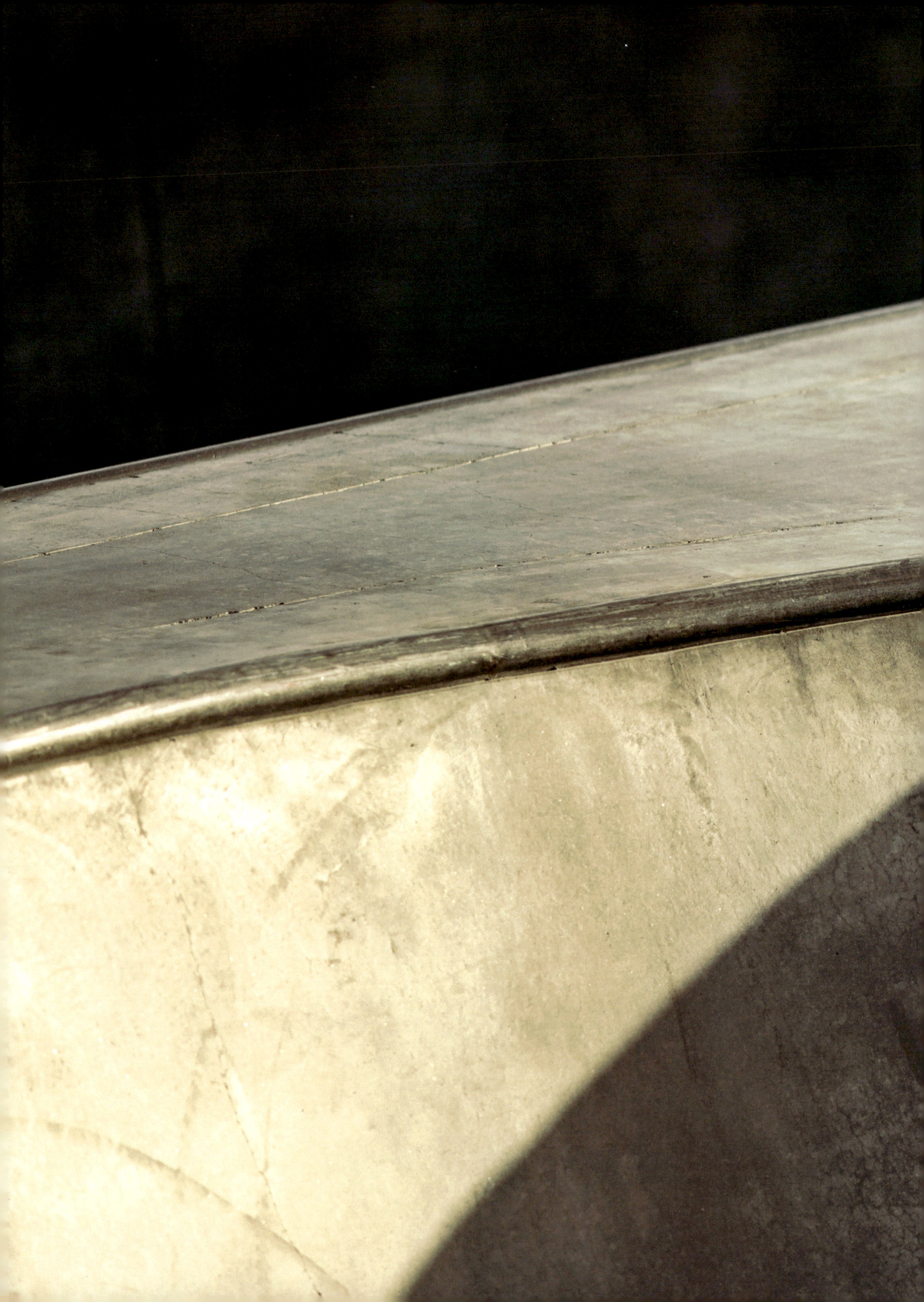

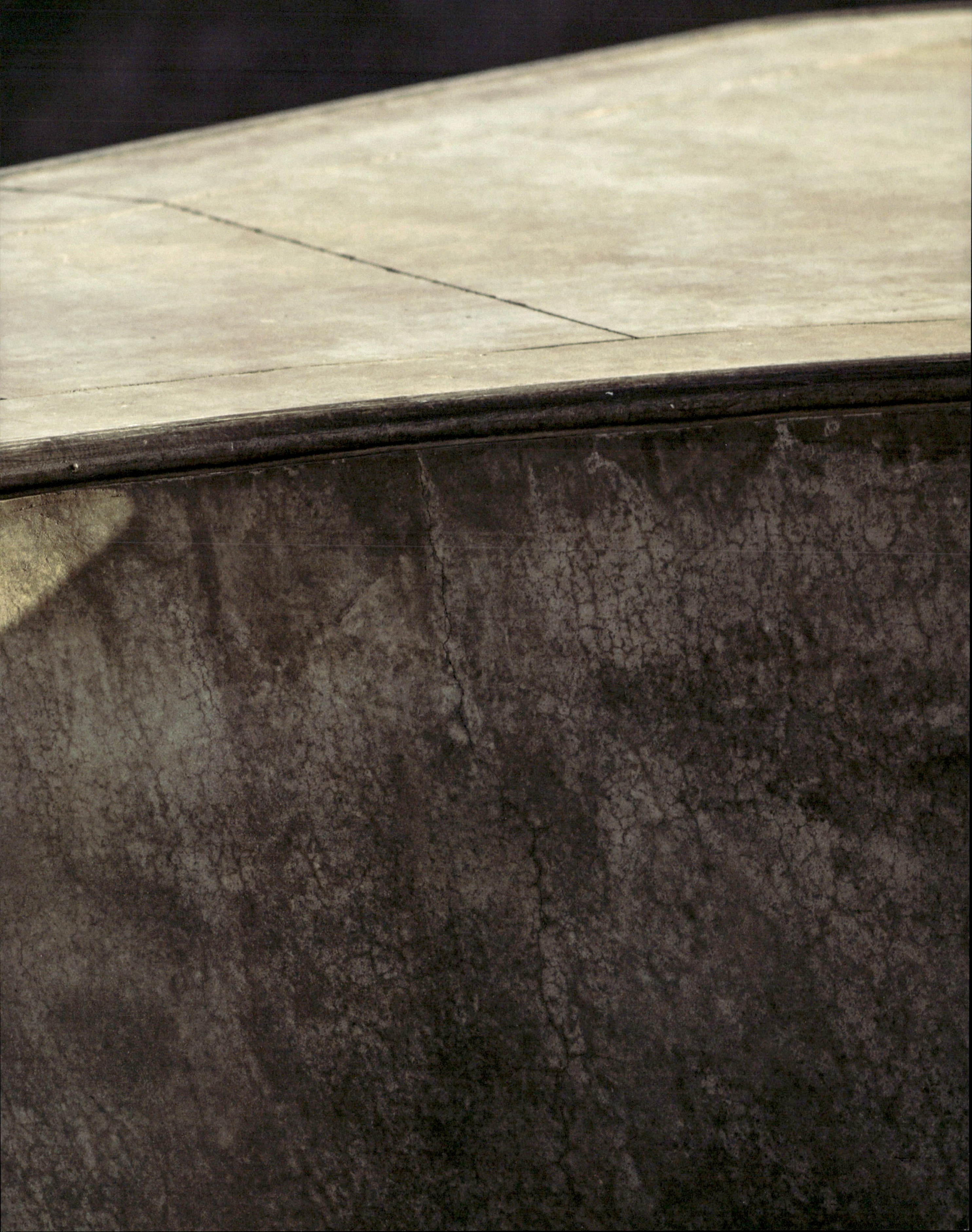

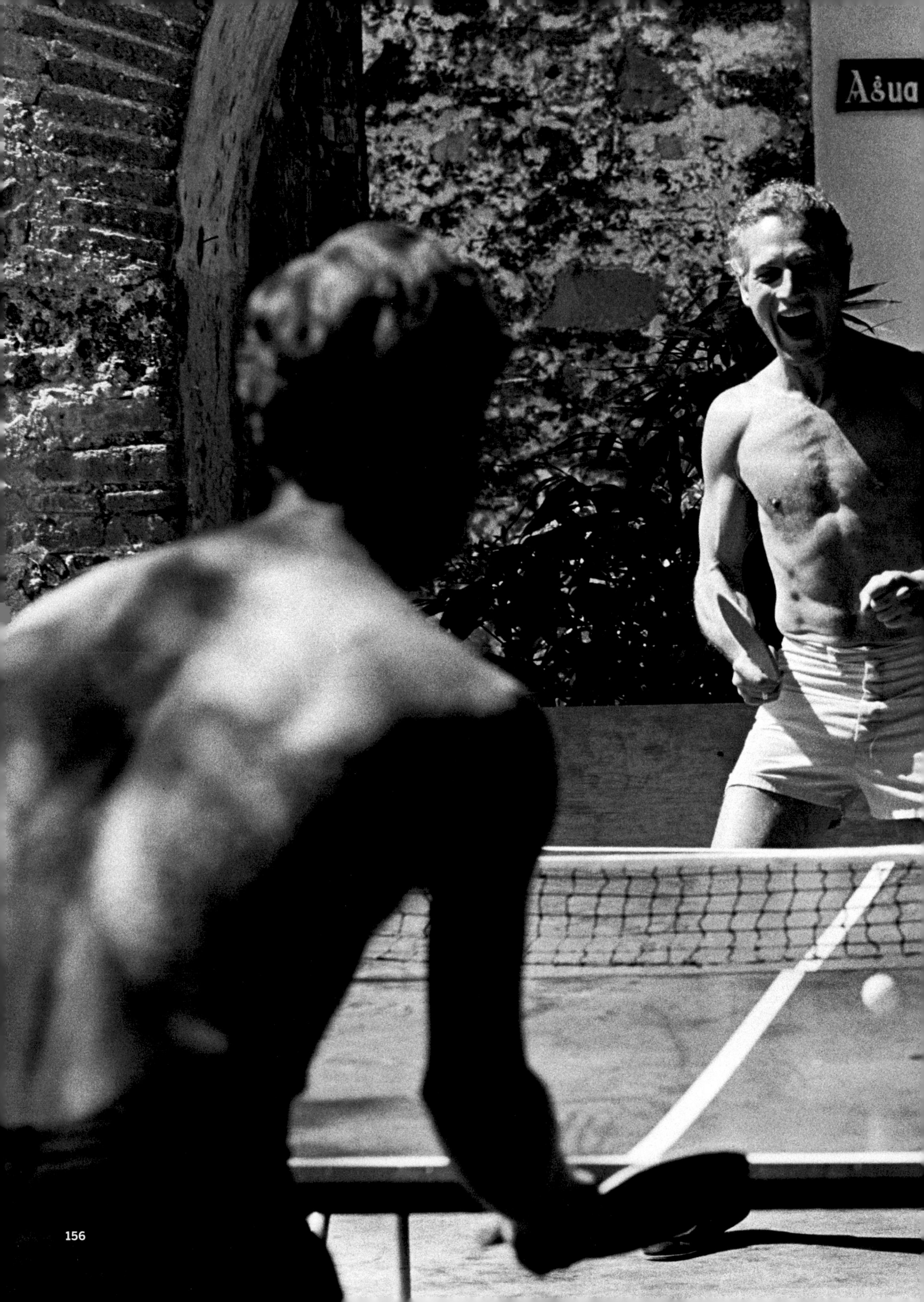
Agua

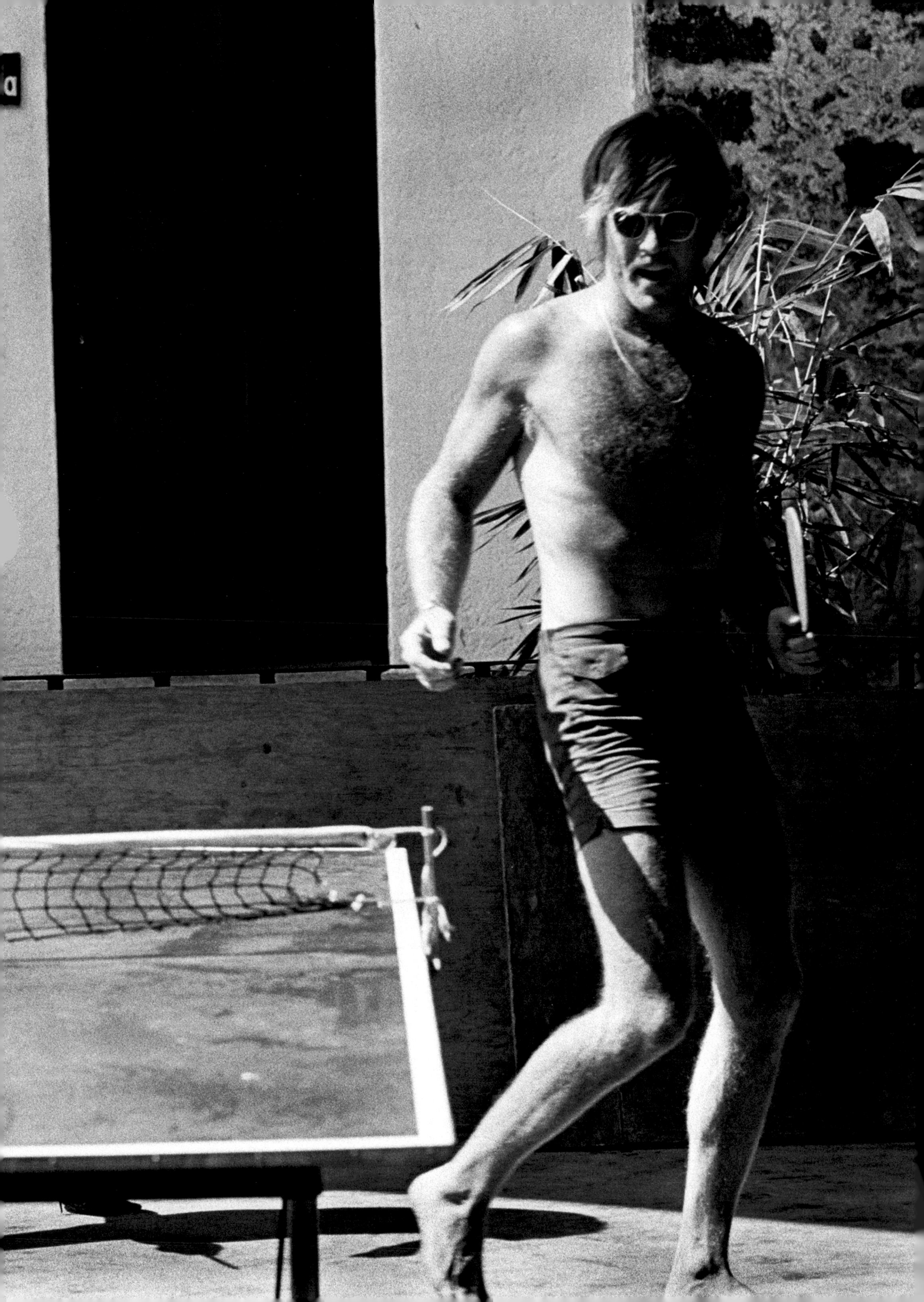

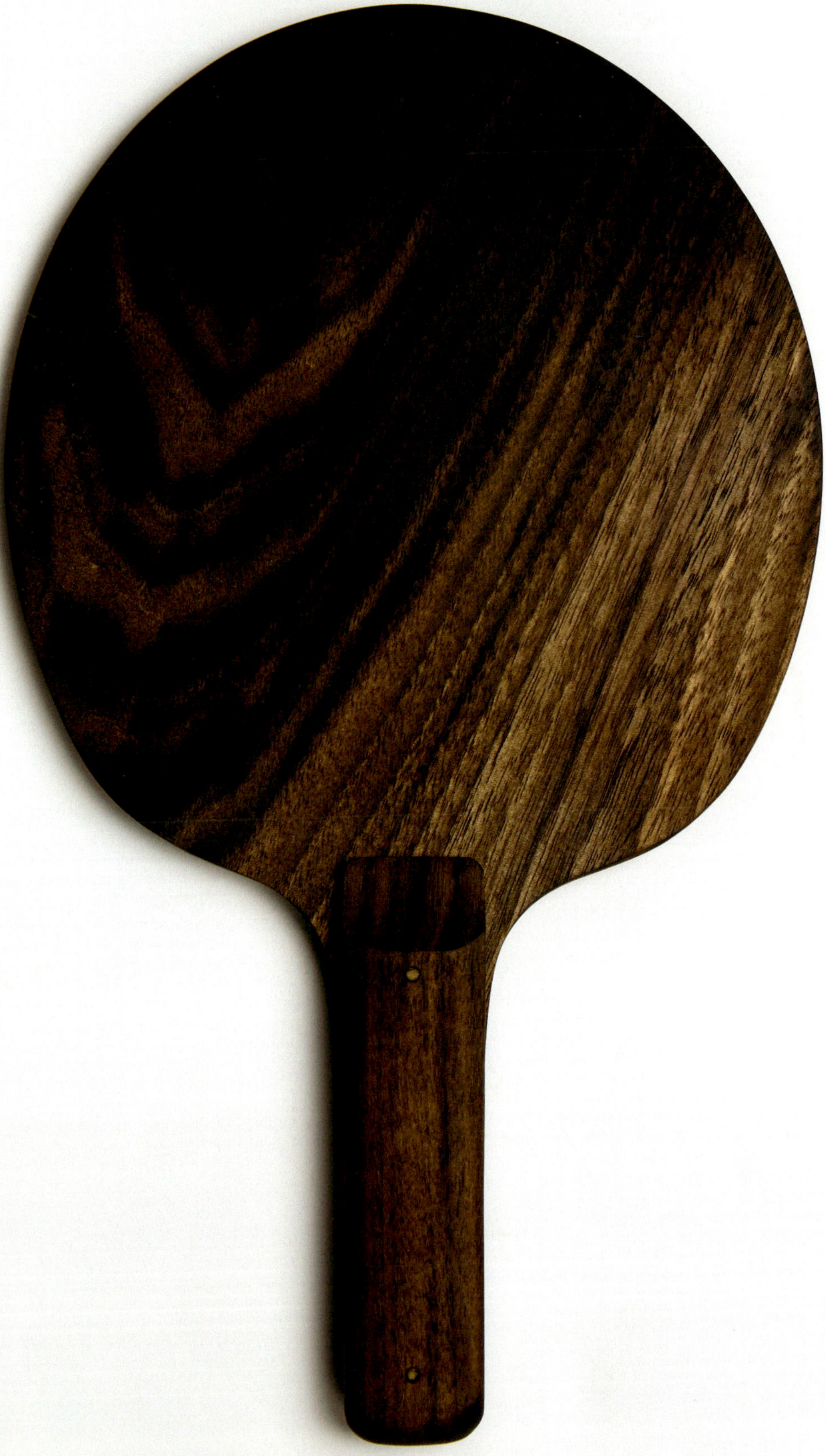

STANGLER WORKS WOODEN PADDLE / $50

Crafted by hand from your choice of walnut or ash wood, this minimalist paddle leaves the bare wood on one side and covers it with cork on the other, giving it a wholly different feel depending on which side you use.

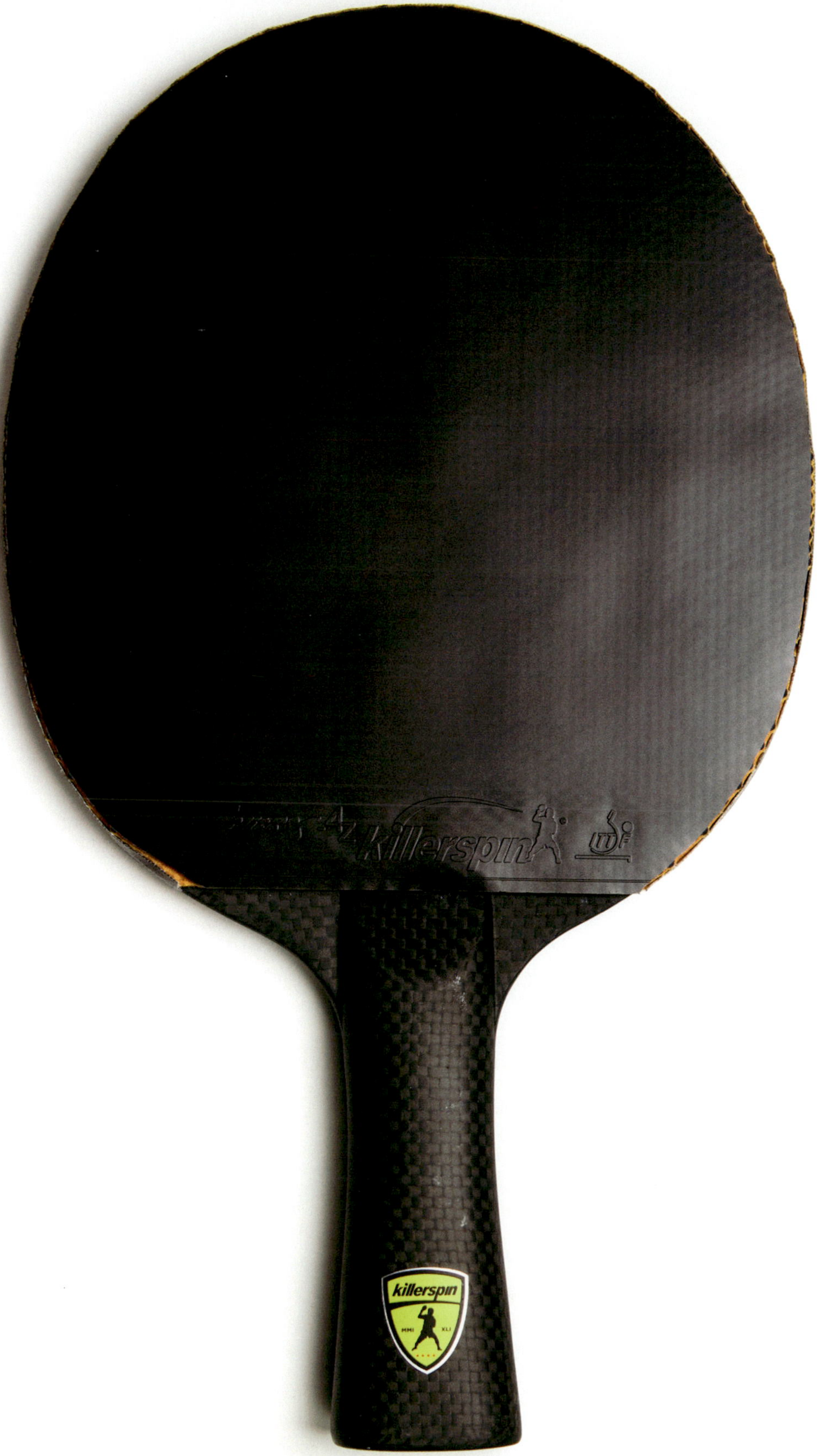

KILLERSPIN STILO7 SVR PADDLE / $350

Made with carbon fiber, this might be the best performing paddle in the world. The carbon handle and core make it both light and extremely comfortable to hold, while the high-quality rubbers give you outstanding spin and feel.

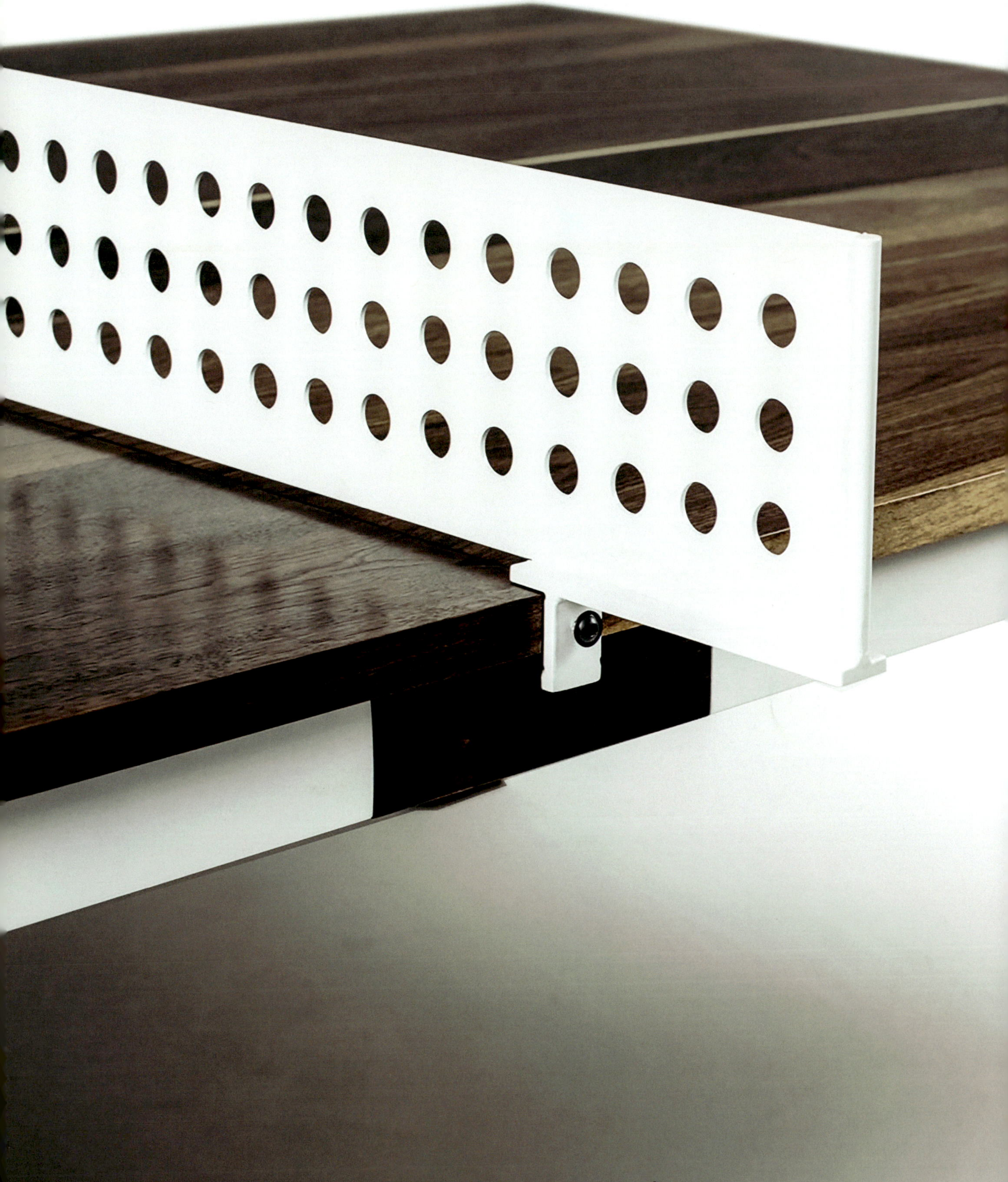

WOOLSEY PING PONG TABLE / $9,000

Ping pong tables take up a huge amount of space. This option pulls double duty as a dinner or office table, with a solid Black Walnut top cut to ITTF specs, a Sugar Maple center line, and a detachable steel net. It's finished with food-safe oil and wax.

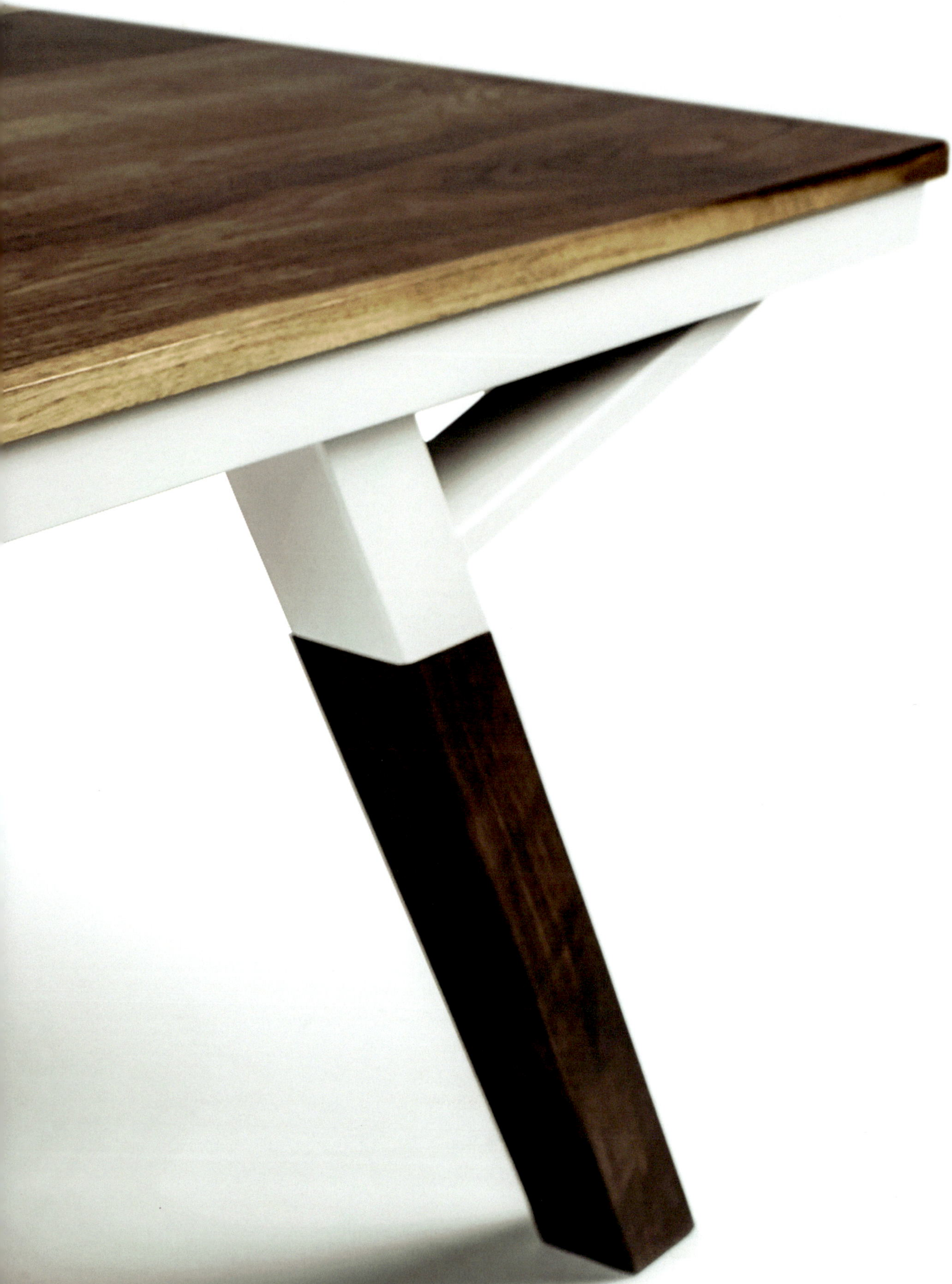

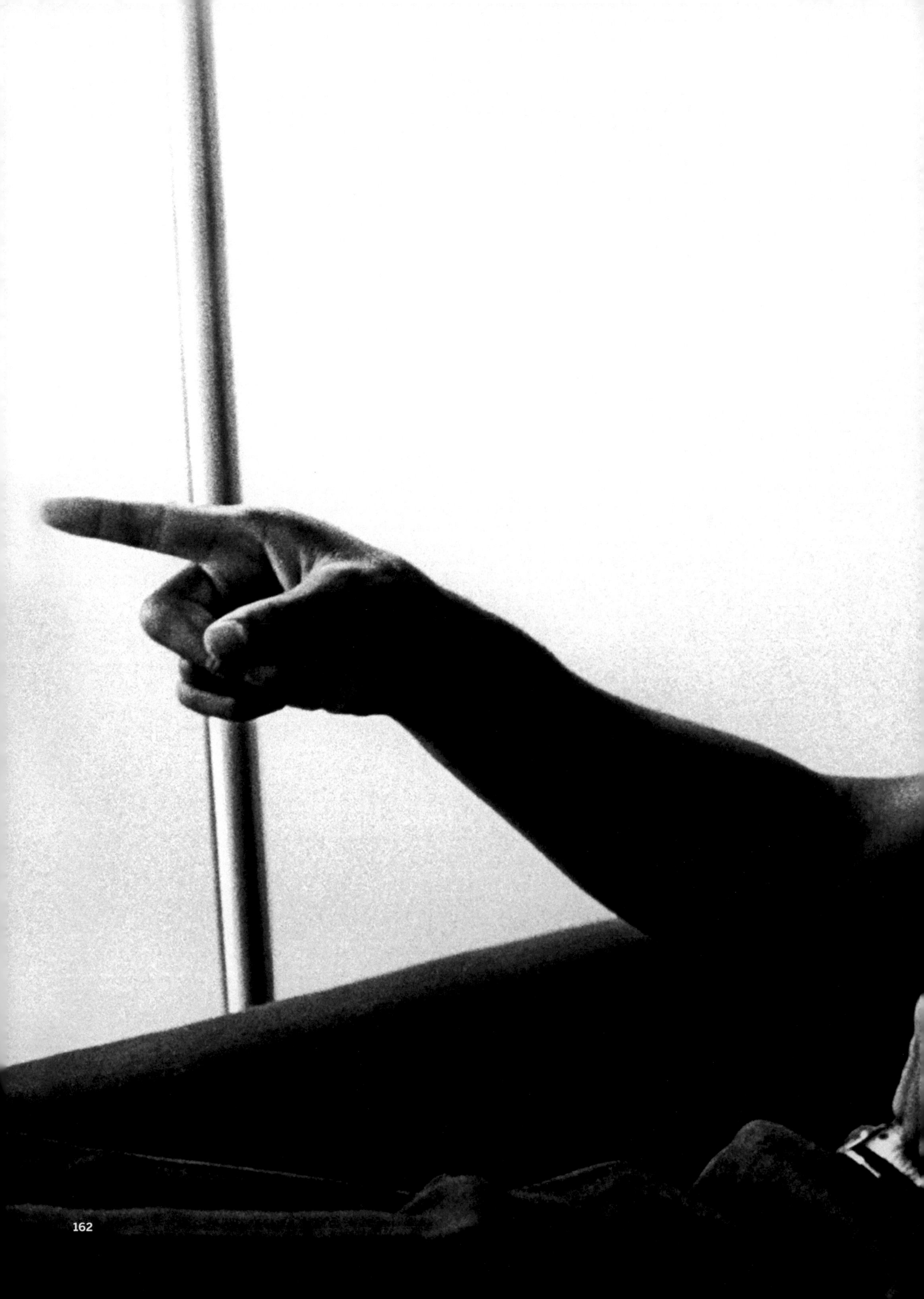

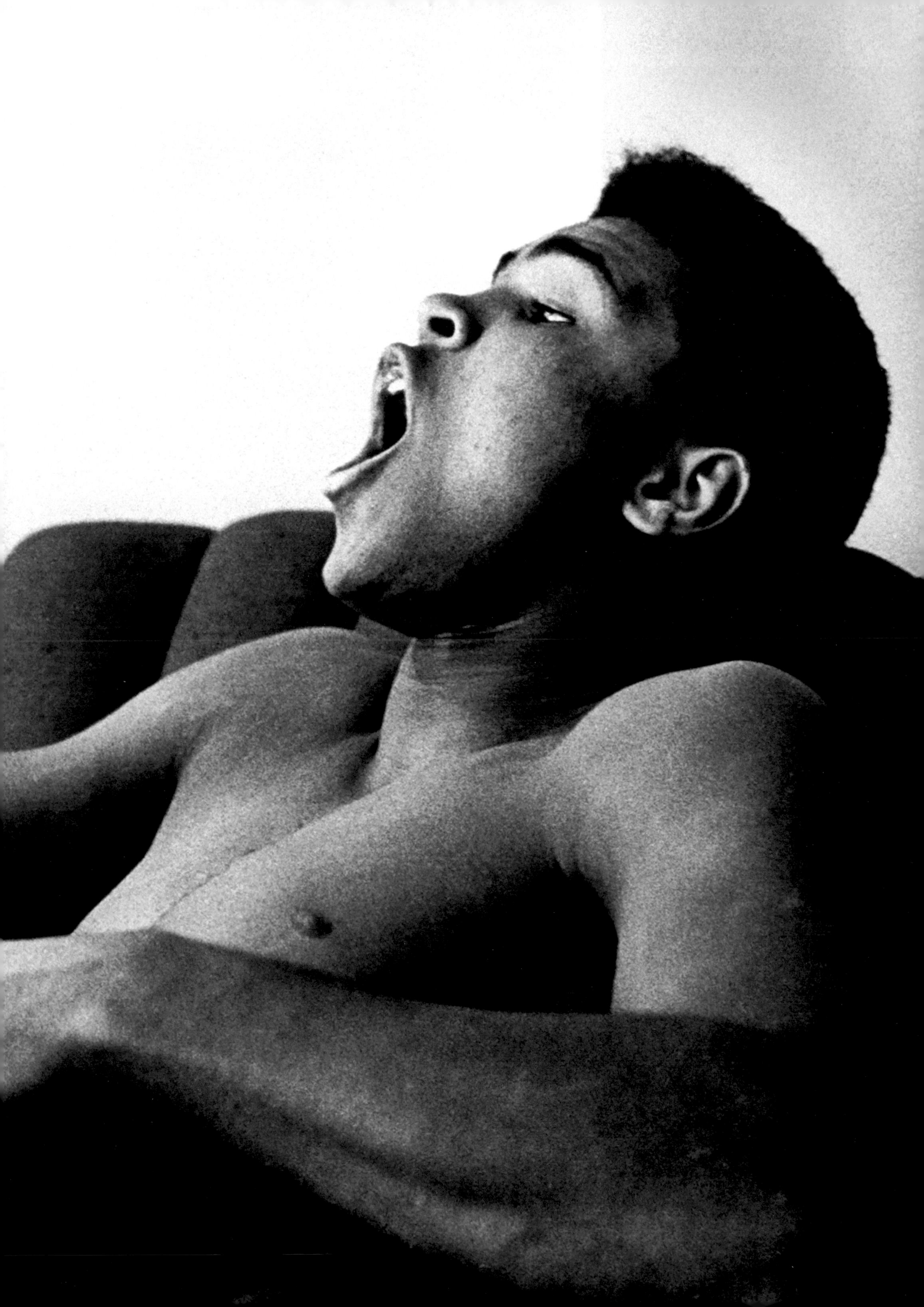

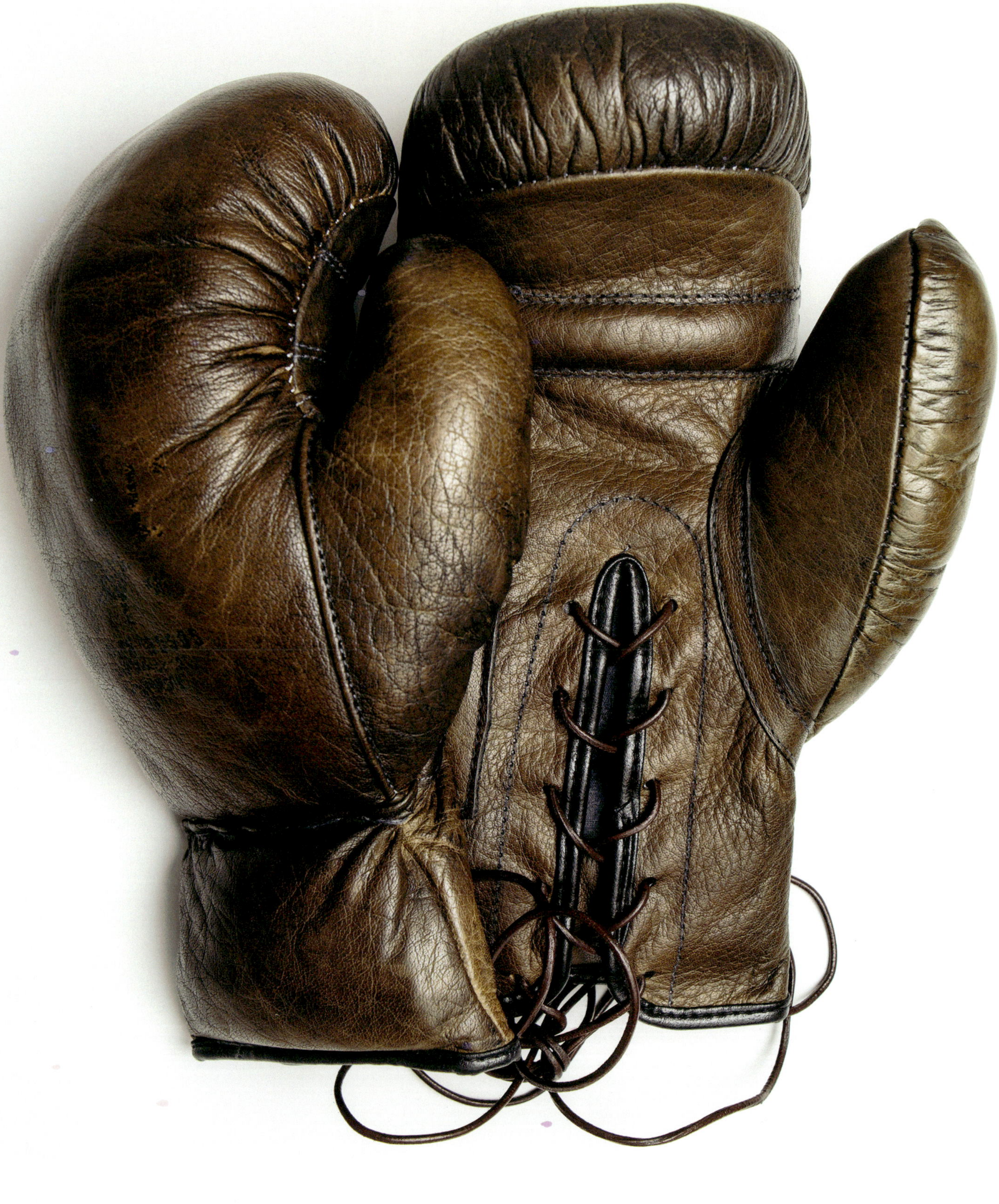

WILL LEATHER GOODS GOLDEN AGE BOXING GLOVES / $120

They're not old. They just look like it. These vintage-inspired gloves are made from leather that's been stained and burnished by hand to create the well-used look. They also have period-correct leather laces and are sized to match early 20th century gloves worn by the likes of Dempsey and Louis.

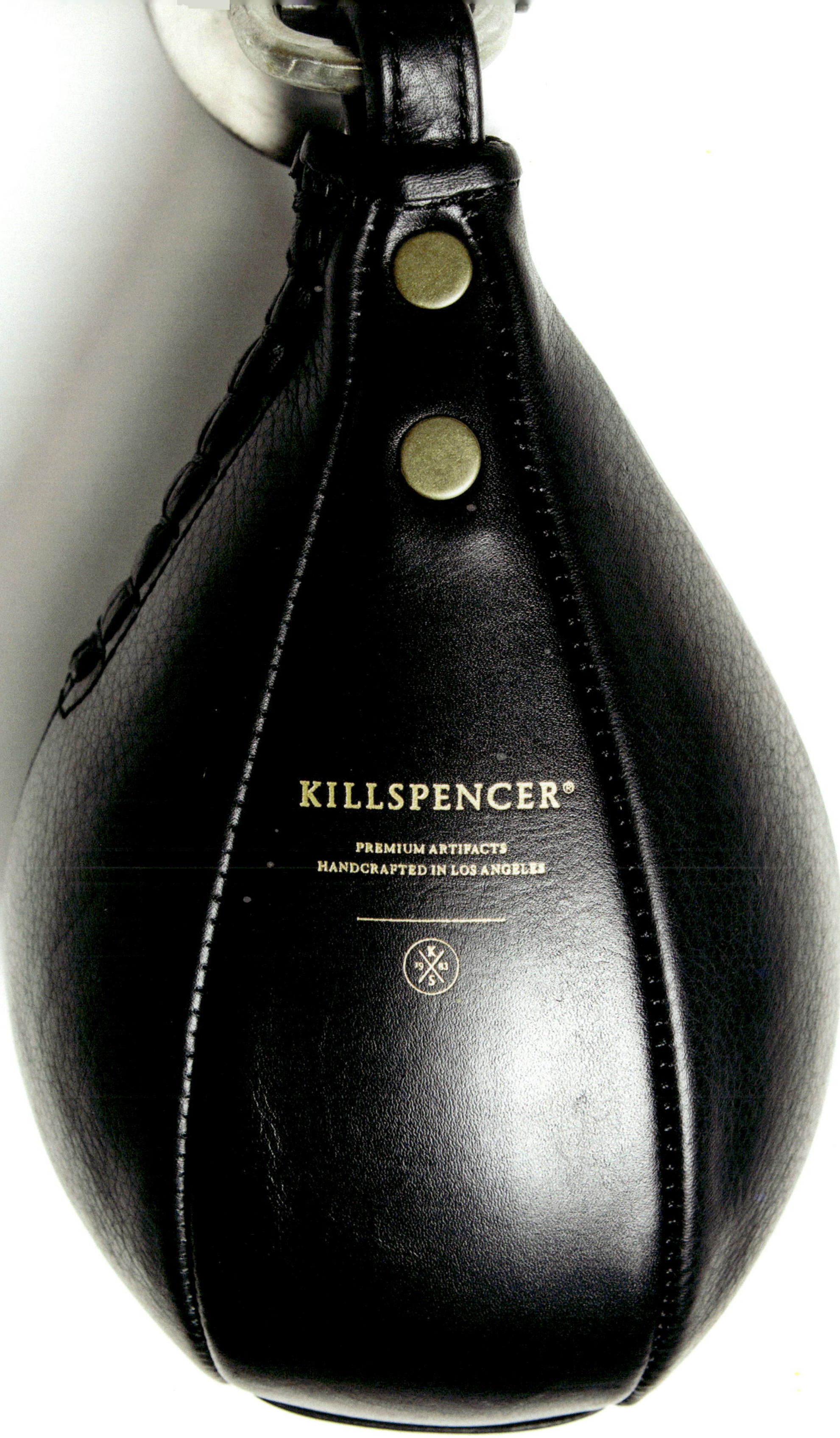

KILLSPENCER SPEED BAG / $225

Boxing is great exercise. Killspencer's speed bag lets you get that excerise without turning your home into a *Rocky* set. Made by hand in Los Angeles from a mix of black bridle and black naked leathers, it has triple stitched seams, antique brass rivets, and waxed linen lacing. Shown with the matching swivel ($69).

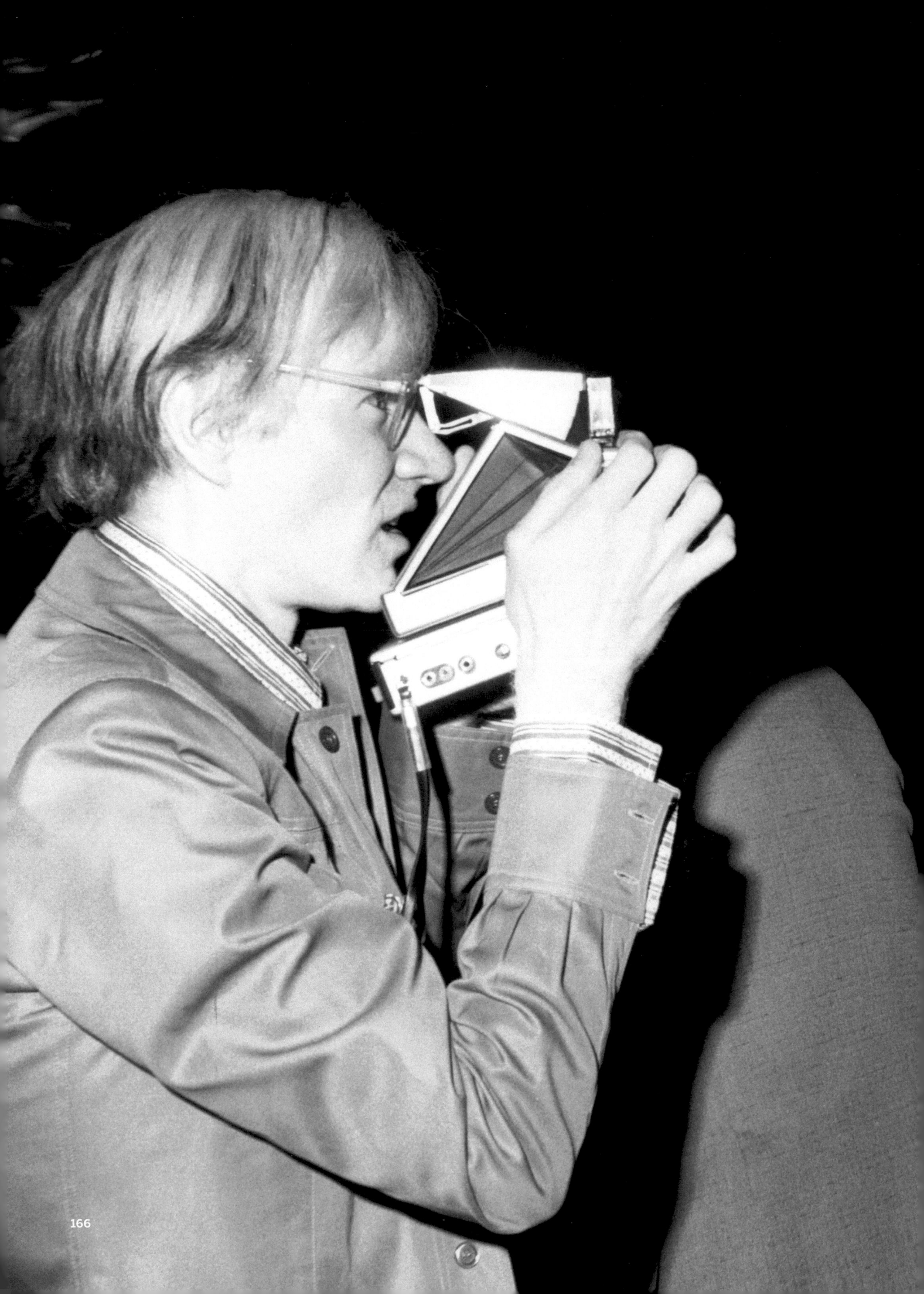

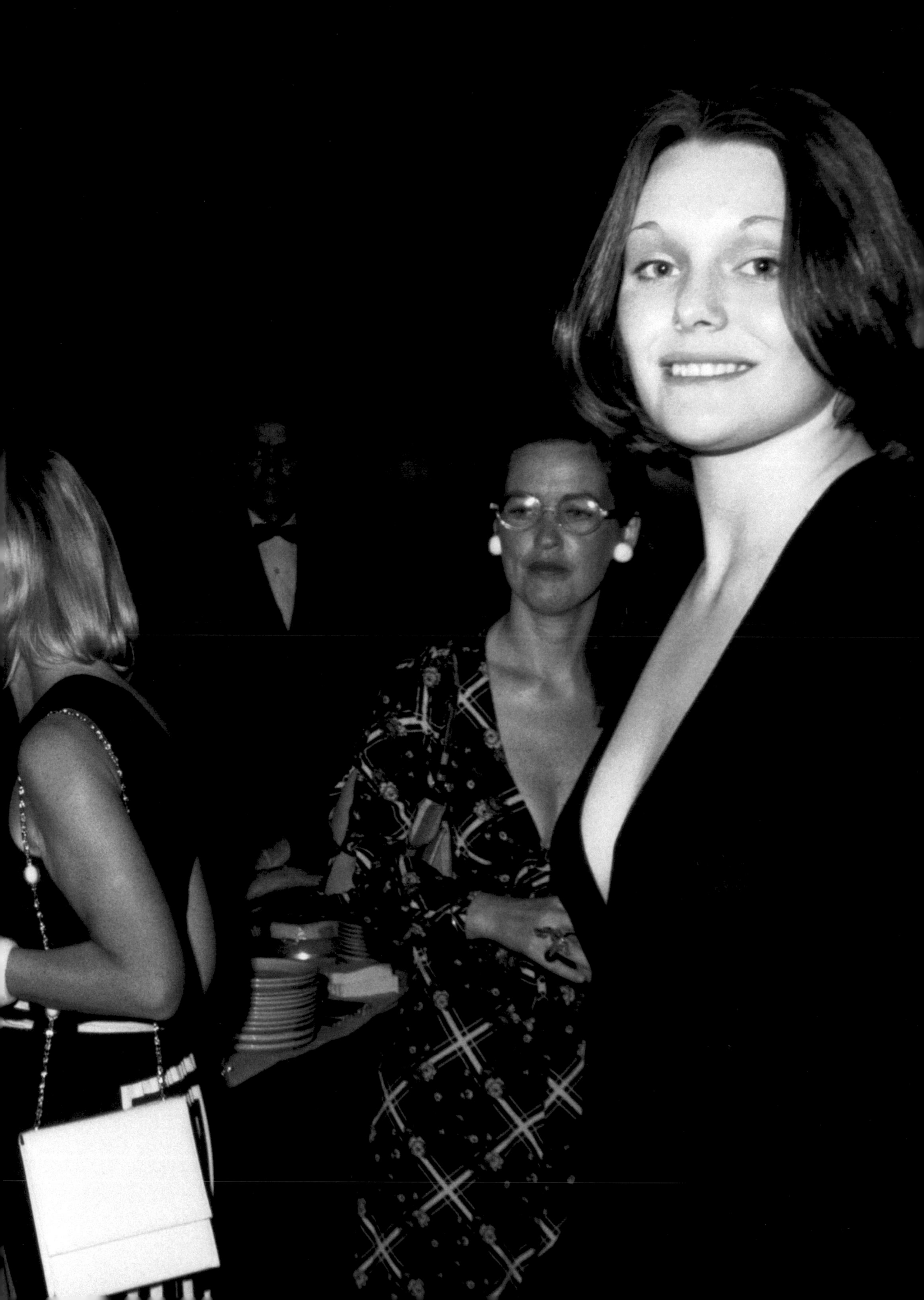

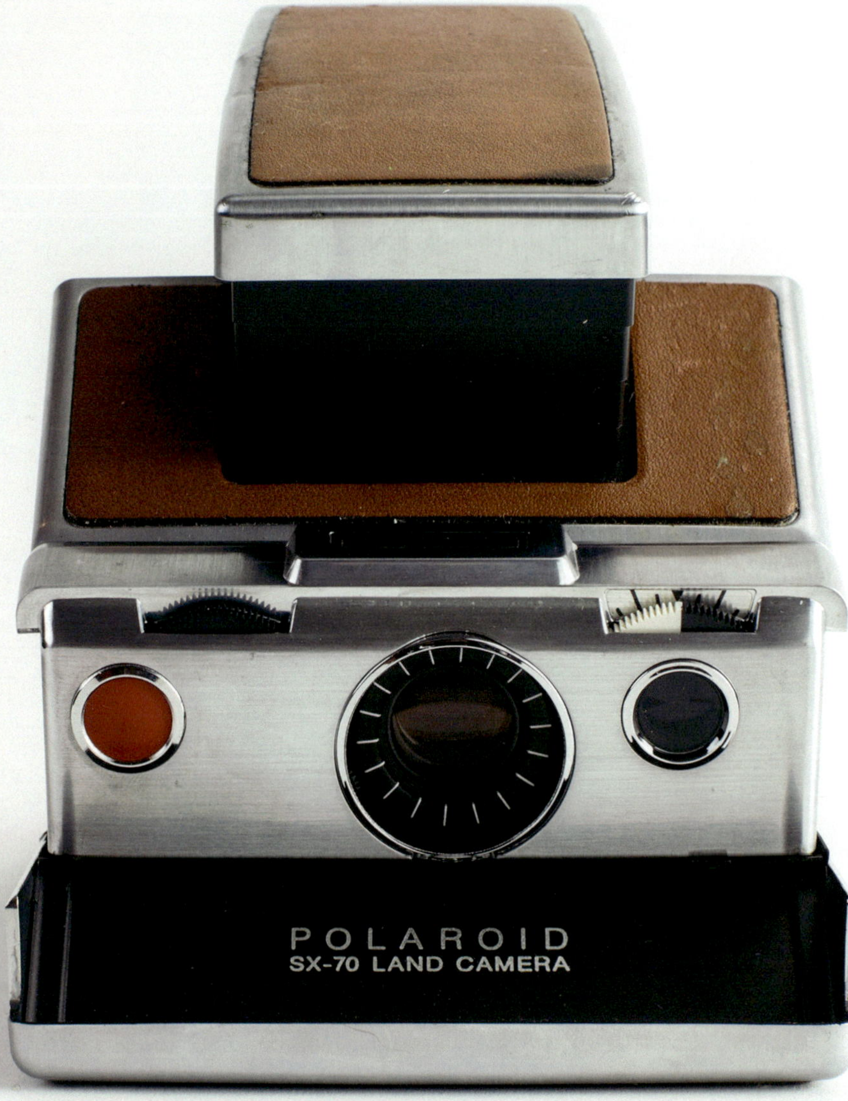

ARTIFACT

POLAROID SX-70 CAMERA / $50

Introduced in 1972, the original SX-70 was a sensation, letting people take photos that ejected automatically and developed quickly, without the residue the Swinger model was known for. Its fold-up design let it be carried anywhere, making it a favorite of artists like Andy Warhol and Ansel Adams.

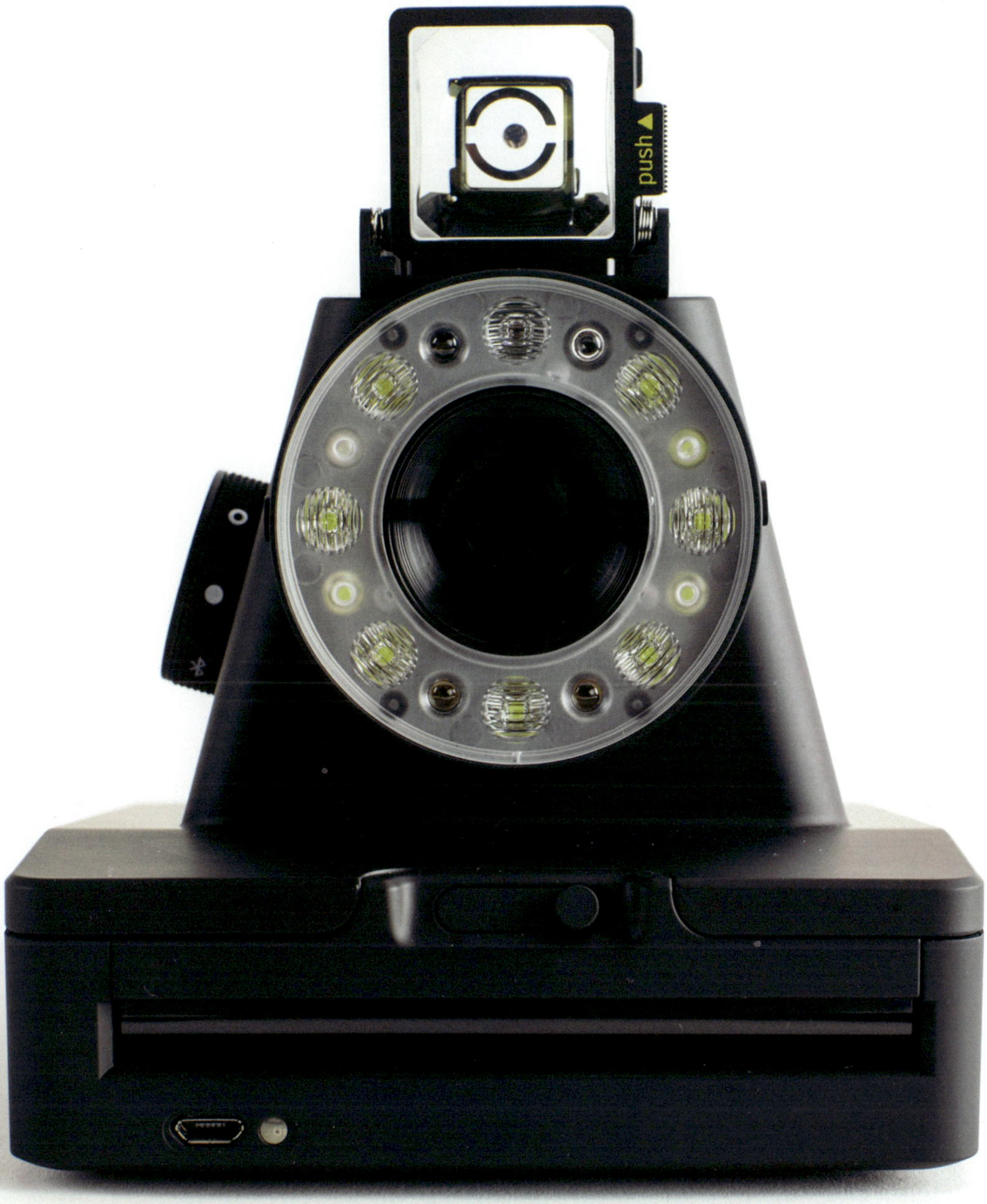

IMPOSSIBLE I-1 CAMERA / $350

When Polaroid quit making film, Impossible decided to start. Now they make their own camera, too. Capable of shooting with either Polaroid 600 or their new I-Type film, the matte black camera has a built-in, portrait-friendly ring flash, an optional manual mode, and can connect to your iPhone for more creative options.

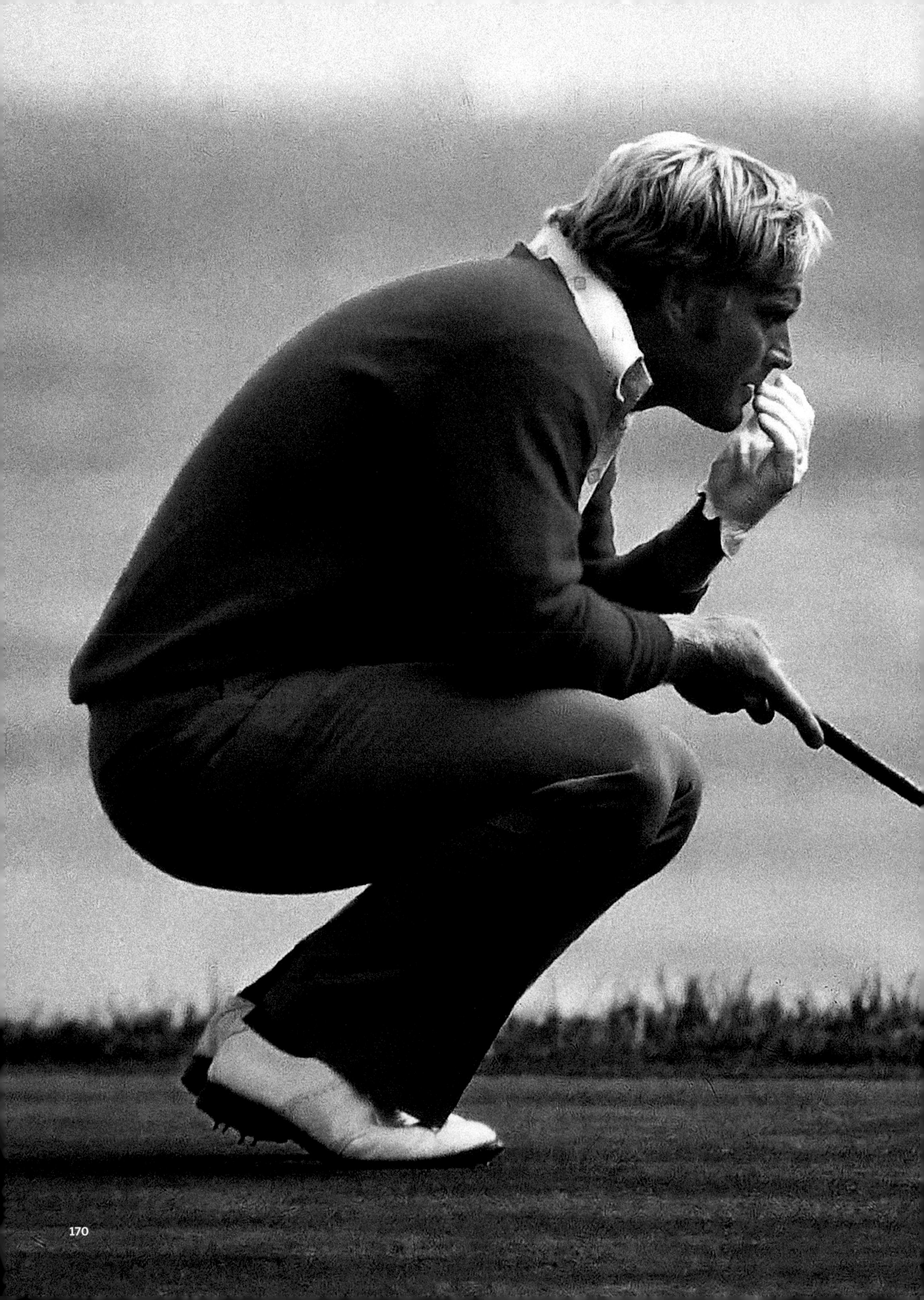

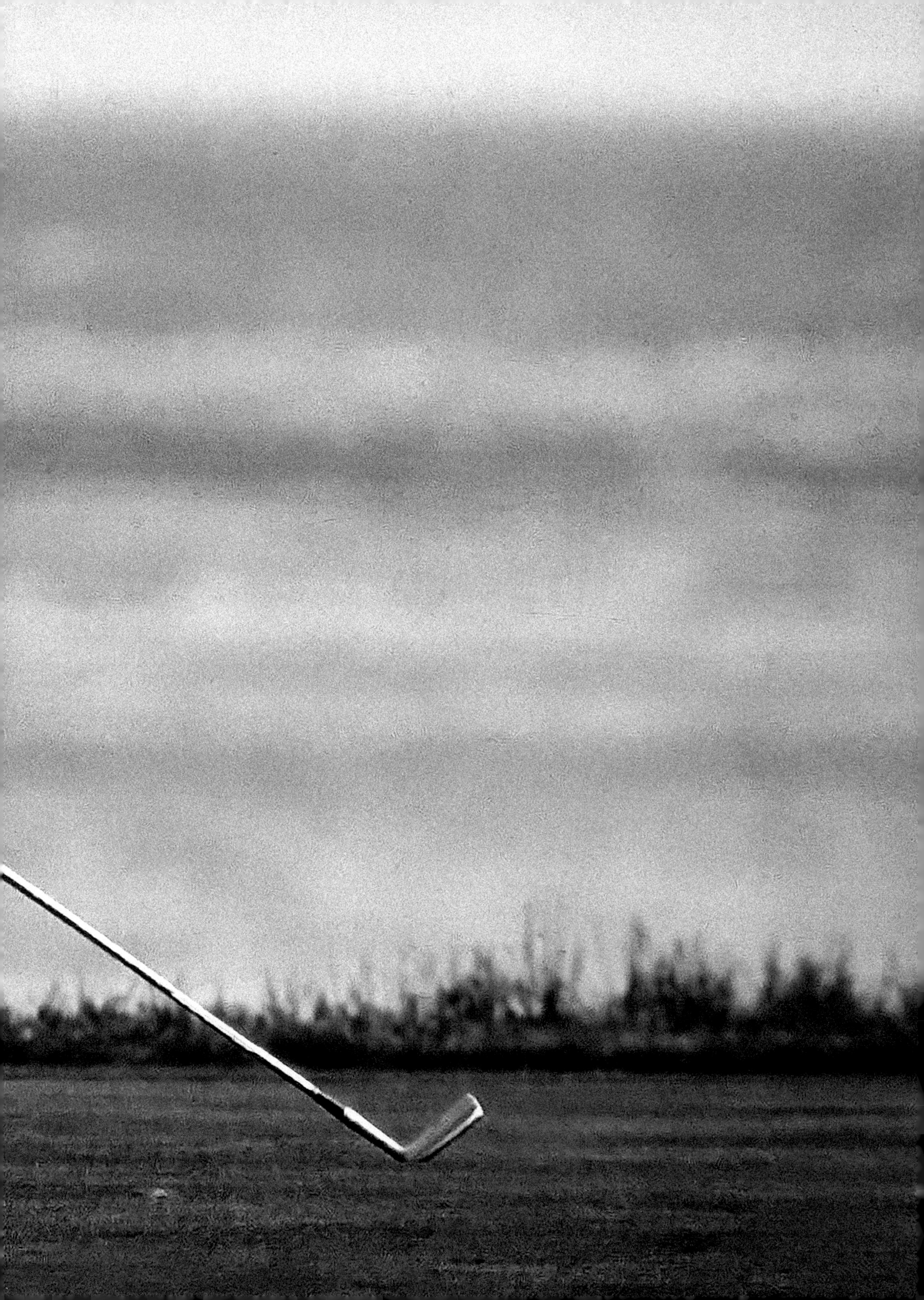

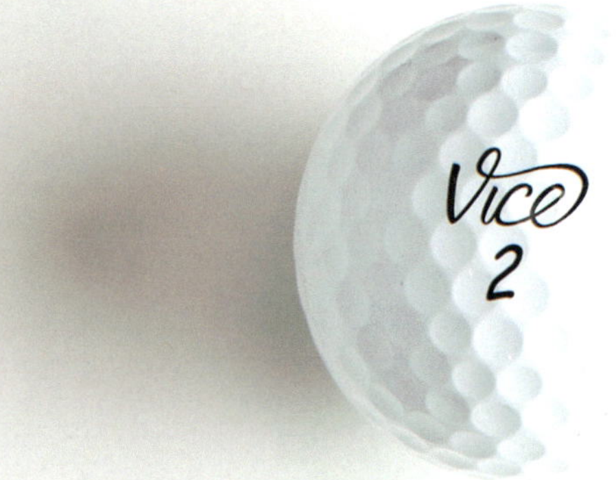

VICE PRO / $35

This three piece ball from the German upstart uses a cast urethane cover for its outstanding feel and ability to stick to the green better than Surlyn covers.

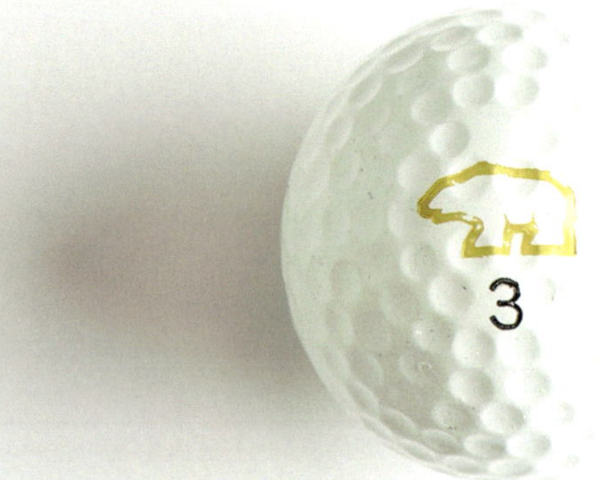

VINTAGE MACGREGOR NICKLAUS GOLDEN BEAR / $10

Jack Nicklaus played MacGregor balls for the majority of his career, so they naturally produced his signature ball, shown here with his iconic Golden Bear logo.

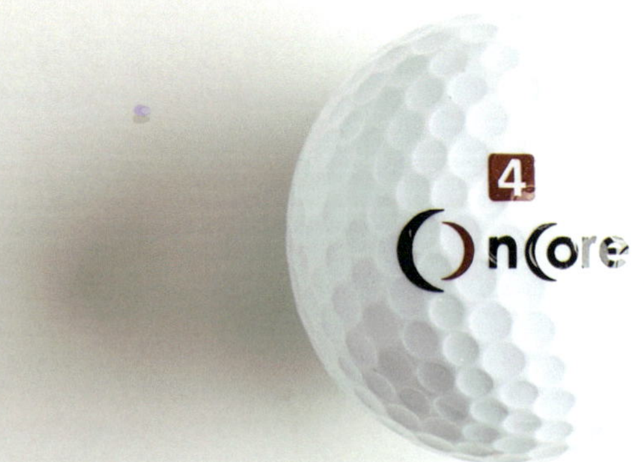

ONCORE CALIBER / $30

Built around a hollow metal core, this ball produces spin similar to a bullet leaving a barrel, resulting in less side spin and straighter flight.

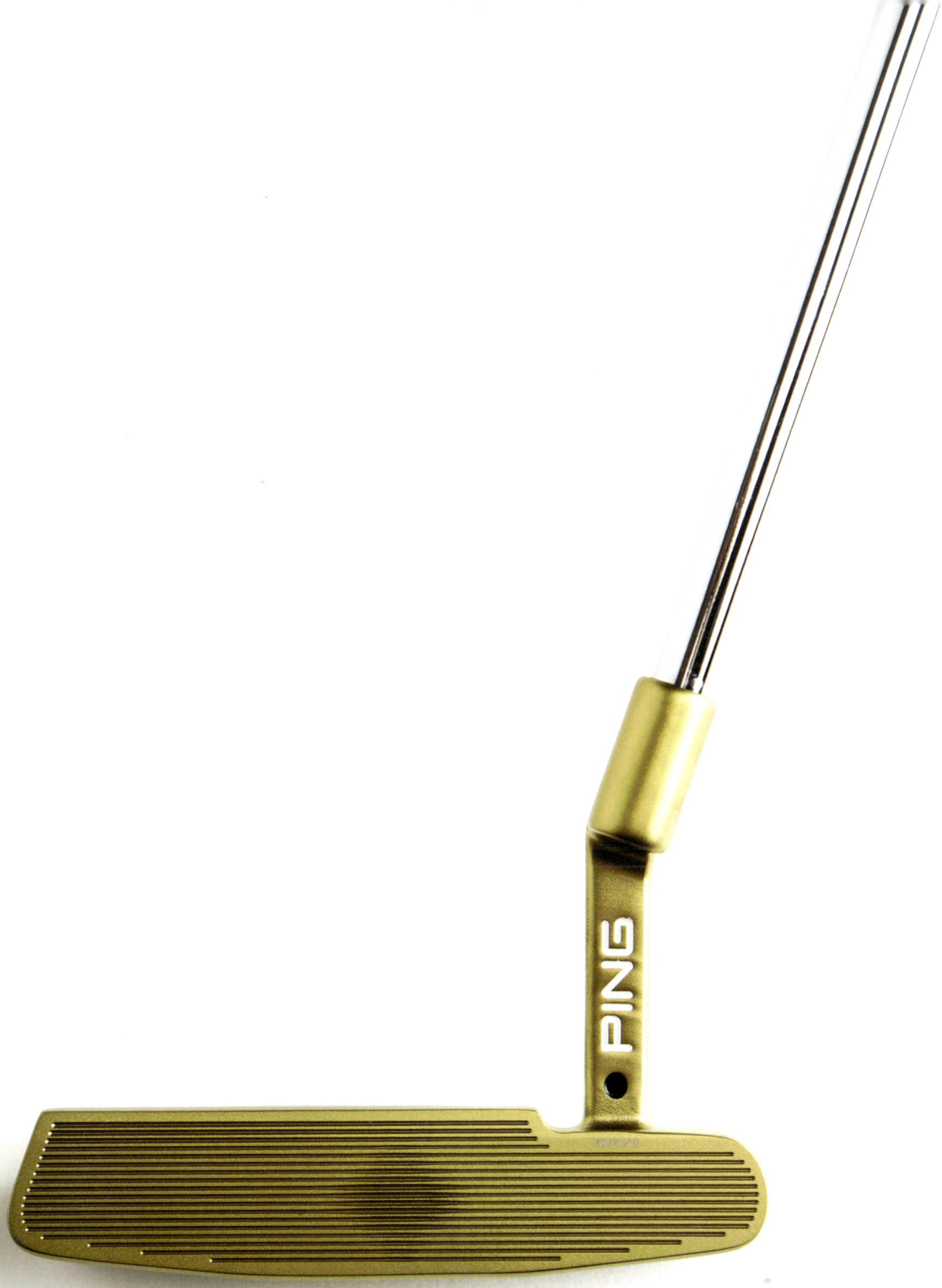

PING TR 1966 ANSER PUTTER / $150

Named for the year Karsten Solheim came up with the design for his most famous putter, this modern version takes the classic, oft-copied silhouette and upgrades it with variable-depth grooves machined into the face that help deliver consistent ball speeds no matter where on the putter you hit it.

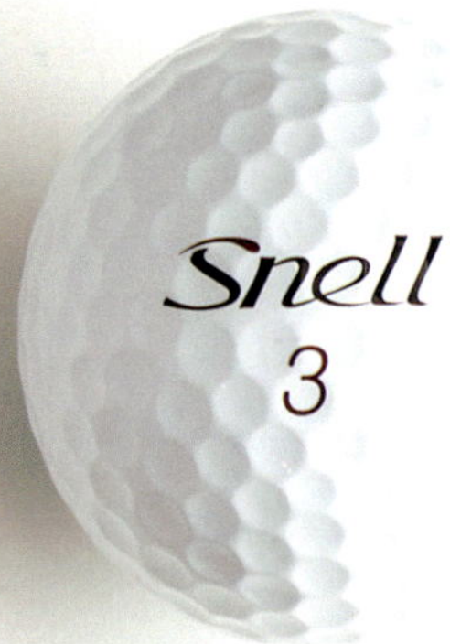

SNELL MY TOUR / $32

Created by Dean Snell, inventor of the Pro V1 (below), this three-piece ball offers tour-level performance and control at an amateur-friendly price.

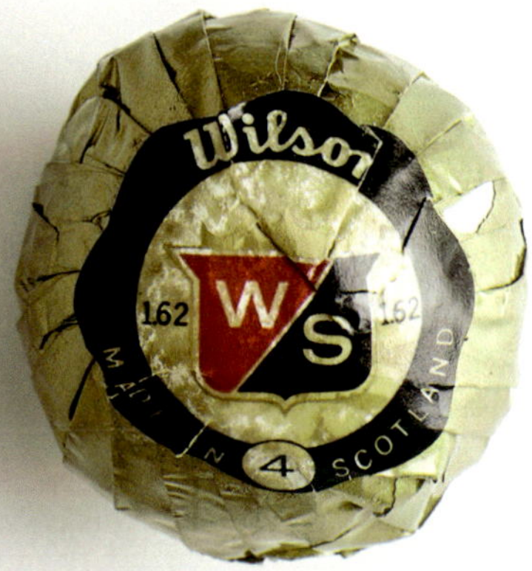

VINTAGE WILSON STAFF 1.62 / $5

Similar to the ball used by Arnold Palmer to win the 1961 Open, it's smaller than the balls used today, letting it fly farther and perform better in windy conditions.

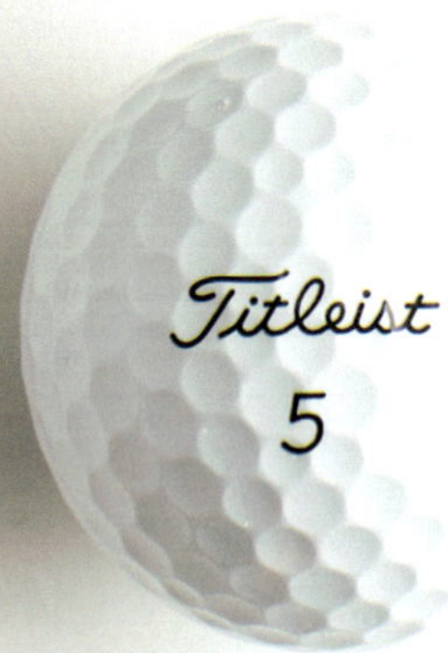

TITLEIST PRO V1 / $50

The gold standard of professional balls, this is the most played ball on tour, yet its mix of length and softness is still suitable for weekend warriors.

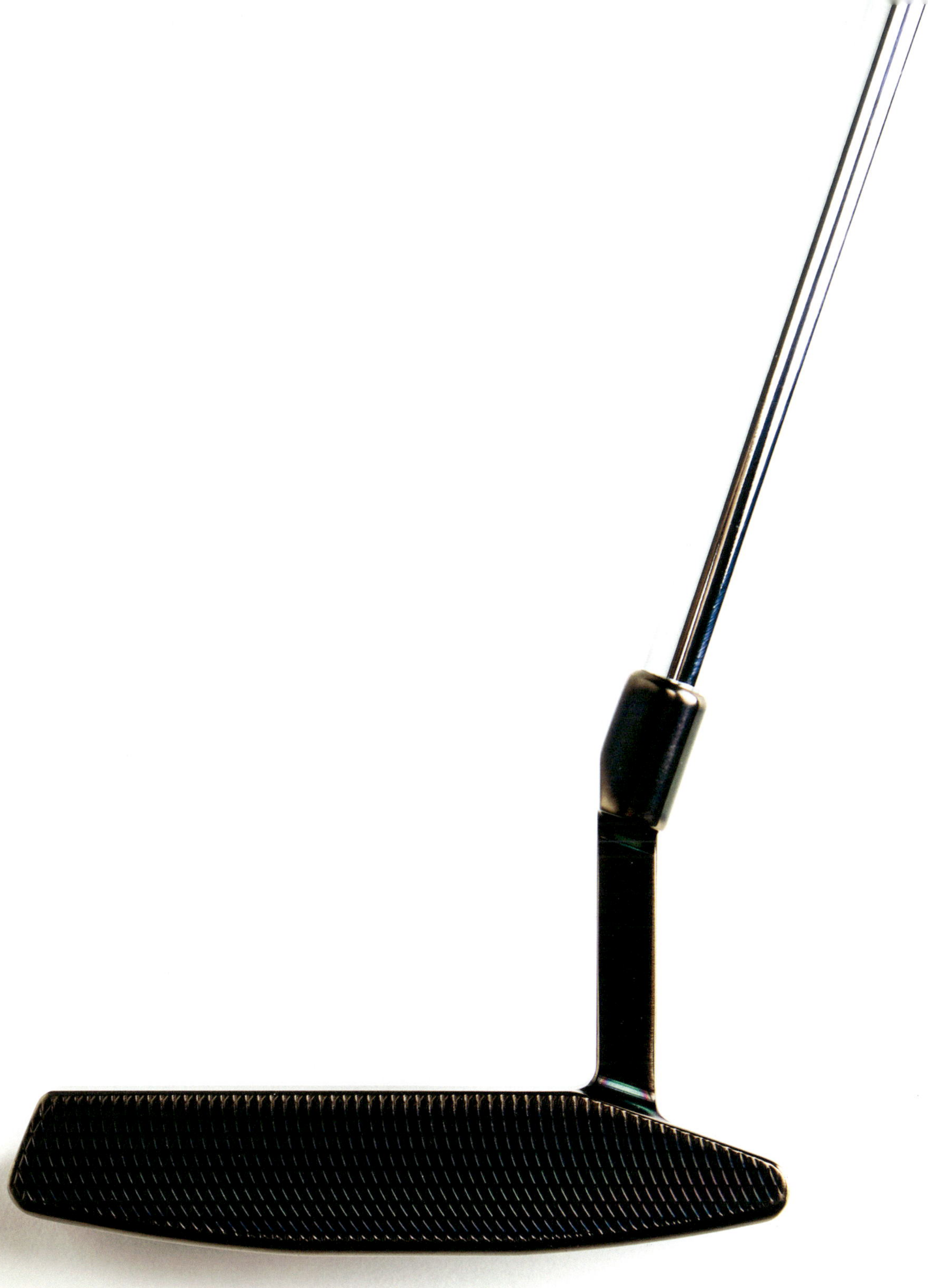

KRONOS AURORA TOUCH PUTTER / $750

Inspired by Japanese pottery glazes, this putter's PVD finish is unique to the company and ages well. It's milled from a single piece of carbon steel, and performance is boosted by a thick topline, a scored face that gives a consistent roll, and a perfectly balanced head that ensures the engraved sightline matches the sweet spot.

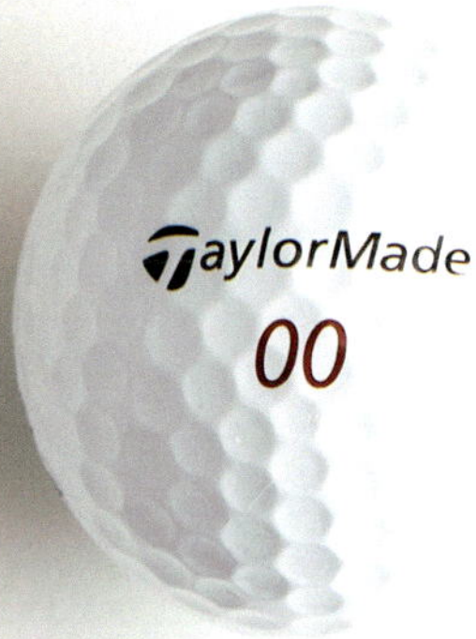

TAYLORMADE PROJECT A / $35

This ball, designed specifically for amateurs, pairs a soft core for added distance with a mantle and cover meant to provide tour-like spin around the green.

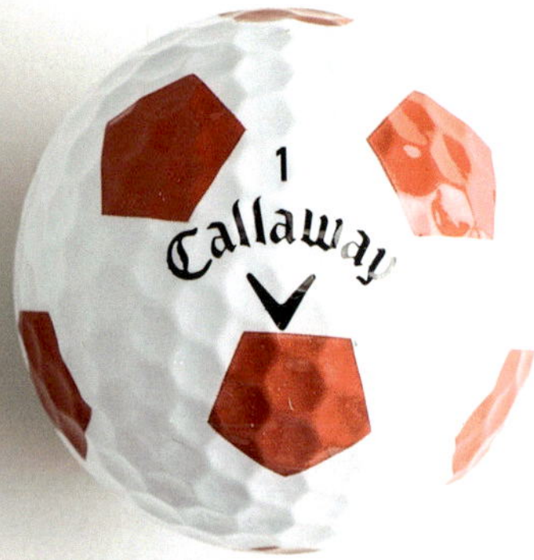

CALLAWAY CHROME SOFT TRUVIS / $40

The Dual SoftFast core of this 4-piece ball delivers both fast ball speeds and low spin off the tee, while its soccer ball-like pattern makes it recognizable in the air.

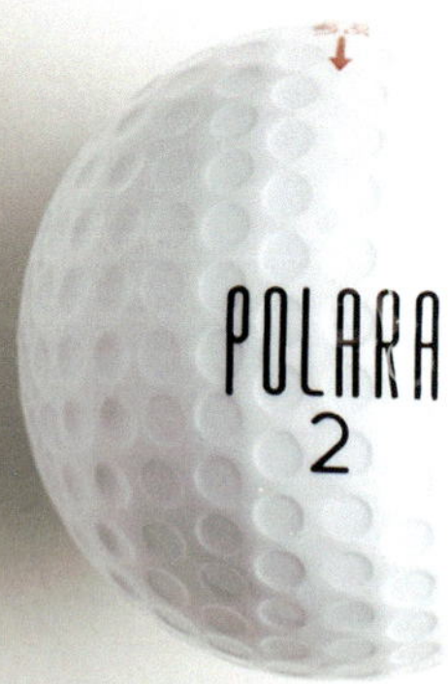

POLARA ULTIMATE STRAIGHT XS / $30

This cheat ball's mix of large and small dimples corrects hooks and slices by up to 75%, ensuring you won't spend your round in the woods. Not legal for tournament play.

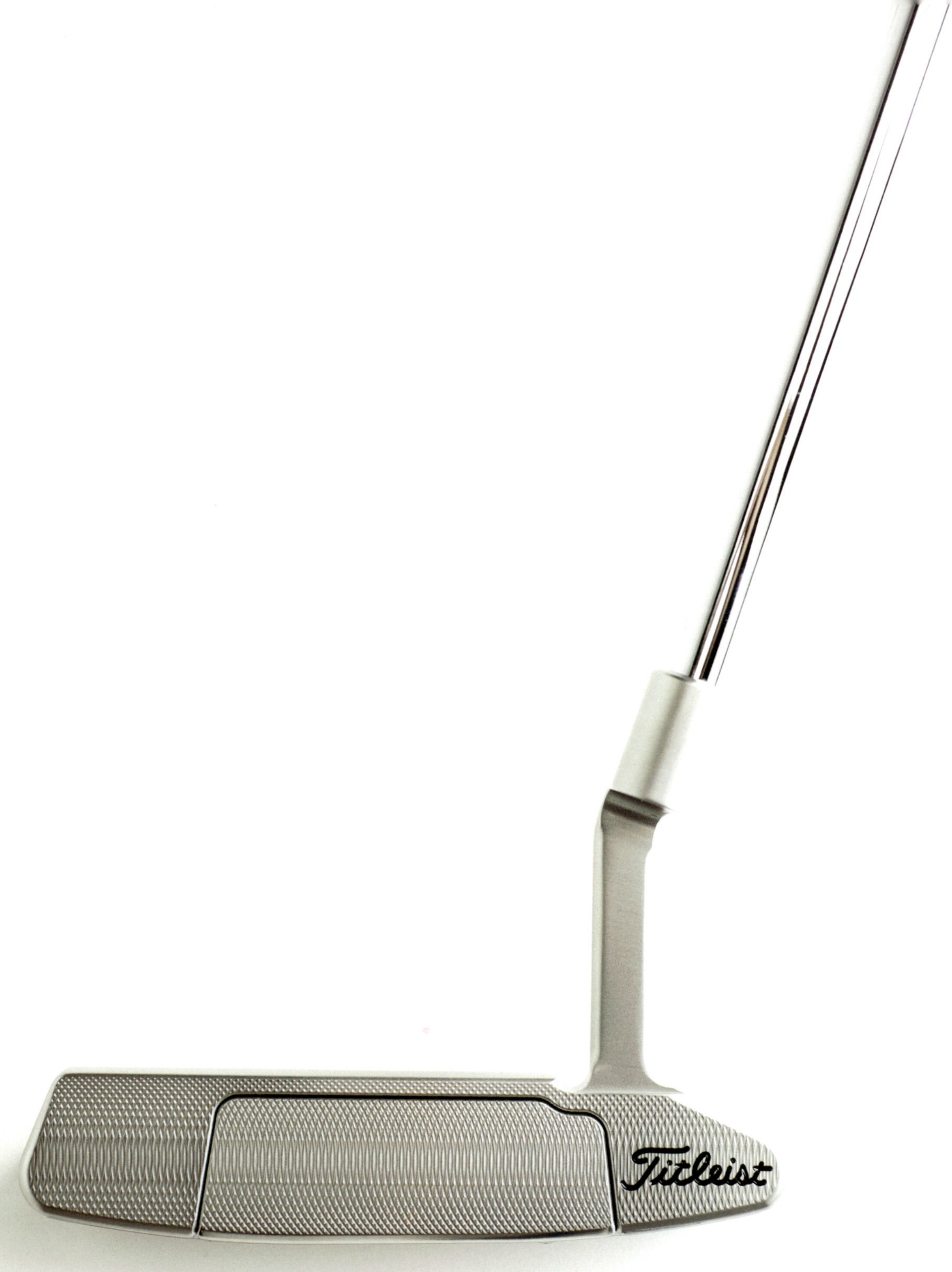

TITLEIST SCOTTY CAMERON SELECT NEWPORT 2 PUTTER / $380

Since 1993, nearly half of all Majors have been won with a Scotty Cameron putter, and a majority of those with a Newport 2. This year's Select improves on prior models with a body-matching stainless steel inlay that sits atop a vibration damper, giving the putter both a solid feel and added forgiveness.

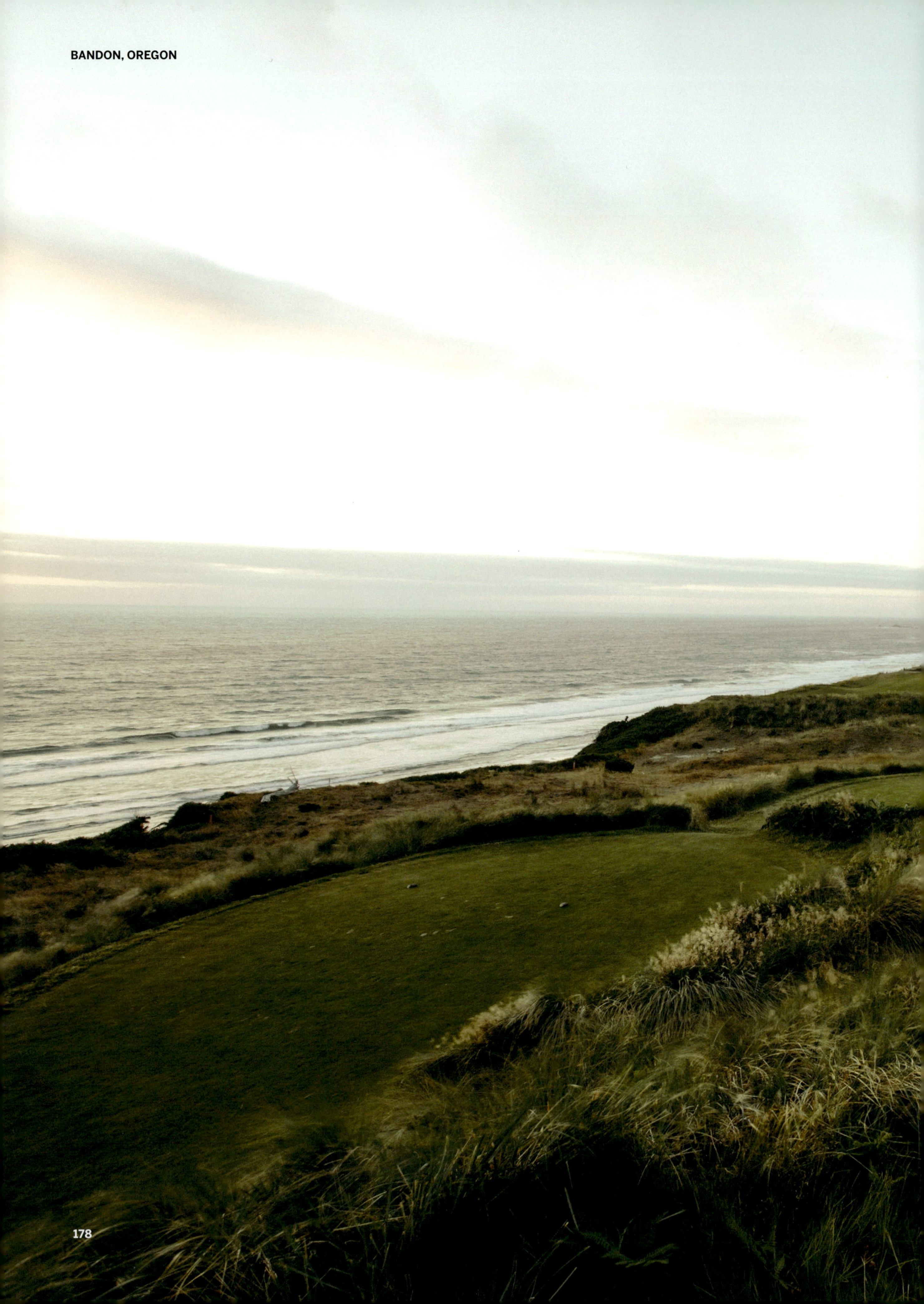

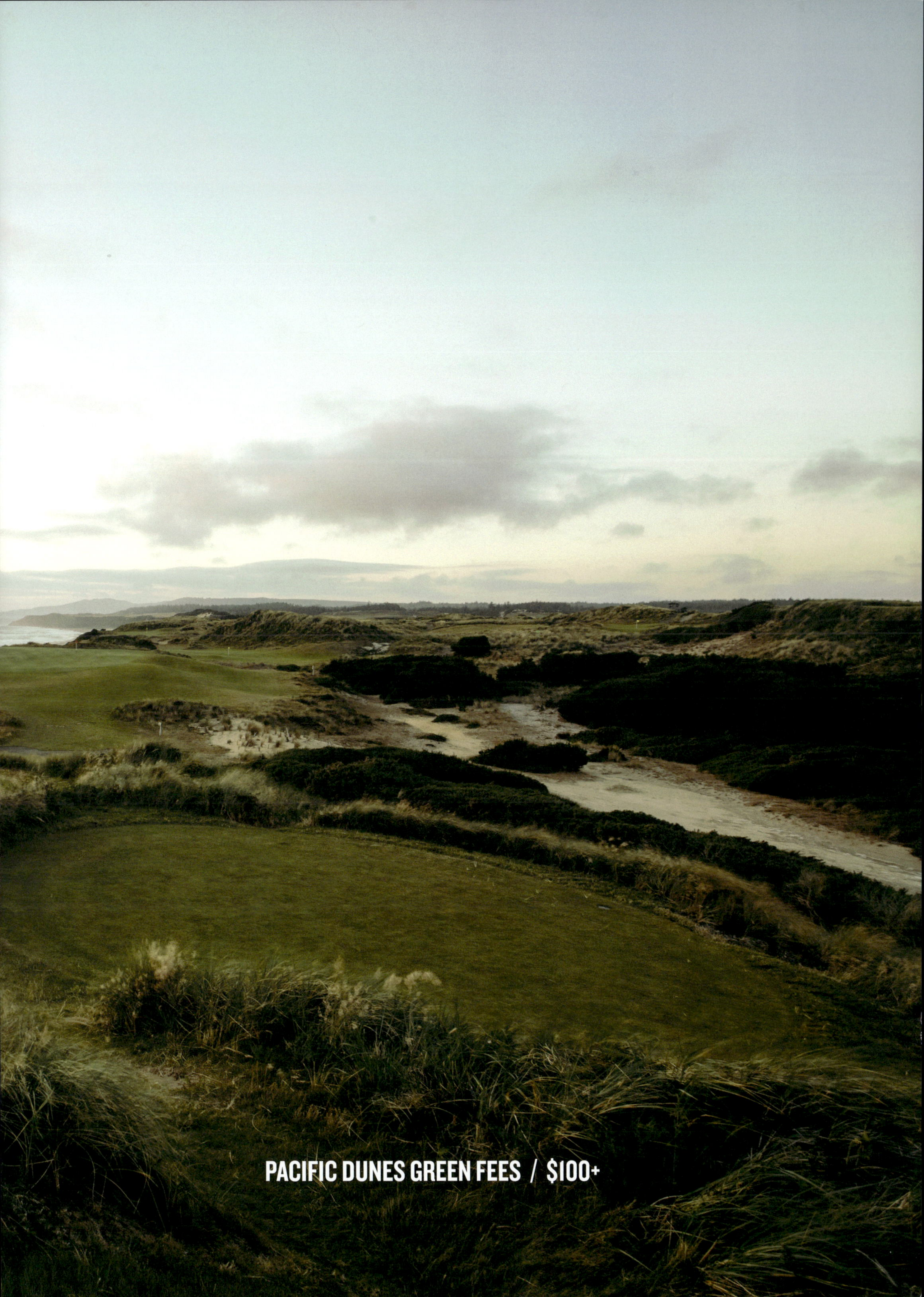
PACIFIC DUNES GREEN FEES / $100+

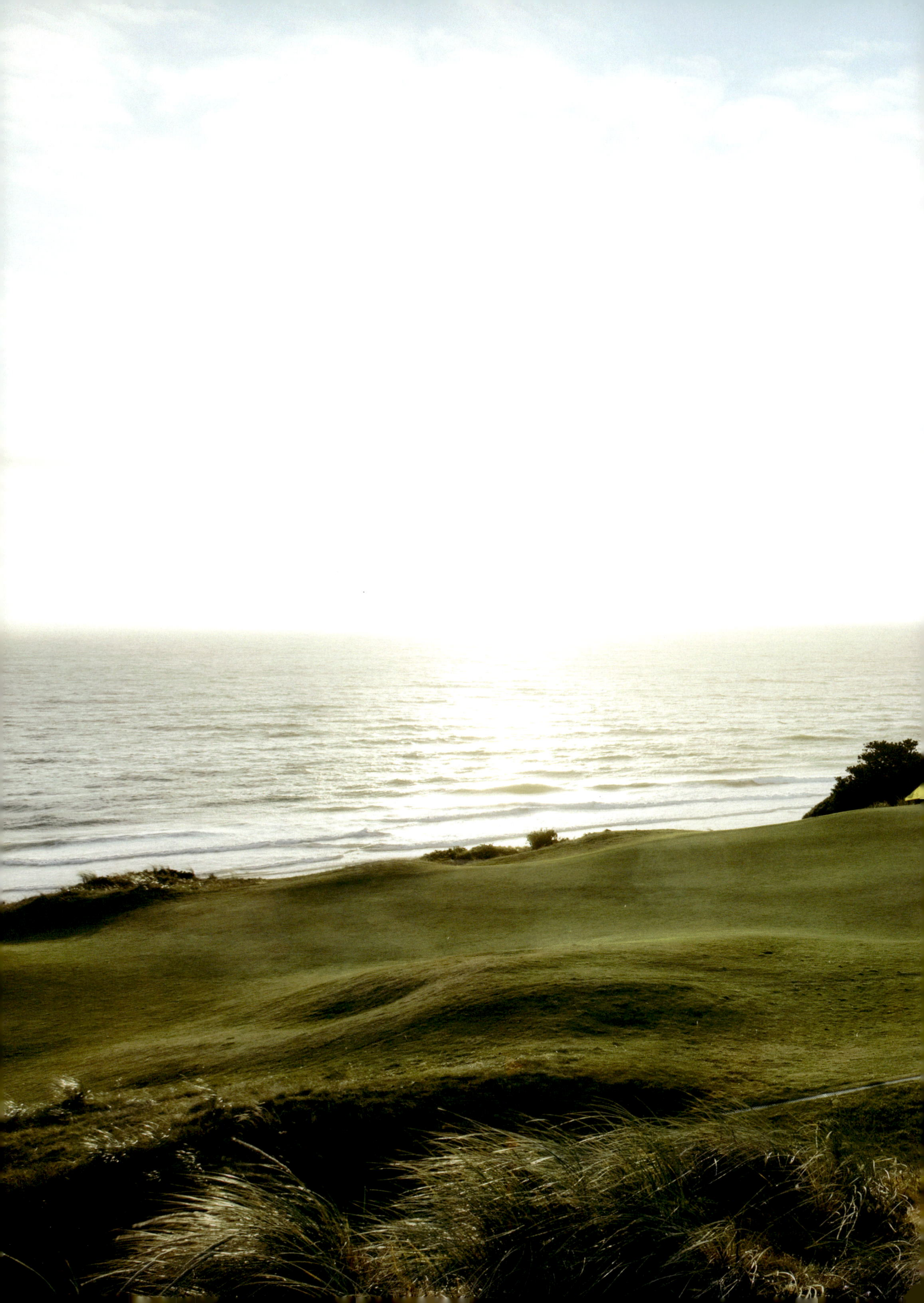

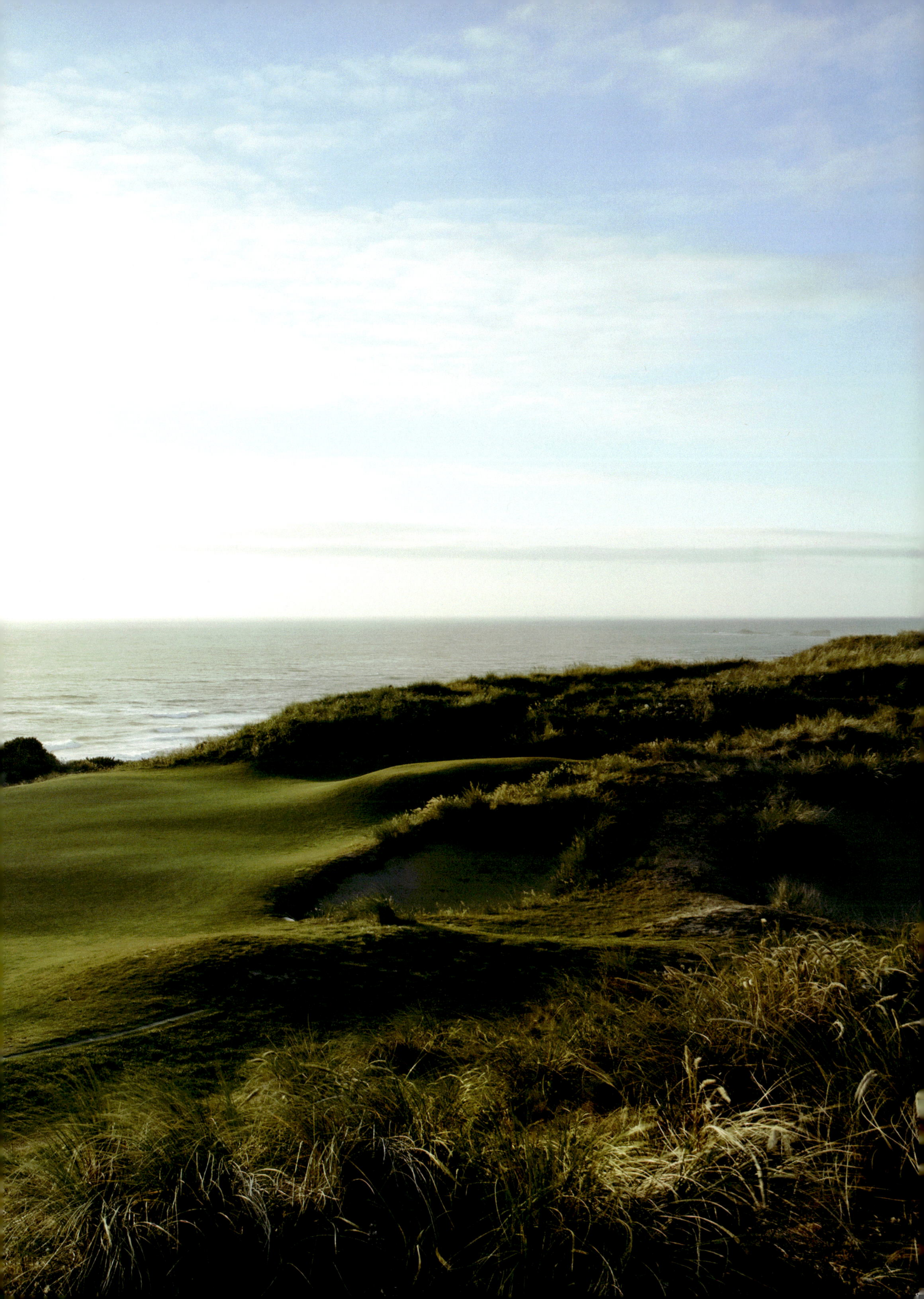

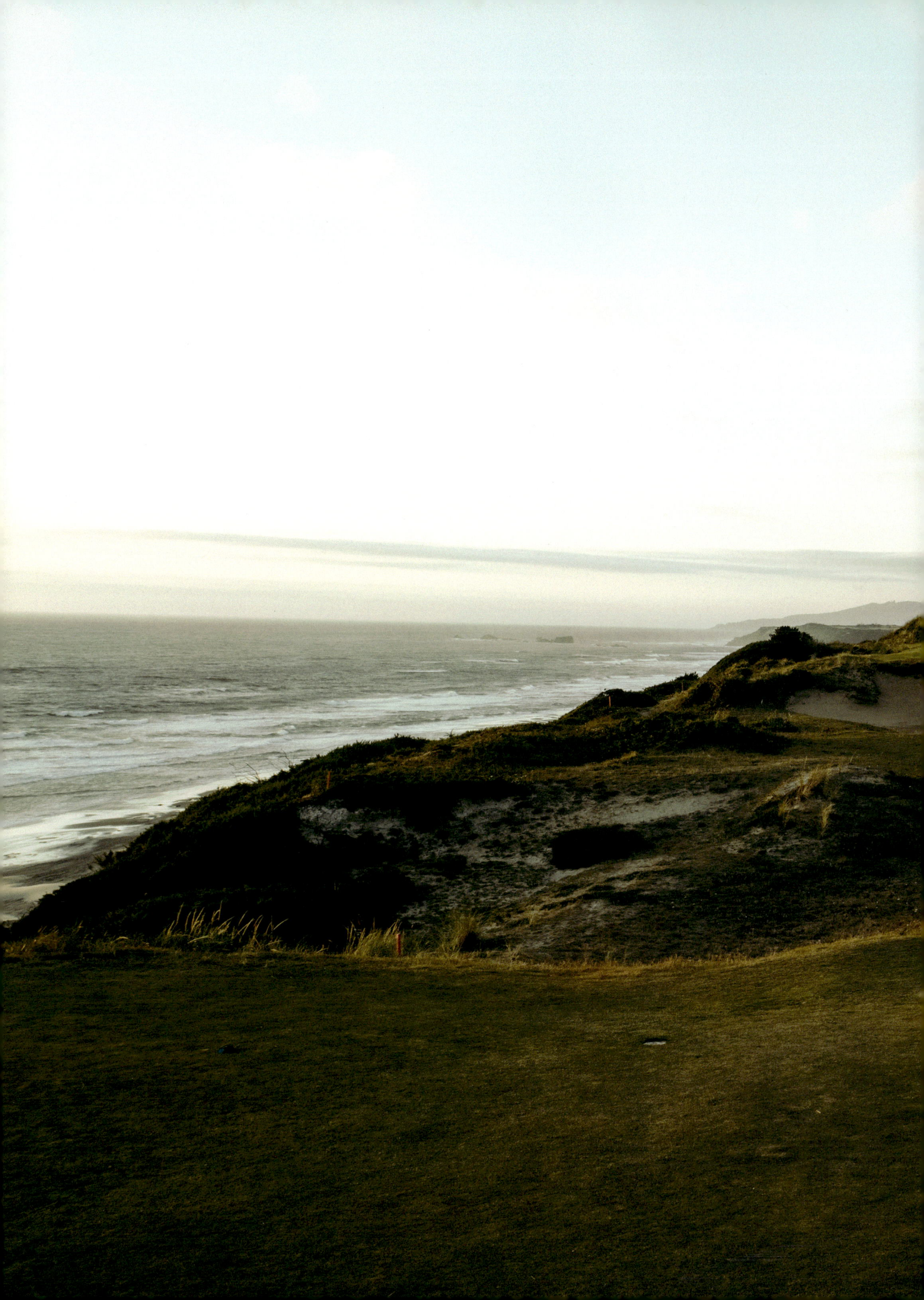

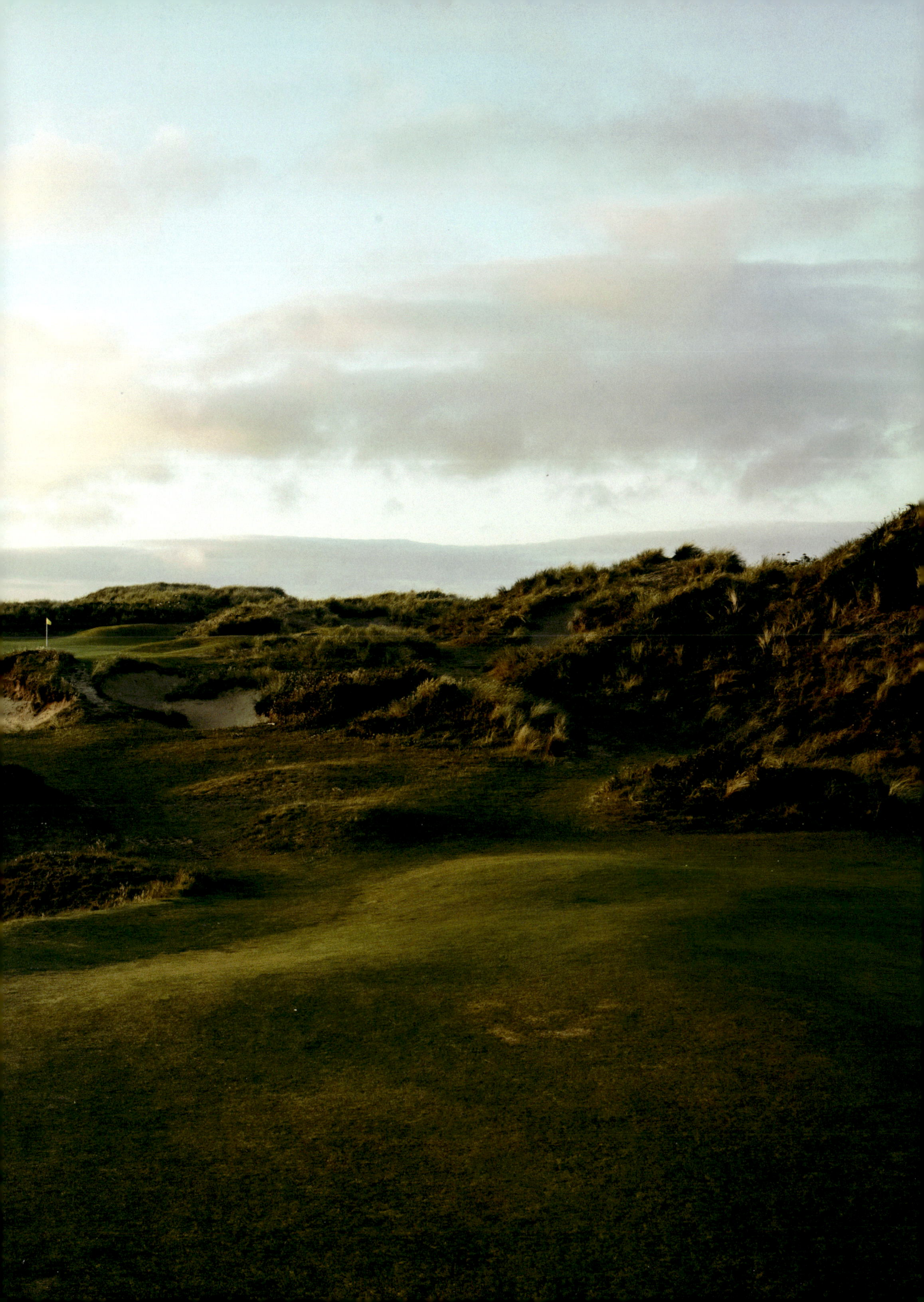

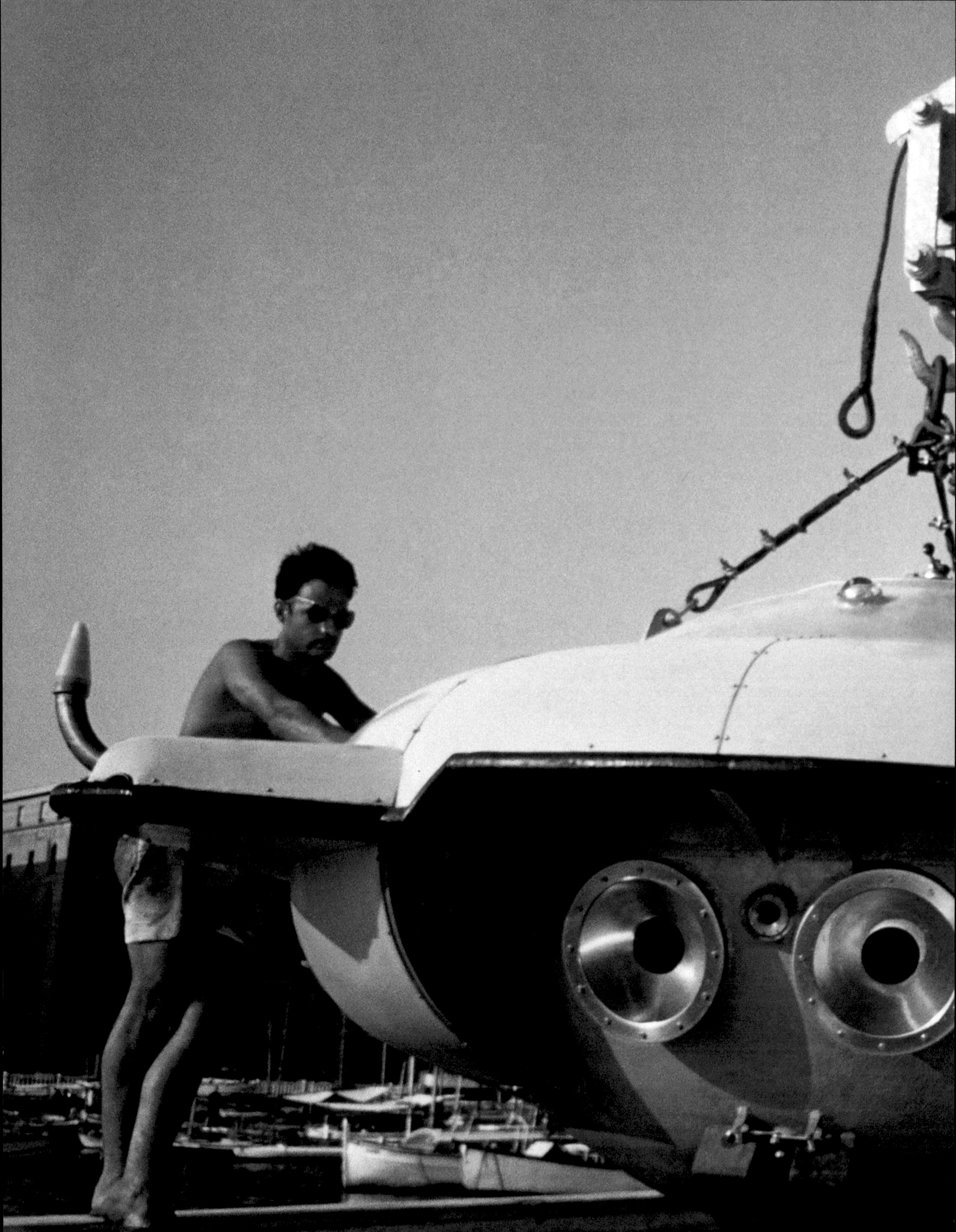

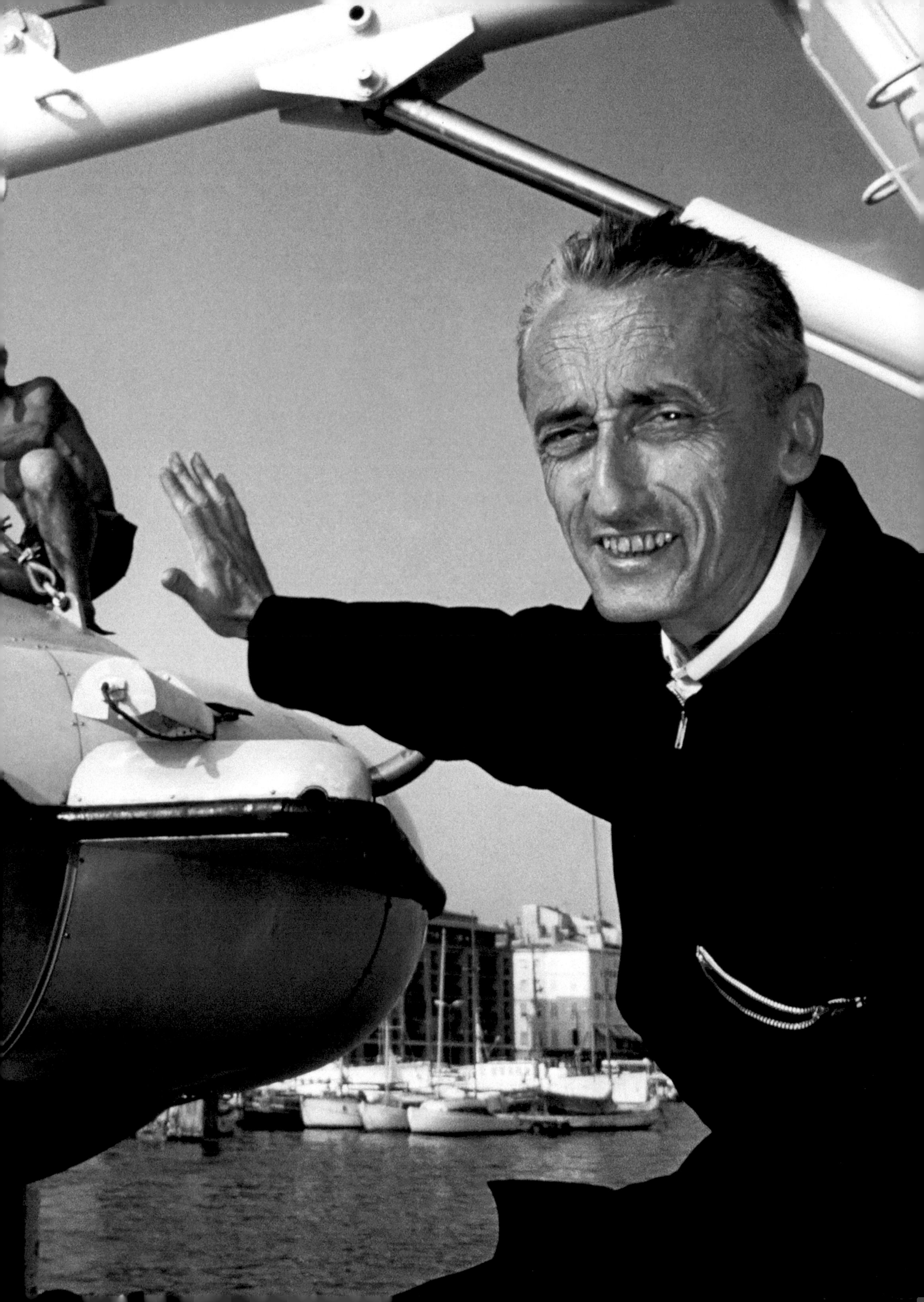

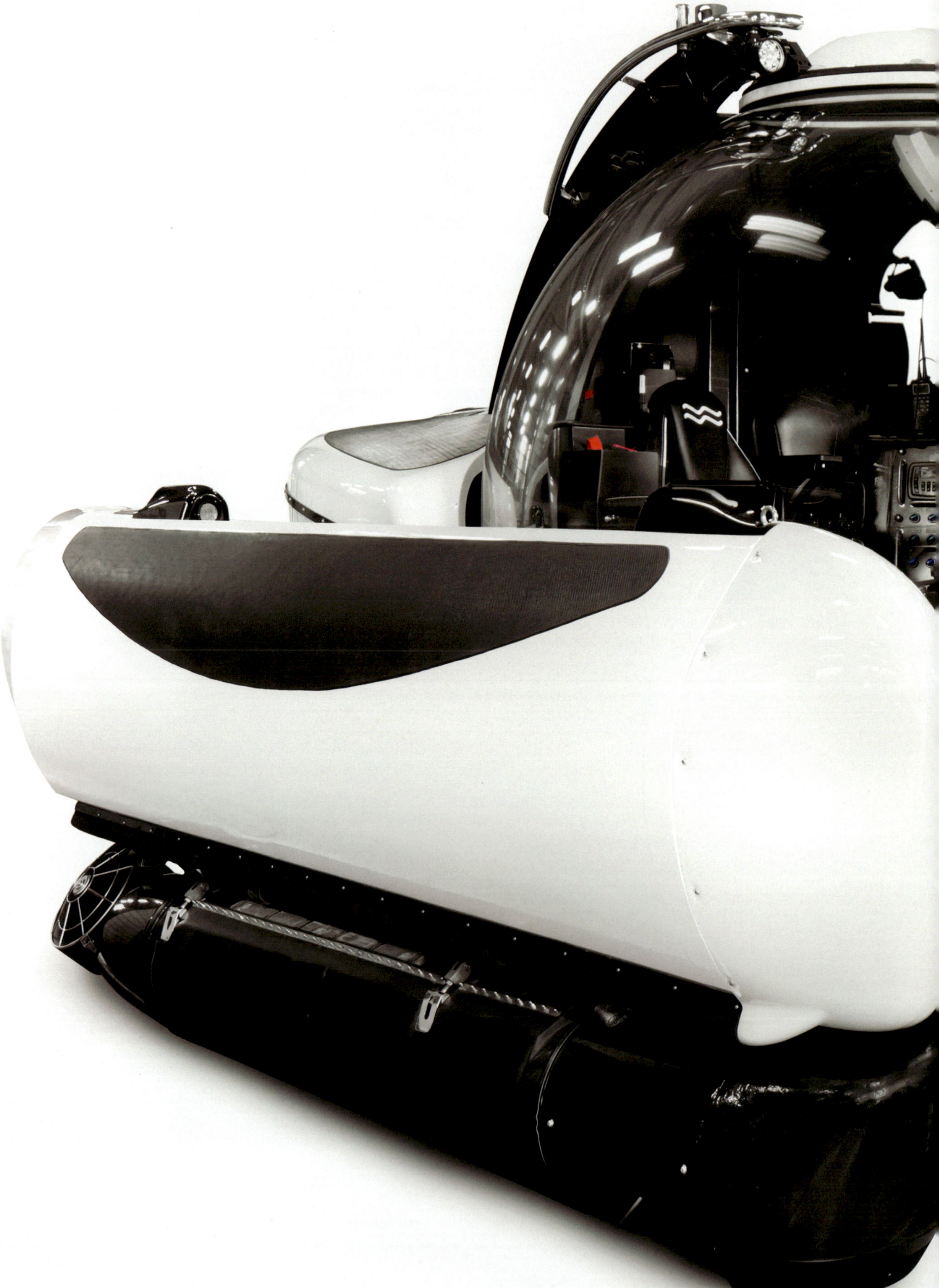

U-BOAT C-EXPLORER 3 / $2.5 MILLION+

A leisurely cruiser it is not. This Red Dot-winning sub is built for serious exploration and research, with a 360° acrylic pressurized hull, a depth rating of 300 meters, and the ability to operate for up to 16 hours at a time. It can also accept a wide range of accessories, including cameras, lighting rigs, robotic arms, and magnetometers.

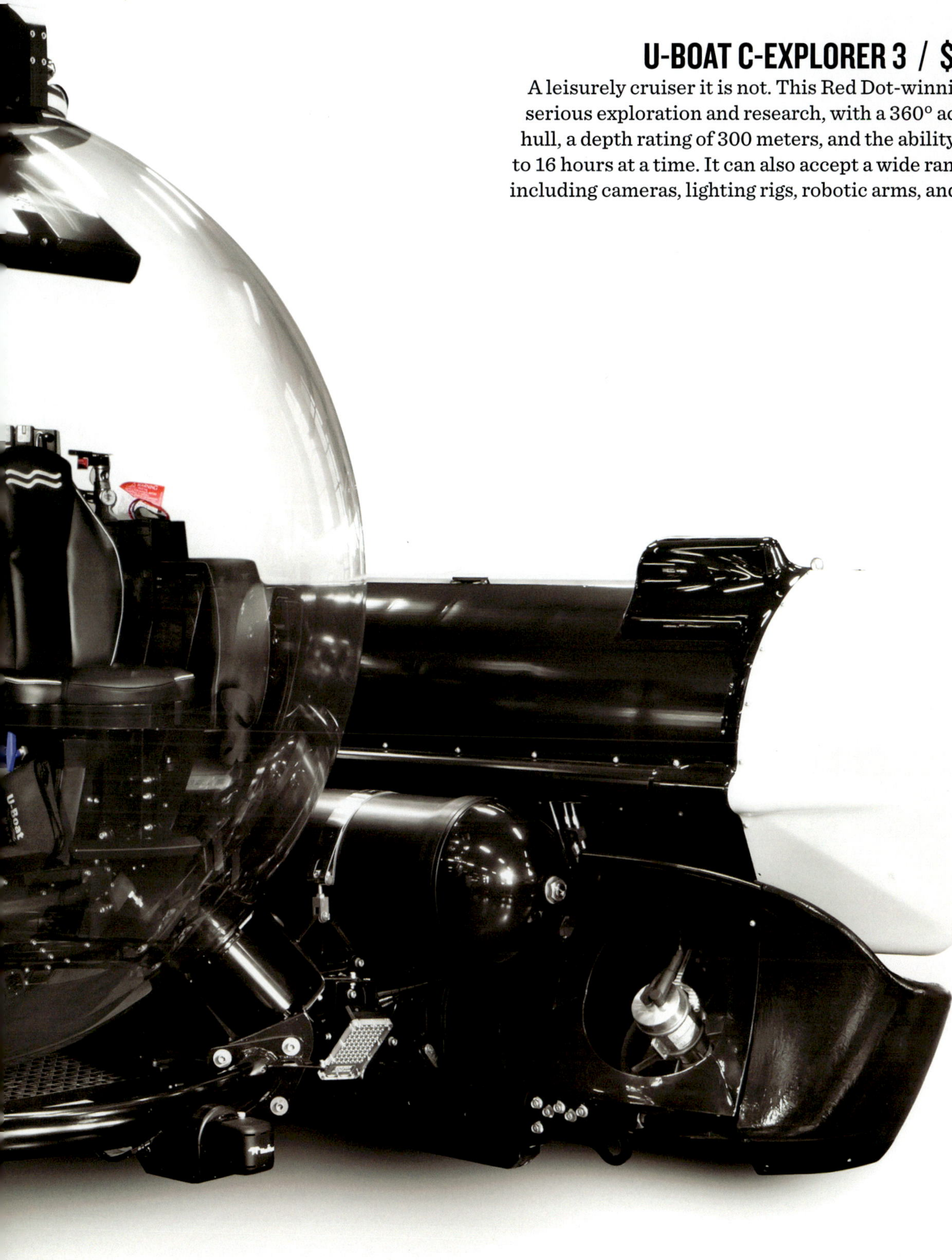

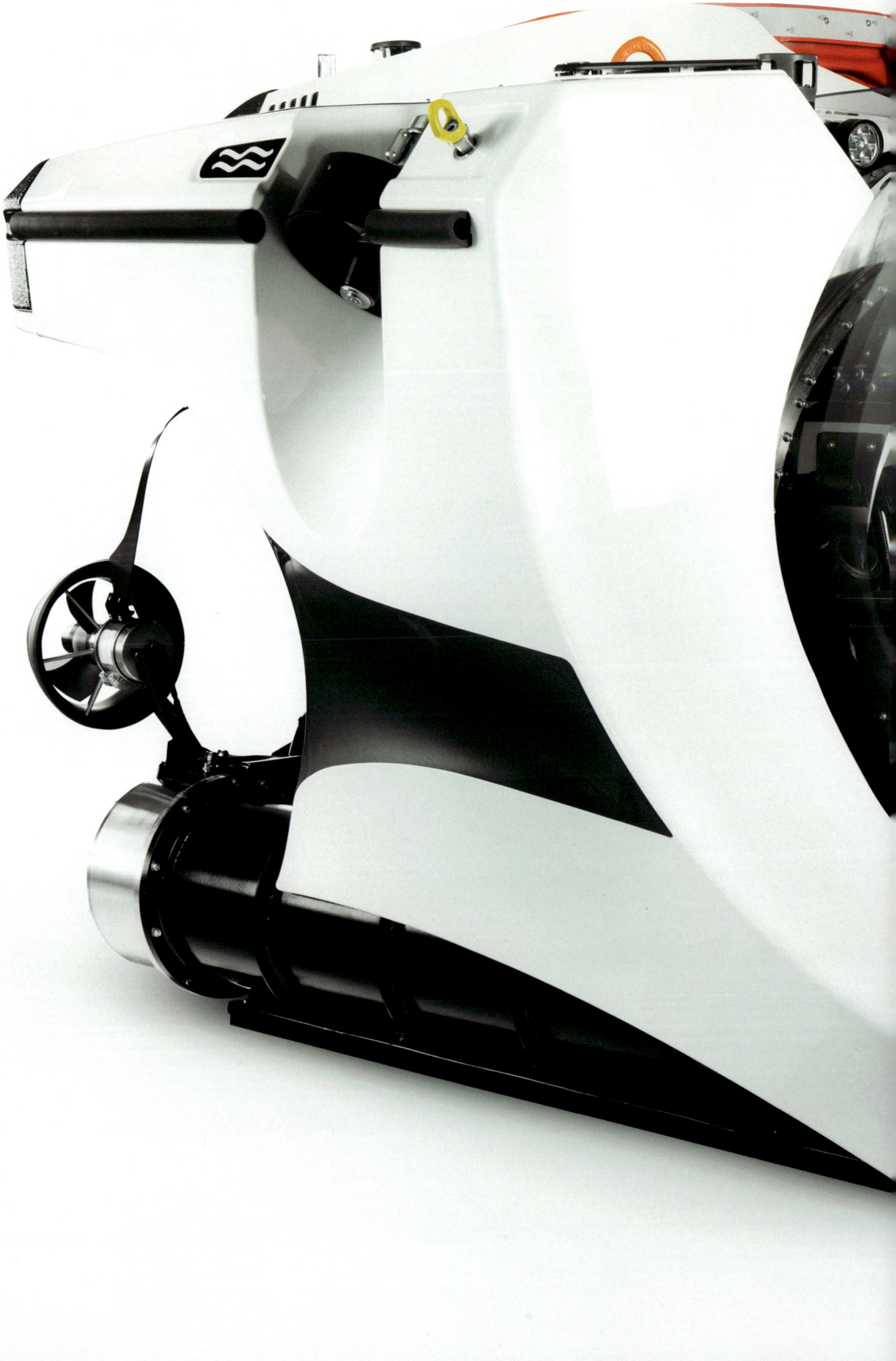

U-BOAT SUPER YACHT SUB 3 / $1.7 MILLION+

The most compact three-person sub ever built, this lightweight machine was designed to fit inside most superyachts. It's available in versions that can go to 100, 200, or 300 meter depths, and has an air-conditioned, pressurized cabin with a seating layout that maximizes the views through the bubble-like acrylic viewport.

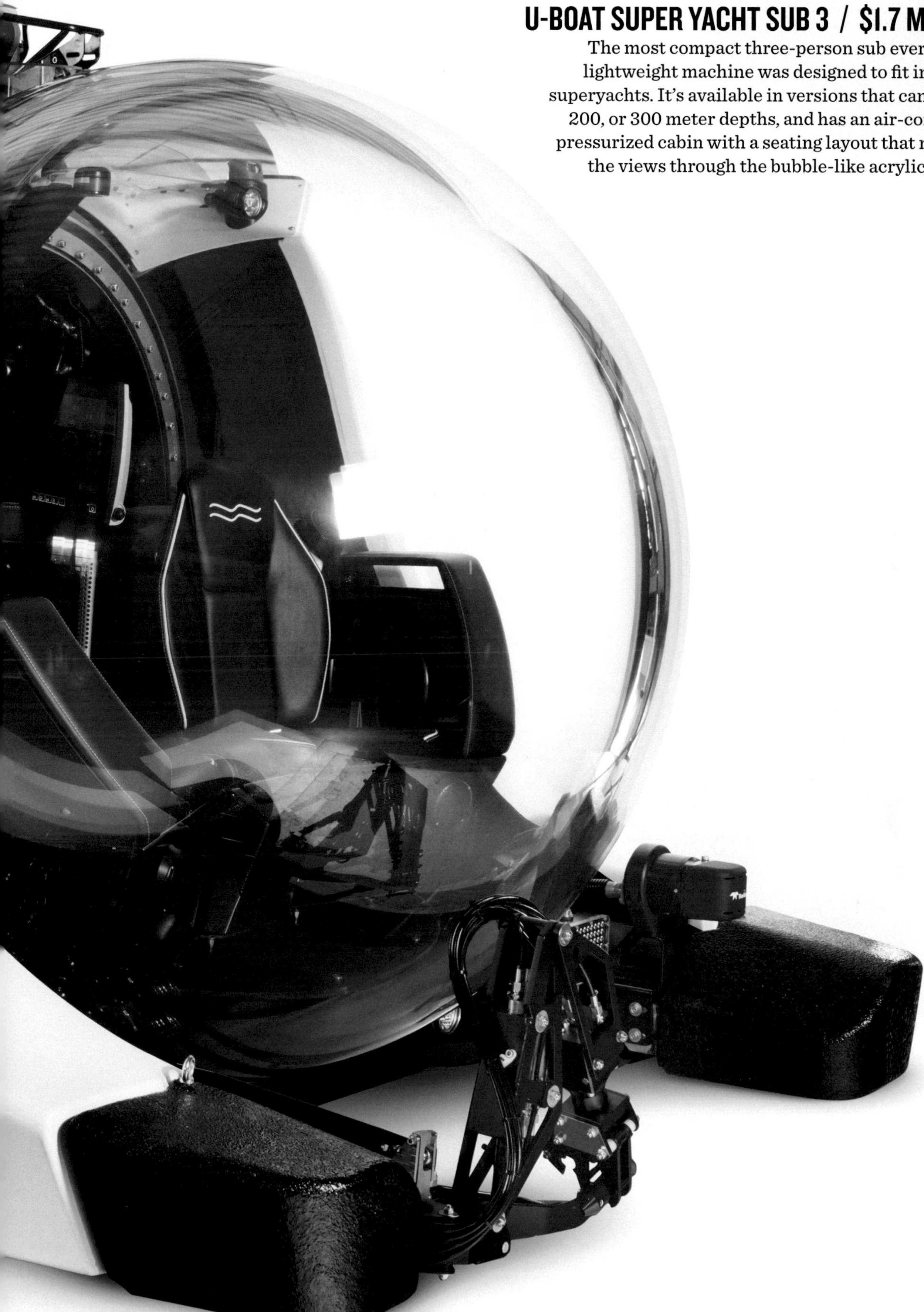

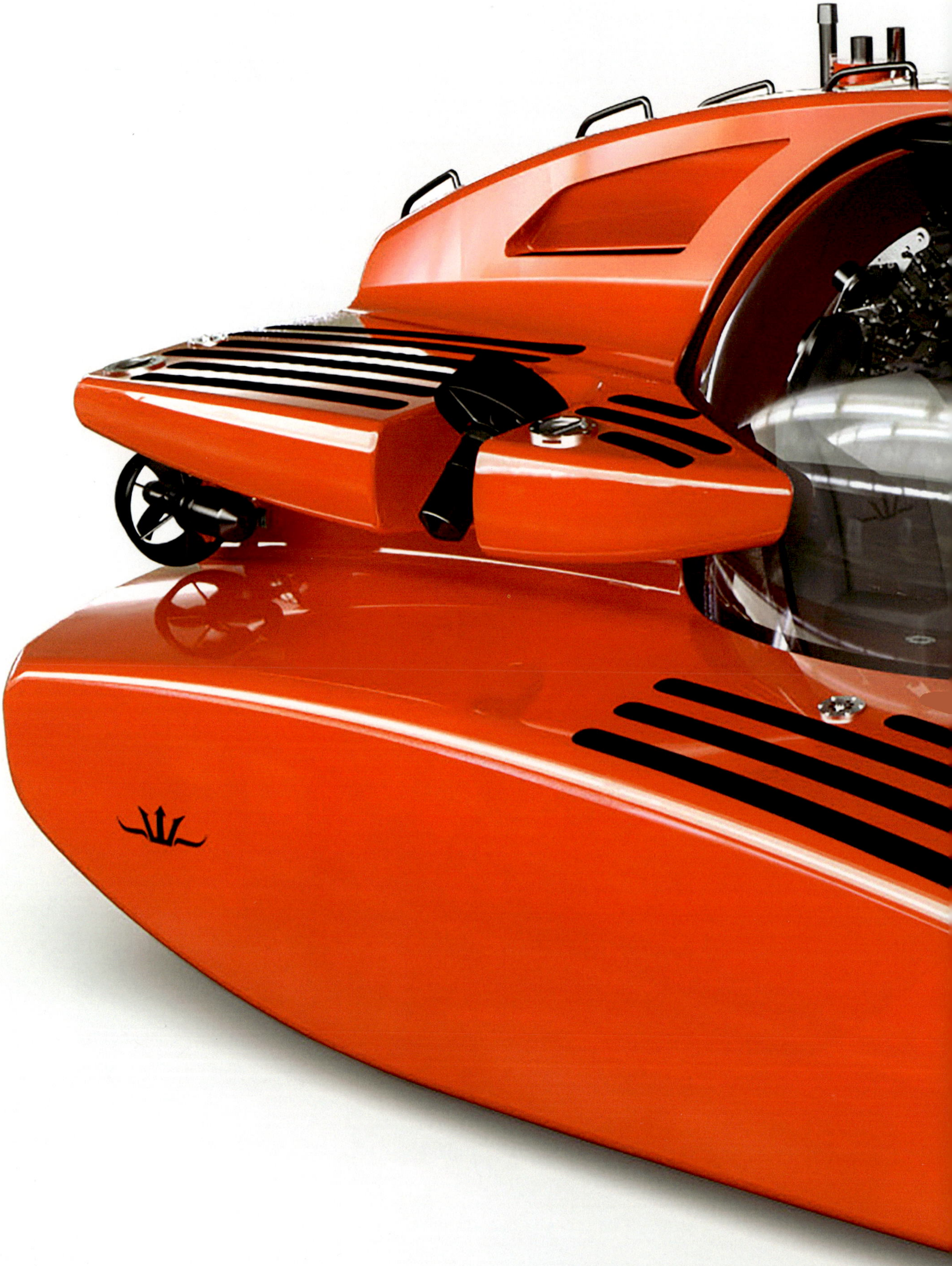

TRITON 6600/2 / $5.5 MILLION

A vast majority of the ocean remains unexplored. This two-person sub might change that. It can reach depths of over a mile, a feat made possible by an optically-perfect acrylic hull that required new manufacturing techniques to build. High-output LEDs light your way when you leave daylight behind, and it can run for up to 12 hours before needing to surface.

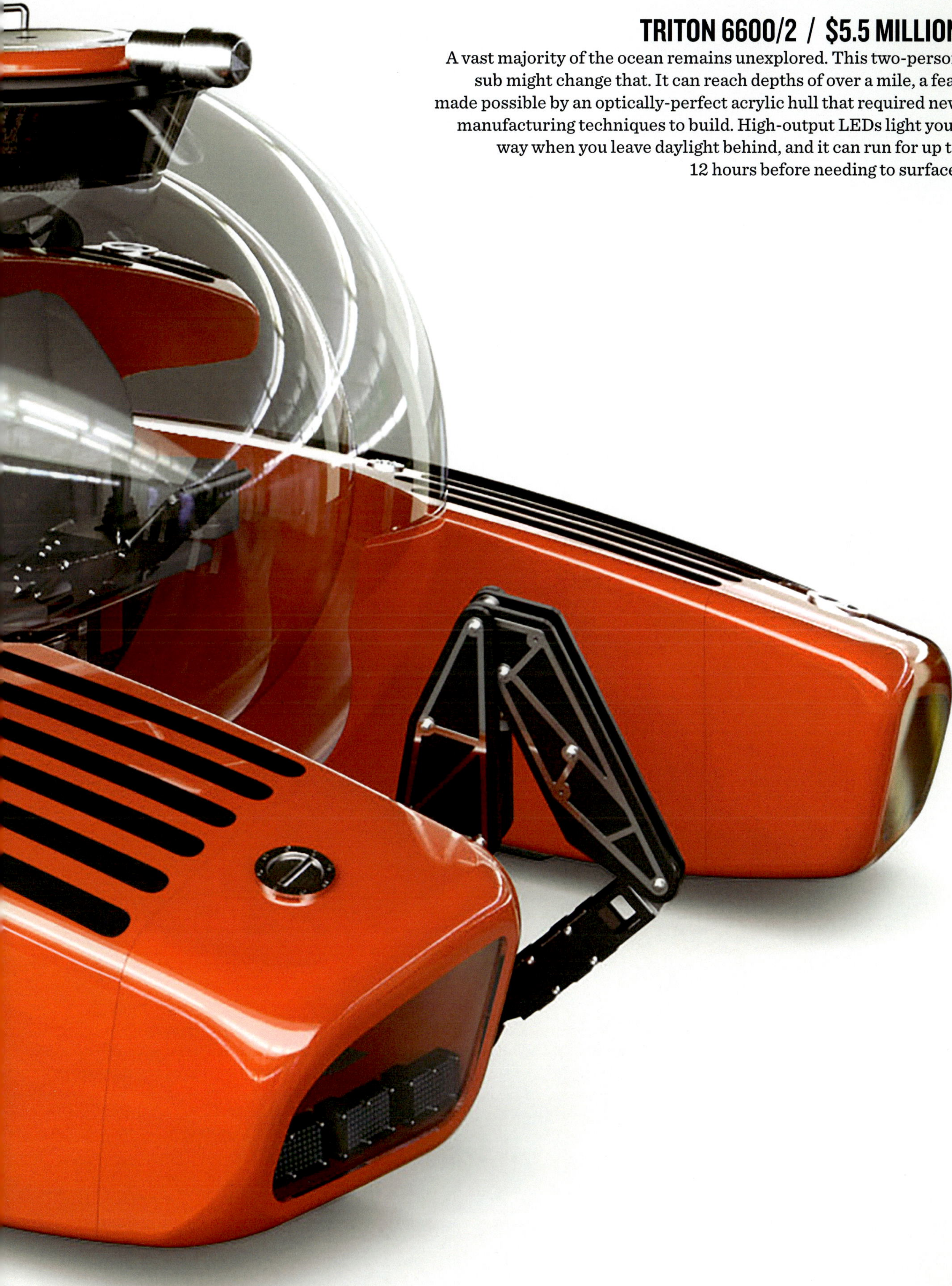

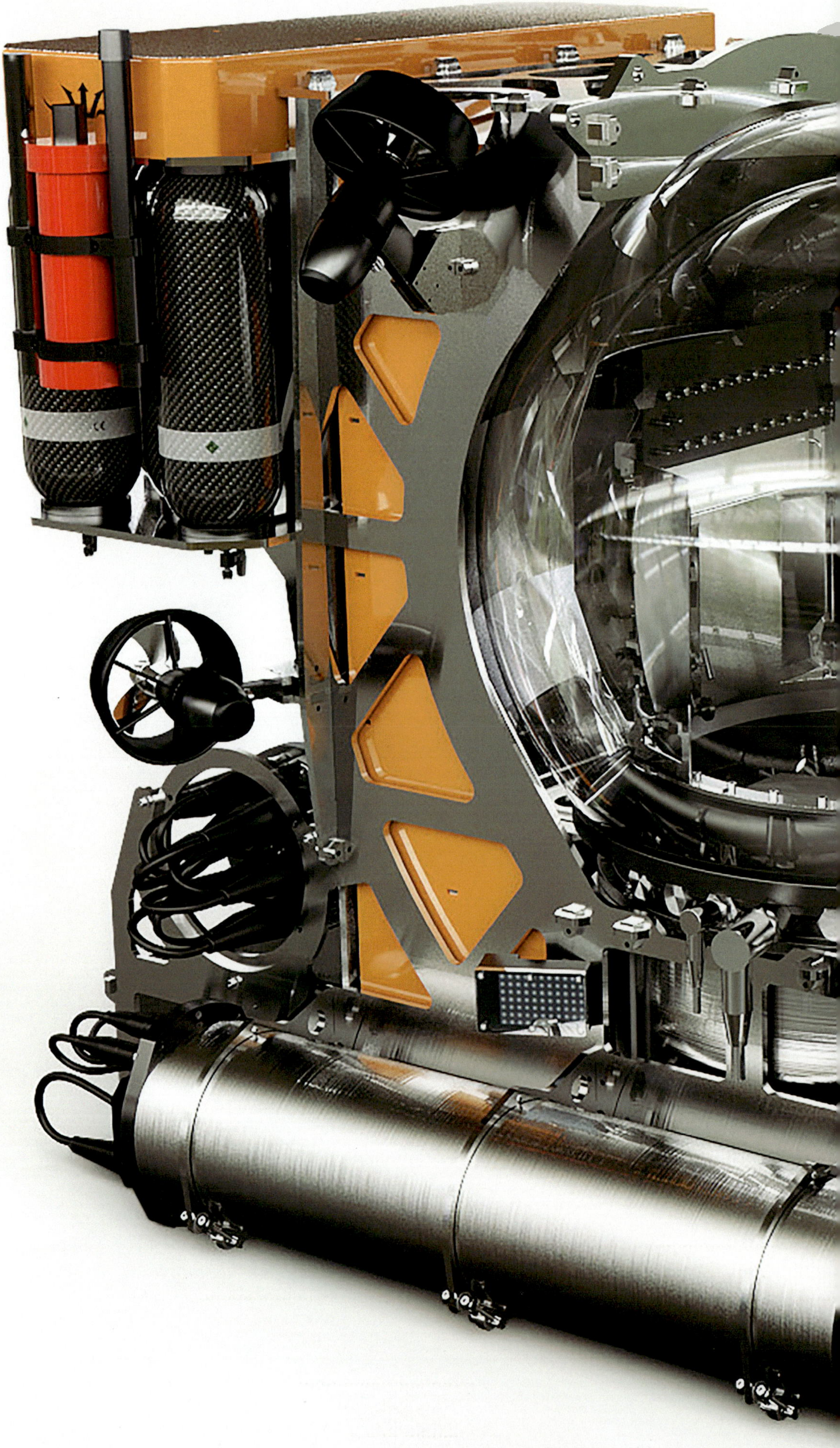

TRITON 3300/1 MD / $1.85 MILLION

This single-operator sub has a robot-like design that's purely focused on utility. It can carry payloads of up to 440 lbs yet still reach over a half mile under the surface, while its modular design makes it easy to swap out different tools and equipment. The six-hour battery gives you plenty of time to get things done.

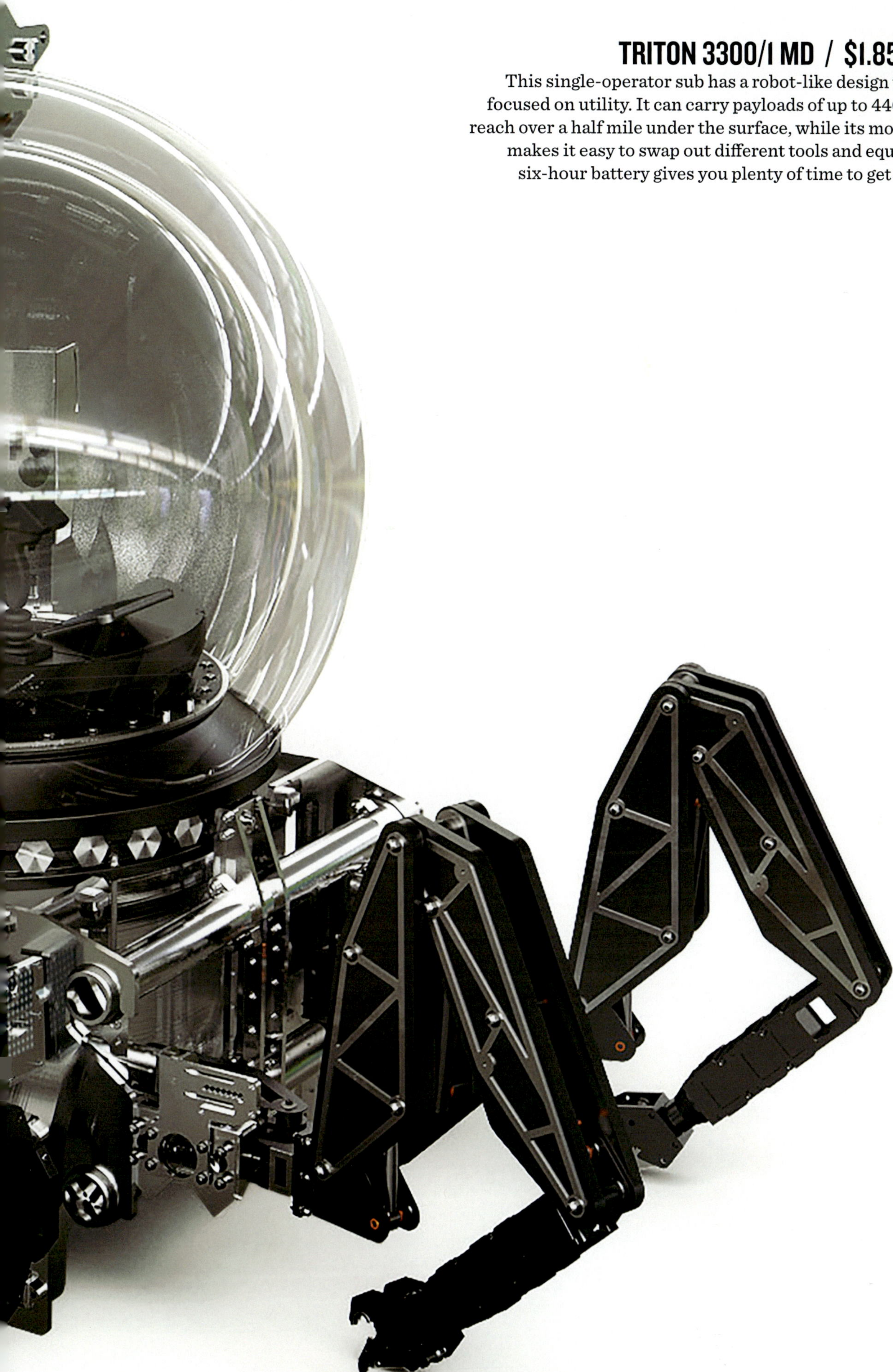

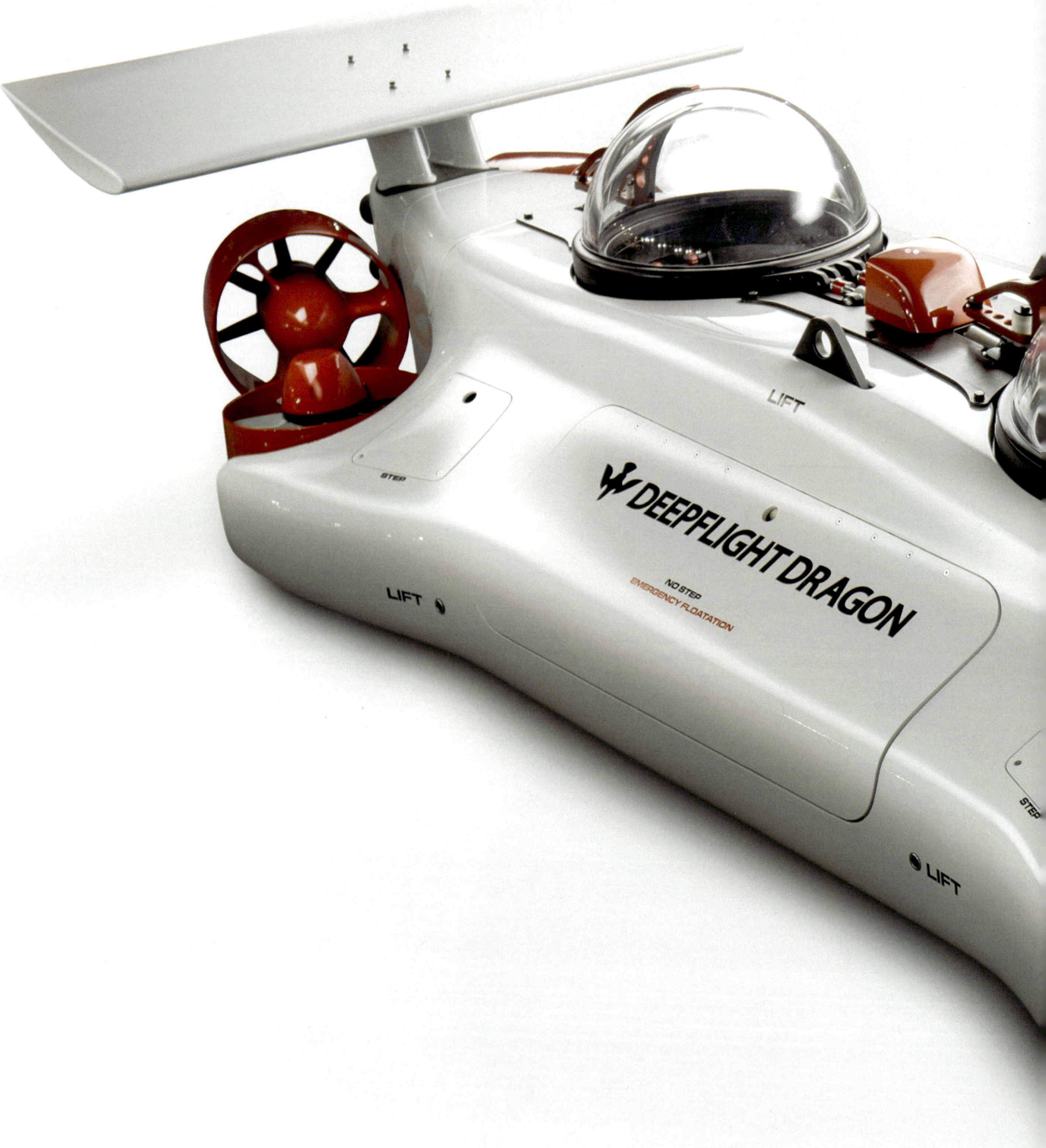
DEEPFLIGHT DRAGON
NO STEP
EMERGENCY FLOATATION
LIFT
LIFT
LIFT
STEP

DEEPFLIGHT DRAGON / $1.5 MILLION

Thanks to four brushless thrusters, this two-person submersible not only lets you fly through the water but also hover next to anything at which you want a better look. It can reach depths of up to 400 feet, has a Dive Manager system that makes it easy to pilot, and fits on many yachts, stowing away in the space meant for your tender.

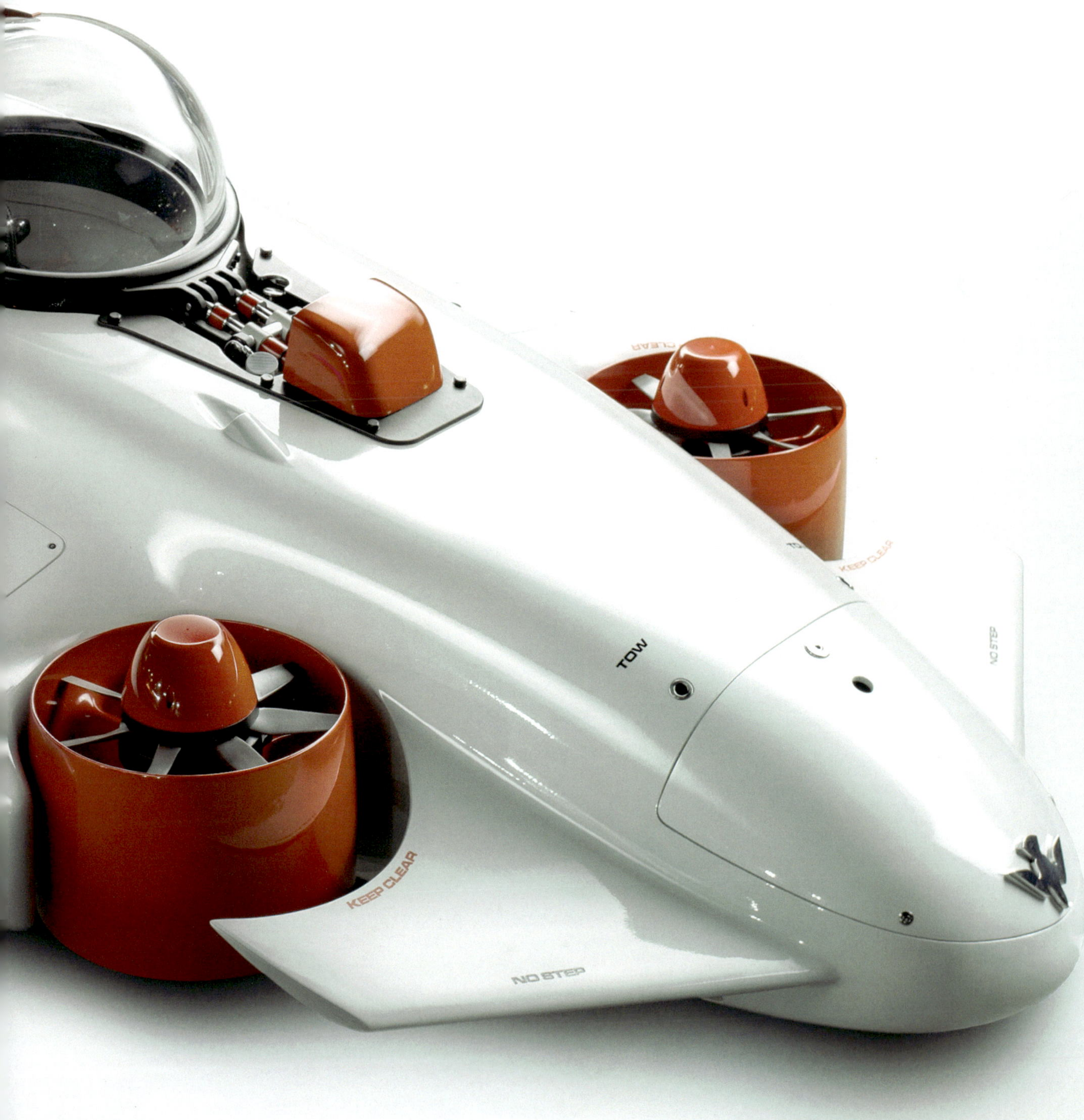

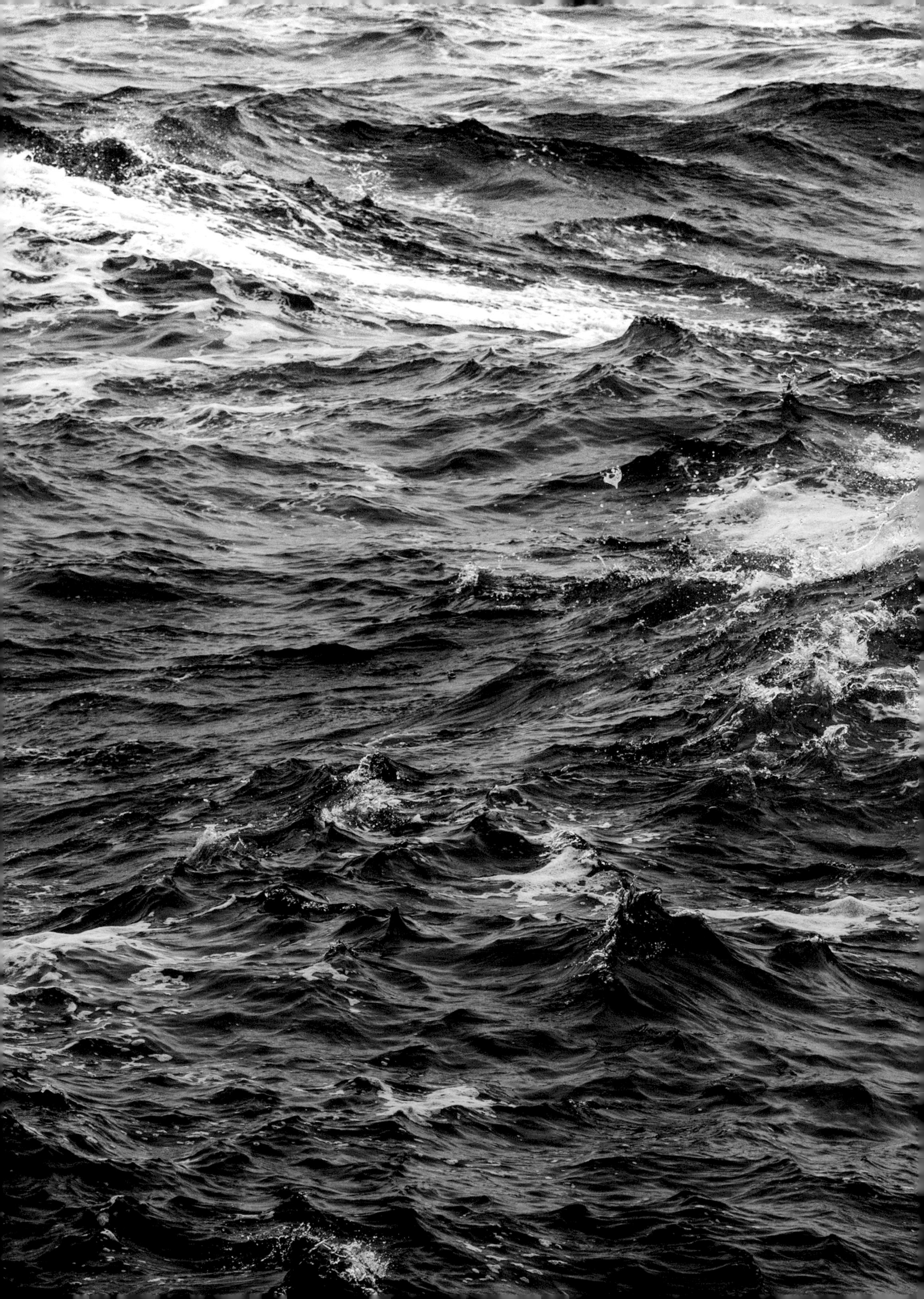

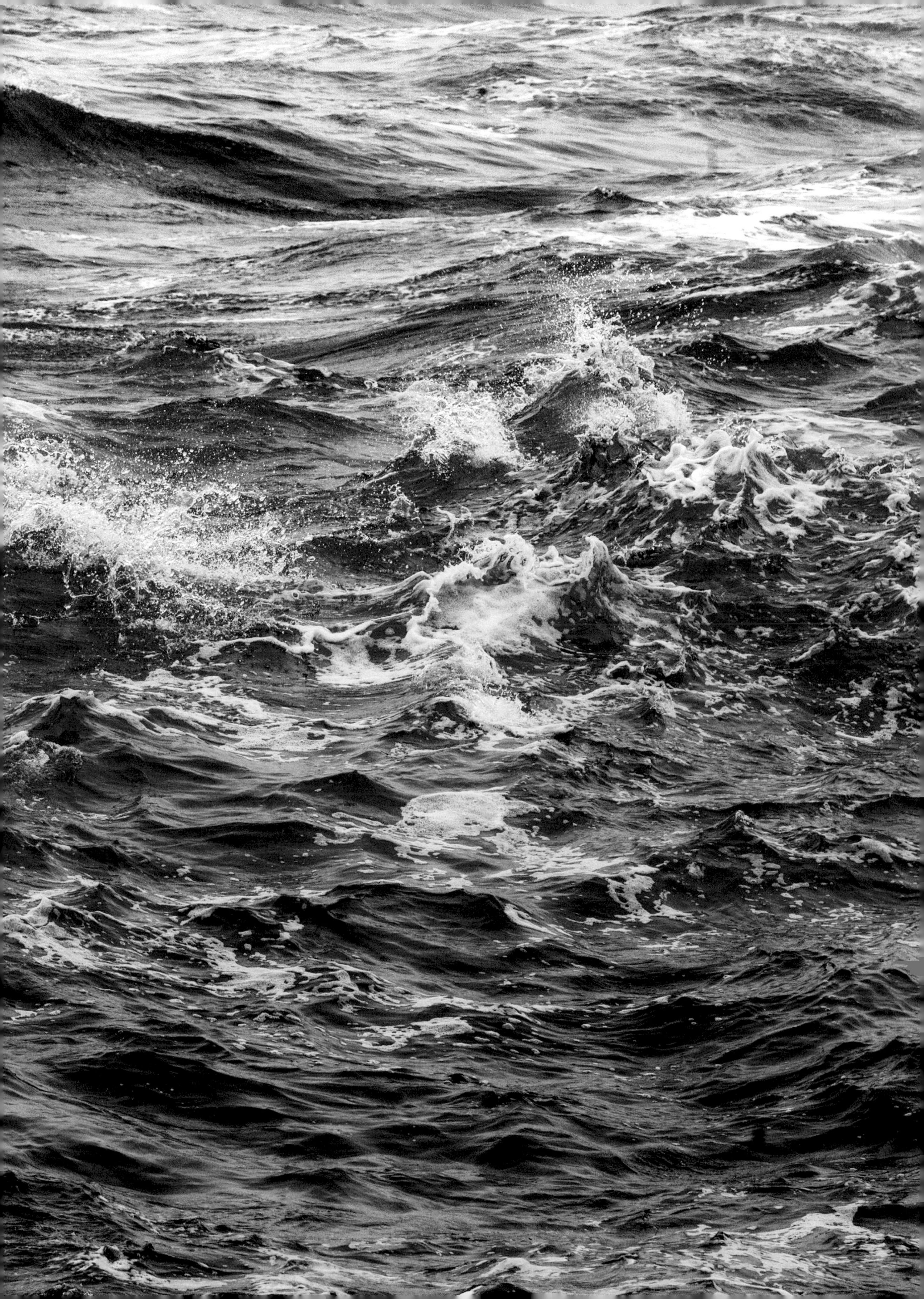

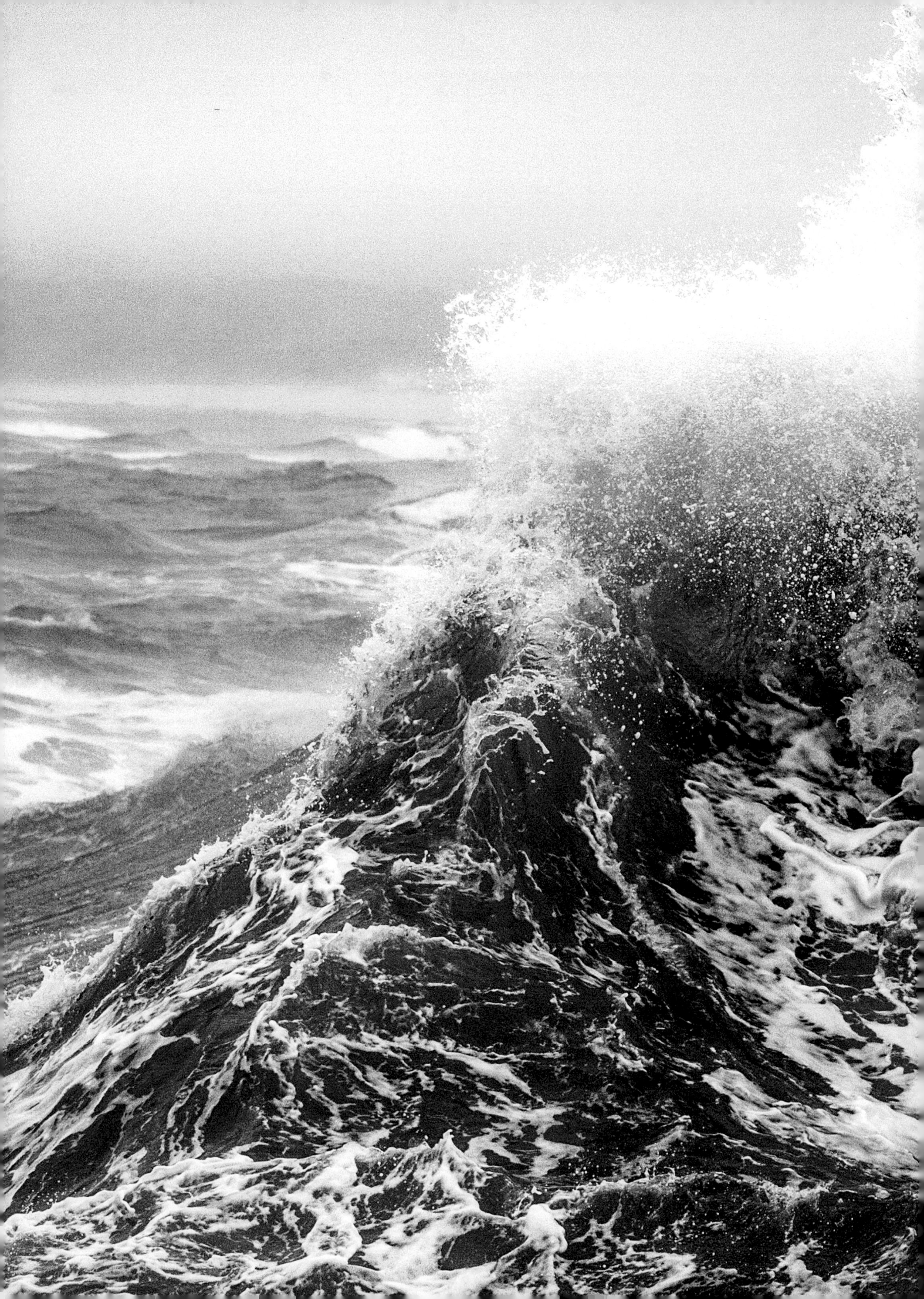

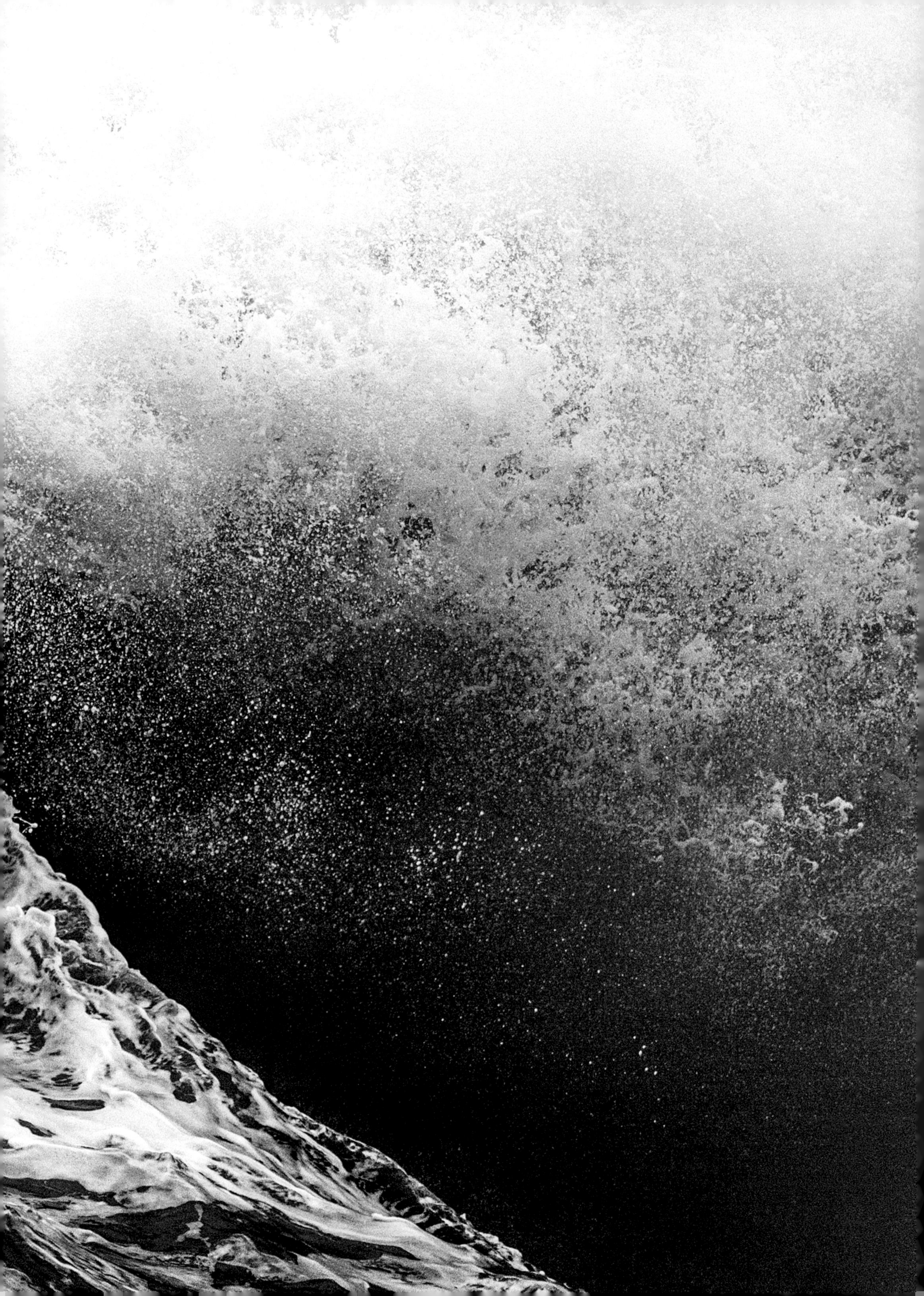

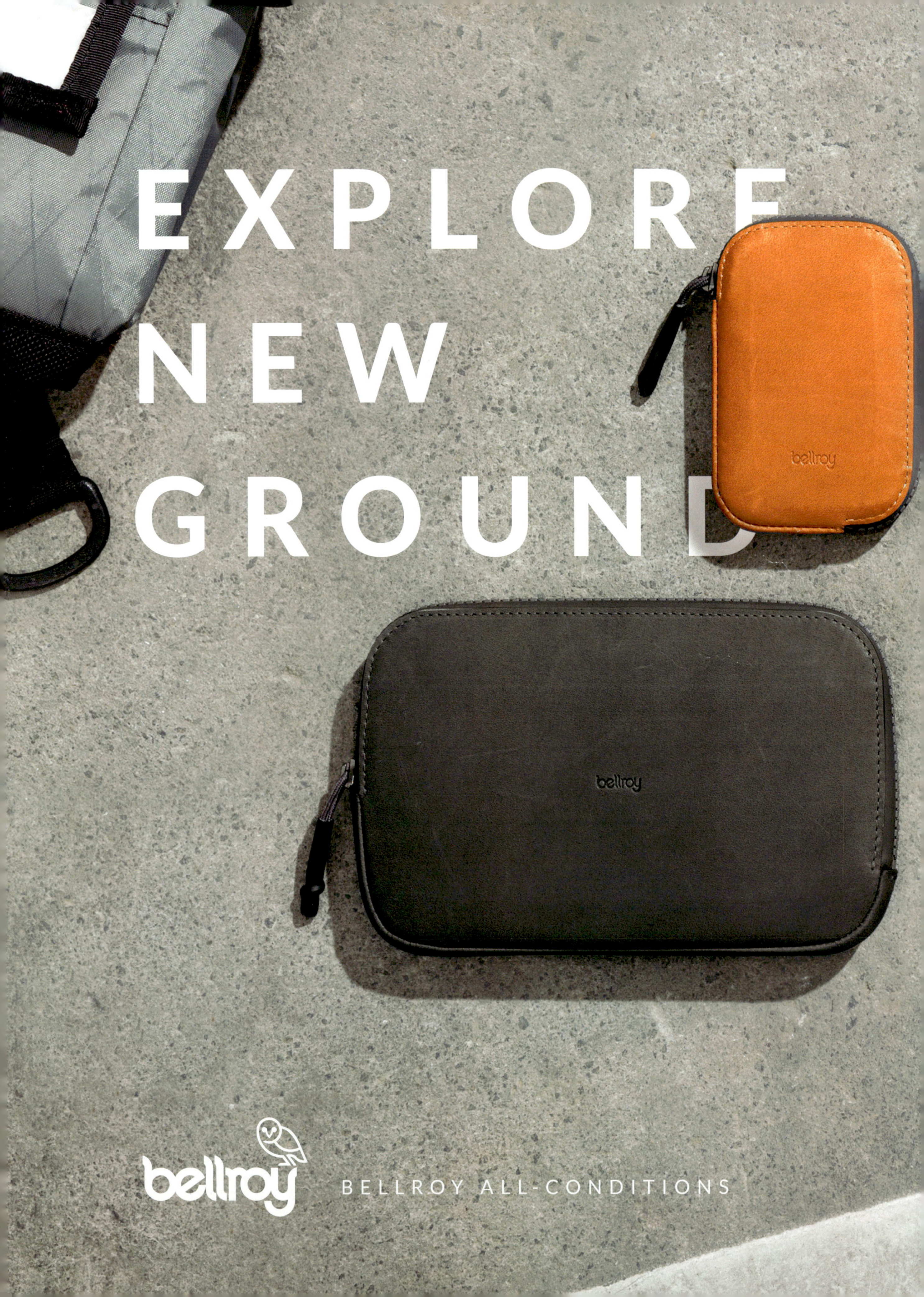
EXPLORE
NEW
GROUND
bellroy
bellroy
bellroy
BELLROY ALL-CONDITIONS